Neglected and Underutilized Plant Species in Horticultural and Ornamental Systems: Perspectives for Biodiversity, Nutraceuticals and Agricultural Sustainability

Neglected and Underutilized Plant Species in Horticultural and Ornamental Systems: Perspectives for Biodiversity, Nutraceuticals and Agricultural Sustainability

Editors

Rosario Paolo Mauro
Carlo Nicoletto
Leo Sabatino

MDPI • Basel • Beijing • Wuhan • Barcelona • Belgrade • Manchester • Tokyo • Cluj • Tianjin

Editors

Rosario Paolo Mauro
Department of Agriculture,
Food and Environment,
University of Catania,
Via Valdisavoia, 5,
95123 Catania, Italy

Carlo Nicoletto
Department of Agronomy,
Food, Natural Resources,
Animals and Environment
(DAFNAE), University of
Padova, 35020 Legnaro, Italy

Leo Sabatino
Department of Agricultural,
Food and Forest Sciences,
University of Palermo,
Viale delle Scienze, Ed. 5,
90128 Palermo, Italy

Editorial Office
MDPI
St. Alban-Anlage 66
4052 Basel, Switzerland

This is a reprint of articles from the Special Issue published online in the open access journal *Horticulturae* (ISSN 2311-7524) (available at: https://www.mdpi.com/journal/horticulturae/special_issues/Neglected_Underutilized_Plant).

For citation purposes, cite each article independently as indicated on the article page online and as indicated below:

LastName, A.A.; LastName, B.B.; LastName, C.C. Article Title. *Journal Name* **Year**, *Volume Number*, Page Range.

ISBN 978-3-0365-4659-9 (Hbk)
ISBN 978-3-0365-4660-5 (PDF)

Contents

About the Editors

Rosario Paolo Mauro

Rosario Mauro: Department of Agriculture, Food and Environment, University of Catania, Via Valdisavoia, 5, 95123 Catania, Italy. Interests: agronomic management in horticulture and floriculture; breeding of horticultural crops; crop physiology; vegetables quality; secondary metabolites.

Carlo Nicoletto

Carlo Nicoletto: Department of Agronomy, Food, Natural Resources, Animals and Environment (DAFNAE), University of Padova, 35020 Legnaro, Italy. Interests: modeling; plant physiology; statistical analysis; genetic diversity; food chemistry; irrigation; plant biology; predictive modeling; plant breeding.

Leo Sabatino

Leo Sabatino: Department of Agricultural, Food and Forest Sciences, University of Palermo, Viale delle Scienze, Ed. 5, 90128 Palermo, Italy. Interests: vegetable production; vegetable grafting; soilless cultivation; nutritional and nutraceutical vegetable fruit quality linked to cultivation condition and practices; biofortification of leafy and fruiting vegetable crops, propagation of ornamental plants.

Editorial

Neglected and Underutilized Plant Species in Horticultural and Ornamental Systems: Perspectives for Biodiversity, Nutraceuticals and Agricultural Sustainability

Rosario Paolo Mauro [1],*, Carlo Nicoletto [2] and Leo Sabatino [3]

[1] Department of Agriculture, Food and Environment, University of Catania, Via Valdisavoia, 5-95123 Catania, Italy
[2] Department of Agronomy, Food, Natural Resources, Animals and Environment (DAFNAE), University of Padova, 35020 Legnaro, Italy; carlo.nicoletto@unipd.it
[3] Department of Agricultural, Food and Forest Sciences, University of Palermo, Viale delle Scienze, Ed. 5, 90128 Palermo, Italy; leo.sabatino@unipa.it
* Correspondence: rosario.mauro@unict.it

Citation: Mauro, R.P.; Nicoletto, C.; Sabatino, L. Neglected and Underutilized Plant Species in Horticultural and Ornamental Systems: Perspectives for Biodiversity, Nutraceuticals and Agricultural Sustainability. *Horticulturae* 2022, 8, 356. https://doi.org/10.3390/horticulturae8050356

Received: 18 April 2022
Accepted: 19 April 2022
Published: 20 April 2022

Publisher's Note: MDPI stays neutral with regard to jurisdictional claims in published maps and institutional affiliations.

Dear Colleagues,

We are pleased to present this reprint of a Special Issue of *Horticulturae*, dedicated to the multifaceted topic of neglected and underutilized plant species (NUS) in horticultural and ornamental systems. Over the last few decades, this topic has received growing attention from the scientific community, due to the fact that it is a possible option that could be used to face the agricultural challenge of producing more foods and services within a framework of greater sustainability. Indeed, in the future, global urbanization processes, climate change and the reduction in natural resources are expected to emphasize the vulnerability of the mainstream agricultural system, which currently satisfies the needs of an ever-increasing world population by leveraging a restricted number of cash crops. In this context, NUS would be able to promote agro-biodiversity, improve the resilience of the agro-ecosystems toward environmental stressors, foster the assumption of nutraceuticals and the diversification of dietary patterns, and provide important local services (environmental, economic, and socio-cultural) as ornamentals or in landscape design. However, there are manifold barriers continue to hamper the reshaping of NUS utilization out from their niche role, many of which stem from our poor knowledge about their biological and technical features.

For the above reasons, this Special Issue aimed to help fill the knowledge gaps concerning NUS in horticultural and ornamental systems, as well as in landscape design, by gathering original research papers, short communications, and review articles dealing with relevant phenomena related to:

✓ Biodiversity and conservation;
✓ Genetics and breeding;
✓ Characterization, propagation, and ecophysiology;
✓ Cultivation techniques and systems;
✓ Landscape protection and restoration;
✓ Product and process innovations;
✓ Biochemistry and composition;
✓ Postharvest factors that affect their end-use quality.

Overall, the Special Issue collected 14 contributions (3 reviews and 11 original research papers).

In their review article, Meena et al. [1] describe the ethnobotany, medicinal and nutritional values, biodiversity conservation and utilization strategies of 19 important and underutilized climate-resilient fruit crops (Indian jujube, Indian gooseberry, lasora, bael, kair, karonda, tamarind, wood apple, custard apple, jamun, jharber, mahua, pilu, khejri, mulberry, chironji, manila tamarind, timroo, and khirni) from arid and semi-arid regions, as they have many advantages in terms of how easy they are to grow and their hardiness and resilience to climate changes compared with major commercially grown crops.

The review of Scarano et al. [2] focuses on both cultivated or spontaneously growing NUS from the Apulia region (southern Italy) that show interesting adaptative, nutritional, and economical potential which can be exploited and properly improved in the future.

Another review article written by Han et al. [3] highlights the importance of *S. aethiopicum* due to its role in crop diversification and in reducing hidden hunger. The authors also stress its nutritive and medicinal benefits, its agricultural sustainability, and future efforts for breeding and the genetic improvement of the solanaceae.

Among the experimental articles, the contribution of Singh et al. [4] reports on three experiments that were carried out to explore the integrated approaches toward nematode control in pomegranate. All of the evaluated genotypes and varieties were found to be susceptible to root knot nematodes, but the severity of an attack varied among them. Hence, more detailed screening is needed in larger populations, although some strategies can be adopted to reduce the attacks.

Amoruso et al. [5] report the effect of salinity (150 mM NaCl) on the growth, quality, and shelf-life of fresh-cut sea fennel grown on a floating system compared to a control condition (9 mM NaCl). The authors found that leaves from plants exposed to salinity had lower NO^{3-}, K^+, and Ca^{2+} contents and an increased Cl^- and Na^+ concentration compared with the control. Sensory quality was similar in both treatments, except that leaves from the NaCl treatment had a salty taste that was easily detected by panelists. This suggests that a saline-nutrient solution applied in hydroponics is a suitable system for sea fennel growth. The product behavior during the postharvest storage was also reported.

Two studies presented by Lamani et al. [6,7] focus on wood apple (*Limonia acidissima* L.), an underutilized fruit-yielding tree native to India and Sri Lanka. In their first paper, the authors analyzed the fatty acid composition, tocopherols, and physico-chemical characterization of wood apple seed oil and the nutritional profile of seed cake, determined using gas chromatography–mass spectrometry (GC-MS). They found interesting results in terms of oleic, alpha-linoleic, and linoleic acid content along with tocochromanols and total phenols. Data on seed oil and cake showed that they are a good source of natural functional ingredients with several health benefits. In the second paper, the authors analyzed the nutritional status of wood apple fruit pulp. They report that the pulp is rich in total carbohydrates, total proteins, oil, fiber, and ash. By using HPLC and GC methods, the article reports a comprehensive characterization of sugars, organic acids, and fatty acids contained in fruit pulp.

In the experiment conducted by Consentino et al. [8], five landraces of *Lagenaria siceraria* L. were subjected to foliar applications of seaweed extracts to improve yield and quality characteristics. The authors report that treated plants produced higher marketable fruit yields, fruit mean masses, young shoot yields, and numbers of young shoots than untreated ones. Relevant increments were also recorded for nitrogen use efficiency and fruits' mineral composition and nutraceutical profile.

Wei et al. [9] investigated the characteristics of suitable habitats for the endangered tree fern, *Sphaeropteris lepifera* (J. Sm. ex Hook.) R.M. Tryon, based on fieldwork, ecological niche modeling, and regression approaches. The ecological niche models indicated several climatic, orographic, and biological features that affect the distribution of *S. lepifera*, thus providing important information for the restoration of this species in the wild.

The possibility of converting *Ginkgo biloba* seeds from an unwanted and unused environmental pollutant into a source of beneficial compounds in Bulgaria was the subject of the article by Tomova et al. [10]. The authors applied various analytical and chromatographic methods to quantify the major constituents and ten biologically active compounds in methanol seed extract. The study revealed that seeds of locally grown Ginkgo trees could be used as a source of biologically active substances, as their composition is similar to that from other geographical areas.

The micro-scale production of microgreens is spreading due to the simplicity of their management, rapid cycle, harvest index, and phytochemical value of the edible product. In this context, two Mediterranean NUS, i.e., purslane (*Portulaca oleracea* L). and borage (*Borago officinalis* L.), offer opportunities to produce nutrient-dense foods as novel vegetable products. For these reasons, Corrado et al. [11] characterized the microgreens of both species, finding that purslane has significant amounts of phenolics and ascorbic acid, and a potential high β-carotene bioavailability, while borage microgreens have a very high fresh yield and a more composite and balanced phenolic profile. Overall, they provide insight into the implementation of NUS market-chains and into the development of added-value food products.

Soil salinization is one of the major threats that affect crop production worldwide. For this reason, NUS could represent an opportunity to find crops tolerant to salt-affected cultivation systems. Starting from this assumption, Alexopoulos et al. [12] studied the effects of increasing salinity in the nutrient solution in dandelion (*Taraxacum officinale* (L.) Weber ex F.H.Wigg.) and common brighteyes (*Reichardia picroides* (L.) Roth) grown under greenhouse conditions. The results revealed that both species are severely affected by high salinity; however, *R. picroides* showed promising results regarding its commercial cultivation under moderate salinity levels, as it exhibited a more effective adaptation mechanism against saline conditions, as evidenced by the higher accumulation of osmolytes such as proline and the higher shoot K content.

Salt stress was also investigated by Sogoni et al. [13]. They studied the effects of NaCl concentration in dune spinach (*Tetragonia decumbens* Mill.) nutrient solution. Dune spinach is an edible, neglected halophyte largely distributed along the coastal regions from southern Namibia to the Eastern Cape. The authors found that this species can be grown and irrigated with brackish water (incorporating up to 50 mM NaCl), as plants showed significant increases in growth parameters, antioxidant power (FRAP essay), along with concentrations of phenolics, nitrogen, phosphorus, and sodium.

Climate change, natural disturbances and human activities are factors that affect plant biodiversity worldwide, meaning the development of conservation and management strategies for the most endangered species is becoming an urgent need. For this reason, Li et al. [14] genotyped 480 individuals of Korean pine (*Pinus koraiensis* (Sieb. et Zucc)) belonging to 16 natural populations present in North-Eastern China by using fifteen polymorphically expressed sequence tag–simple sequence repeat (EST-SSR) markers to evaluate their genetic diversity, population structure, and differentiation. The results provide new genetic information for future genome-wide association studies (GWAS), marker-assisted selection (MAS), and genomic selection (GS) in natural *P. koraiensis* breeding programs. These findings can improve conservation and management strategies for this valuable species.

Informed Consent Statement: Not applicable.

Data Availability Statement: Not applicable.

Conflicts of Interest: The authors declare no conflict of interest.

References

1. Meena, V.S.; Gora, J.S.; Singh, A.; Ram, C.; Meena, N.K.; Pratibha; Rouphael, Y.; Basile, B.; Kumar, P. Underutilized Fruit Crops of Indian Arid and Semi-Arid Regions: Importance, Conservation and Utilization Strategies. *Horticulturae* **2022**, *8*, 171. [CrossRef]
2. Scarano, A.; Semeraro, T.; Chieppa, M.; Santino, A. Neglected and underutilized plant species (Nus) from the apulia region worthy of being rescued and re-included in daily diet. *Horticulturae* **2021**, *7*, 177. [CrossRef]
3. Han, M.; Opoku, K.N.; Bissah, N.A.B.; Su, T. Solanum aethiopicum: The nutrient-rich vegetable crop with great economic, genetic biodiversity and pharmaceutical potential. *Horticulturae* **2021**, *7*, 126. [CrossRef]
4. Singh, A.; Kaul, R.K.; Khapte, P.S.; Jadon, K.S.; Rouphael, Y.; Basile, B.; Kumar, P. Root Knot Nematode Presence and Its Integrated Management in Pomegranate Orchards Located in Indian Arid Areas. *Horticulturae* **2022**, *8*, 160. [CrossRef]
5. Amoruso, F.; Signore, A.; Gómez, P.A.; Martínez-Ballesta, M.D.C.; Giménez, A.; Franco, J.A.; Fernández, J.A.; Egea-Gilabert, C. Effect of Saline-Nutrient Solution on Yield, Quality, and Shelf-Life of Sea Fennel (*Crithmum maritimum* L.) Plants. *Horticulturae* **2022**, *8*, 127. [CrossRef]

6.   Lamani, S.; Anu-Appaiah, K.A.; Murthy, H.N.; Dewir, Y.H.; Rihan, H.Z. Fatty acid profile, tocopherol content of seed oil, and nutritional analysis of seed cake of wood apple (*Limonia acidissima* L.), an underutilized fruit-yielding tree species. *Horticulturae* **2021**, *7*, 275. [CrossRef]

7.   Lamani, S.; Anu-Appaiah, K.A.; Murthy, H.N.; Dewir, Y.H.; Rikisahedew, J.J. Analysis of Free Sugars, Organic Acids, and Fatty Acids of Wood Apple (*Limonia acidissima* L.) Fruit Pulp. *Horticulturae* **2022**, *8*, 67. [CrossRef]

8.   Consentino, B.B.; Sabatino, L.; Mauro, R.P.; Nicoletto, C.; De Pasquale, C.; Iapichino, G.; La Bella, S. Seaweed extract improves Lagenaria siceraria young shoot production, mineral profile and functional quality. *Horticulturae* **2021**, *7*, 549. [CrossRef]

9.   Wei, X.; Harris, A.J.; Cui, Y.; Dai, Y.; Hu, H.; Yu, X.; Jiang, R.; Wang, F. Inferring the potential geographic distribution and reasons for the endangered status of the tree fern, sphaeropteris lepifera, in lingnan, china using a small sample size. *Horticulturae* **2021**, *7*, 496. [CrossRef]

10.   Tomova, T.; Slavova, I.; Tomov, D.; Kirova, G.; Argirova, M.D. Ginkgo biloba seeds—An environmental pollutant or a functional food. *Horticulturae* **2021**, *7*, 218. [CrossRef]

11.   Corrado, G.; El-Nakhel, C.; Graziani, G.; Pannico, A.; Zarrelli, A.; Giannini, P.; Ritieni, A.; De Pascale, S.; Kyriacou, M.C.; Rouphael, Y. Productive and morphometric traits, mineral composition and secondary metabolome components of borage and purslane as underutilized species for microgreens production. *Horticulturae* **2021**, *7*, 211. [CrossRef]

12.   Alexopoulos, A.A.; Assimakopoulou, A.; Panagopoulos, P.; Bakea, M.; Vidalis, N.; Karapanos, I.C.; Petropoulos, S.A. Impact of salinity on the growth and chemical composition of two underutilized wild edible greens: Taraxacum officinale and reichardia picroides. *Horticulturae* **2021**, *7*, 160. [CrossRef]

13.   Sogoni, A.; Jimoh, M.O.; Kambizi, L.; Laubscher, C.P. The impact of salt stress on plant growth, mineral composition, and antioxidant activity in tetragonia decumbens mill.: An underutilized edible halophyte in South Africa. *Horticulturae* **2021**, *7*, 140. [CrossRef]

14.   Li, X.; Zhao, M.; Xu, Y.; Li, Y.; Tigabu, M.; Zhao, X. Genetic diversity and population differentiation of pinus koraiensis in China. *Horticulturae* **2021**, *7*, 104. [CrossRef]

horticulturae

MDPI

Article

Root Knot Nematode Presence and Its Integrated Management in Pomegranate Orchards Located in Indian Arid Areas

Akath Singh [1,*], Ramesh Kumar Kaul [1], Pratapsingh Suresh Khapte [2], Kuldeep Singh Jadon [1], Youssef Rouphael [3], Boris Basile [3,*] and Pradeep Kumar [1]

[1] ICAR-Central Arid Zone Research Institute, Jodhpur 342003, India; rk.kaul@icar.gov.in (R.K.K.); kuldeep.jadon@icar.gov.in (K.S.J.); pradeep.kumar4@icar.gov.in (P.K.)
[2] ICAR-National Institute of Abiotic Stress Management, Baramati 413115, India; pratapsingh.khapte@icar.gov.in
[3] Department of Agricultural Sciences, University of Naples Federico II, 80055 Portici, Italy; youssef.rouphael@unina.it
[*] Correspondence: Akath.Singh@icar.gov.in (A.S.); boris.basile@unina.it (B.B.)

Abstract: Nematodes are a serious problem across pomegranate-growing areas worldwide, but the severity is higher in light sandy soils of arid regions. The present study was carried out to explore the integrated approaches for the control of nematodes in pomegranate. Three different experiments were carried out during 2017–2020 to (a) delineate nematode abundance in major pomegranate areas, (b) screen pomegranate genotypes against nematode, and (c) assess the efficacy of integrated management for the control of root knot nematode in pomegranate. The survey results revealed that none of the pomegranate orchards were found to be free from nematode infestation. Moreover, the majority of the orchards (78%) showed moderate incidence (10.1 to 40%) of infestation. A significant yield reduction (40.2%) and a decrease in fruit size was observed in nematode-affected trees. Pattern of cuticular markings in the perineal area of the mature female confirmed the occurrence of *Meloidogyne incognita* only in all the surveyed orchard of pomegranate. All the evaluated genotypes and varieties were found susceptible to root knot nematodes, but the severity of the attack varied among them. Hence, more detailed screening is needed on a larger population. Nematode population (number of galls g^{-1} root) can be minimized significantly with the combined applications of Carbofuran at 20 g + Fluensulfone at 20 g per plant or Neemcake 500 g + *Paecilomyceslilacinus* at 25 mL + Carbofuran at 20 g + Fluensulfone at 20 g per plant in April and August.

Keywords: root knot nematode; *Punica granatum*; bioagents; nematicides; neemcake

Citation: Singh, A.; Kaul, R.K.; Khapte, P.S.; Jadon, K.S.; Rouphael, Y.; Basile, B.; Kumar, P. Root Knot Nematode Presence and Its Integrated Management in Pomegranate Orchards Located in Indian Arid Areas. *Horticulturae* **2022**, *8*, 160. https://doi.org/10.3390/horticulturae8020160

Academic Editor: Giovanni Bubici

Received: 29 November 2021
Accepted: 9 February 2022
Published: 12 February 2022

Publisher's Note: MDPI stays neutral with regard to jurisdictional claims in published maps and institutional affiliations.

1. Introduction

Pomegranate (*Punica granatum* L.) is an underutilized crop that can be cultivated in several climatic conditions (Mediterranean, subtropical, and tropical), and this indicates its wide adaptability. Low water requirement, good response to modern horticultural practices, high economic returns, and great global demand have made this species one of the most popular commercial cash crops at a global scale [1]. Global production of pomegranates is estimated around 6.3 million metric tons (MT) from an area of 556 thousand hectares [2]. At global level, India is a leading producer (3186 thousand MT) followed by China (1600 thousand MT), Iran (1100 thousand MT), Turkey (220 thousand MT), the USA (210 thousand MT), Afghanistan (150 thousand MT), and Spain (53.18 thousand MT) [2]. The international export market is estimated to be around 362.6 thousand MT, and India has recorded a constant increase in foreign earning by exporting 80,547.74 MT worth USD 92.46 million in 2019–2020 [3]. Recent export trends depict a high amenability in supply–demand of Indian pomegranate in international market with increased prices. This ambiance provides opportunities to stretch the growing area of pomegranate from its traditional belt, viz., Central and Southern India to arid and semi-arid northwestern Indian states such as Gujarat and Rajasthan. The quality of the fruits produced in these regions also matches the export specifications with attractive color, soft seeds, and low acidity [4].

Pomegranate is attacked by several non-insect pests. Among these, root knot nematode has emerged as a major threat to sustainable production of pomegranate in these regions [5]. Although the infestation of nematodes is a serious problem throughout its growing areas worldwide, the severity of their attack is higher in arid climates and light sandy soils [6]. Root knot nematodes can be spread by water or by soil or farm equipment or through infested planting materials [7,8]. Substrate mixture used during seedling preparation may also harbor these nematodes [9]. Their feeding leads to impaired root functions such as nutrient and water uptake due to gall formation, and consequently plants become progressively sick (yellow aspect, tip drying, and stunted growth) and may be exposed to secondary infections of fungi, bacteria, and nutritional deficiencies that can even cause plant death [8]. These nematodes are responsible for 30 to 40% yield losses with poor quality fruits in the current season [6,10]. The association of different species and population density of root knot nematode has not been well documented in affected pomegranate orchards thusfar; however, Holland et al. [9] reported that *Meloidogyne incognita* and *M. javanica* are the main species attacking this tree crop species. Darekar et al. [11] reported 10 different species of plant parasitic nematodes associated with pomegranate orchards in the Maharashtra state. The root knot nematode *M. incognita* is one of the main species causing considerable yield losses in pomegranate [10,11]. Application of nematicides has remained the most common short-term management strategy against root knot nematode [12]. Several horticultural practices such as intercropping of marigold, inoculation of various bio-formulations with neem or castor cakes, etc., have been recommended, but their effectiveness in controlling nematodes varies depending on the growing conditions [13,14]. Presently, there is barely any information on the nematode tolerant/resistant genotypes of pomegranate, which can be used as rootstocks or as a source of resistance genes to be used in future breeding programs [15]. In other fruit crops such as grape, citrus, and stone fruits, there are several successful examples of steady solutions to different abiotic and biotic problems including nematodes. Nematode abundance, knowledge of nematode species, use of resistant/tolerant genotypes, organic amendments, and use of biocontrol agents can be important components of an integrated nematode management approach. In light of the above, the present investigation included three experiments that were carried out during 2017–2020 at the ICAR-Central Arid Zone Research Institute, Jodhpur, to delineate nematode abundance in pomegranate orchards, screen tolerant genotypes, and study integrated management strategies of nematodes in field conditions.

2. Materials and Methods

2.1. Collection, Isolation, and Identification of Nematode

A field survey was conducted during 2017–2018 in two major pomegranate-growing districts of Rajasthan, India, i.e., Barmer and Jodhpur, in order to ascertain the nematode abundance and to identify their dominant species. In each district, two locations comprising four villages were considered for sample collection. The selection of these fields was based on visual symptoms of nematode infestation, viz., plants with pale green or yellowish leaves, wilting symptoms, reduced growth, and drying twigs (Figure 1).

The number of infected plants in each orchard was recorded to calculate the percentage of nematode incidence. A total of 25 orchards with an area of at least one hectare each (715 trees/ha) with these characteristics and a minimum age of seven years were selected randomly in each location. The percentage of nematode incidence in each orchard was calculated with the following equation:

Nematode Incidence (%) = No. of plants infected/Total No. of plants × 100

Figure 1. Selection of orchards according to visual symptoms of plants: (**A**) showing symptoms of nematode infestation; (**B**) healthy plant without nematode infestation.

On the basis of the Nematode Incidence, we categorized the severity of the infestation of surveyed orchards as severe (>40%), moderately severe (10–40%), and mild (0–10%).

Soil and root samples were collected from five randomly selected plants in each orchard and mixed thoroughly to prepare composite samples. One hundred composite samples were brought to the laboratory for assessing the nematode abundance and identification of nematode species for further studies. The egg masses were isolated from soil and root samples and kept in fresh water for hatching. These eggs were obtained from a pure culture established from single egg mass for identification according to the characteristics of the perineal pattern of matured females as previously described [16,17].

2.2. Screening of Genotypes

An experiment was conducted under protected conditions (fan and pad greenhouse with air temperature and relative humidity maintained in the range of 25–32 °C and 40–65%, respectively) during 2018–2019 and 2019–2020 to study the response of several genotypes and varieties of pomegranate to root knot nematode. Sufficient quantity of infected soil from a previously surveyed nematode-affected pomegranate orchard was collected for mass culturing of root knot nematode in earthen pots. The nematode was sub-cultured for mass multiplication by removing egg masses from the mother culture plants. Egg masses were collected and kept in fresh water for hatching. These eggs were obtained from a pure culture established from single egg masses of previously identified species according to the characteristics of the perineal pattern [16,17] and reared under protected conditions. The sterilized soil mixture (sand/farmyard manure in a 3:1 proportion) was used to fill disinfected earthen pots (with a diameter of 30 cm). Three-month-old pomegranate saplings were grown in 10 earthen pots and were maintained for three months under protected conditions for mass multiplication of nematode. Simultaneously, around 500 seedlings of 27 genotypes, namely, IC-318790, IC-318723, IC-318728, IC-318754, Wonderful, Joyti, Tabesta, Yercawd-1, Jodhpur collection, Kabuli yellow, Co-white, Ruby, KRS, Kasturi, Amali dana, Bhagwa, CAZRI-Sel., Phule Arakta, P-26, G-137, P-23, Jodhpur Red, Ganesh, Mridula, Dholka, Basein Seedless, and Jalore Seedless, were raised in the nursery. Four 6-month-old saplings of each genotype were shifted in small earthen pots (1 kg capacity) containing soil mixed with well-rotted farmyard manure (FYM) in a 3:1 proportion. Once seedlings were well established, freshly hatched second-stage nematode juveniles were inoculated to each pot (3000 J2/pots). This experiment was carried out in a completely randomized design under protected conditions. Observations on height and girth of seedlings, as well as the number of root knot galls g^{-1} root, were recorded after completing three growth cycles of nematode, i.e., nine months after inoculation of seedlings.

2.3. Integrated Management of Nematode

A field experiment was also conducted during 2019–2020 and 2020–2021 in a naturally root knot nematode-infested pomegranate orchard to ascertain the effect of different integrated approaches for the control of nematodes. The experiment was carried out in a 7-year-old pomegranate orchard (3 ha; 715 trees/ha) of own-rooted trees of the Bhagwa variety. This cultivar was selected because it is the most cultivated pomegranate variety in the area where this experiment was carried out (more than 95% of the total area cultivated with this tree crop). Growth, yield, and the number of root knot galls g^{-1} root were recorded before applying the treatment. The granular nematicide (carbofuran and Fluensulfone), bioagents (*Paecilomyces lilacinus* and *Trichoderma harzianum*), and organic amendment (neem cake) were assessed as soil applications for the control of root knot nematode in pomegranate. All the treatments were applied twice in April and August, each year. The experiment was conducted in a randomized block design with three replications and 13 treatments (Table 1) on a total of 78 homogenous trees (6 trees per treatment).

Table 1. Description of treatments applied for integrated management of root knot nematode.

Treatment ID	Treatment Description
T1	Carbofuran at 40 g/plant
T2	Fluensulfone at 40 g/plant
T3	Carbofuran at 40 g + Fluensulfone at 40 g
T4	Carbofuran at 20 g + Fluensulfone at 20 g
T5	Neemcake at 500 g + *P. lilacinus* at 25 mL + Carbofuran at 40 g
T6	Neemcake at 500 g + *T. harzianum* at 50 g + Carbofuran at 40 g
T7	Neemcake at 500 g + *P. lilacinus* at 25 mL + Fluensulfone at 40 g
T8	Neemcake at 500 g + *T. harzianum* at 50 g + Fluensulfone at 40 g
T9	Neemcake at 500 g + *P. lilacinus* at 25 mL + Carbofuran at 40 g + Fluensulfone at 40 g
T10	Neemcake at 500 g + *T. harzianum* at 50 g + Carbofuran at 40 g + Fluensulfone at 40 g
T11	Neemcake at 500 g + *P. lilacinus* at 25 mL + Carbofuran at 20 g + Fluensulfone at 20 g
T12	Neemcake at 500 g + *T. harzianum* at 50 g + Carbofuran at 20 g + Fluensulfone at 20 g
T13	Control

2.4. Observations Recorded

2.4.1. Growth, Yield, and Quality Parameters

Plant growth in terms of plant height and canopy spread (measured in east–west and north–south directions) were recorded in November (middle of fruiting season). On each picking, fruit yield was recorded, and the average cumulative yield per tree was calculated. On each picking, 20 randomly selected fruits from all the tree canopy sides were sampled and weighed with a top pan digital balance, and their average weight was calculated. Fruit size (fruit length and width) was measured on the same fruits with a digital vernier calliper. Fruit juice was extracted using a mechanical juicer and grinder followed by squeezing and filtering with a muslin cloth. The quantity of extracted juice was measured with a graduated measuring cylinder, and fruit juice content was expressed as percentage on a fresh fruit weight basis. Total soluble solids (TSS) of the juice were determined in the laboratory using a digital hand held refractometer (Model: Brix 54, Bellingham + Stanley Ltd., Tunbridge Wells, Kent, United Kingdom), which was calibrated using distilled water before measurements. The acidity of fruit juice was determined with the titration method described by Ranganna [18].

2.4.2. Nematode Population

In each treatment, 500 g composite soil-root samples were collected at 30–60 cm depth from four sides of a tree, each 45–60 cm apart from the tree trunk, at the beginning and at the end of the experiment. Soil from 15 different locations within each site wascollected and mixed thoroughly, and composite samples were brought to the laboratory to count the root knot nematode population, root galls, and egg masses. The soil samples were processed by Cobb's decanting and sieving method [19], as previously described [20,21]. Randomly, 5–10 mature females were separated from infested roots using needles and forceps, teased with the stereoscopic binocular microscope to create perineal patterns for identification and confirmation of the root knot species in collected samples [16,17]

2.5. Statistical Analysis

Data of the different measured parameters were subjected to analysis of variance (ANOVA), followed by the Tukey HSD test ($p < 0.05$) for mean comparison. Before running the ANOVAs, the assumptions for normality of data distribution and homogeneity of variance were tested using the Shapiro–Wilk test and Levene's test, respectively. Counts of the number of galls per gram of root, number of eggs per egg mass, and number nematodes per 200cc soil were analyzed with the general linear model procedure using the negative binomial model, which best fitted the data, followed by the least-squares means (with *p*-value adjustment using Tukey method) for multiple comparison. All analyses were performed using SAS 9.2 (SAS Ins., Cary, NC, USA). A cluster analysis of different genotypes was conducted using the statistical package JMP Pro 10 (SAS Institute Inc., Cary, NC, USA). To study the similarity among the genotypes in terms of tolerance to the root knot nematode, we conducted a hierarchical cluster analysis (Ward linkage method) using the data of root knot galls g^{-1} root.

3. Results and Discussion

3.1. Growth, Yield, and Fruit Quality of Nematode-Infested Pomegranate

The survey results revealed that none of the pomegranate orchard was found free from nematode infestation in both the investigated districts. Thirteen percent of the surveyed orchards had severe incidence of nematodes (>40.0% incidence), 78% had moderate incidence (10–40%), and 9% had a mild nematode incidence (<10%) (Table 2). In this part of the Indian arid region, commercial cultivation of pomegranate is relatively recent, and nematode infestation is a new emerging problem. This may be a reason for the moderate severity level of nematode we found in this study. However, the severity of nematode incidence was higher in both the locations in Barmer district compared to Jodhpur (Figure 2). Nematodes multiply faster in light soil when available soil moisture is constant and temperature is high [9]. In the region where this survey was carried out, pomegranate is planted under high density with regular supply of water through drip system in light-sandy soils. These factors may explain the high incidence of root knot nematode in this region, as also suggested by the findings of Dasgupta and Gaur [7].

Table 2. Total number of surveyed pomegranate orchards in each district and location and their distribution in three categories of severity of root knot nematode incidence: mild (nematode incidence of 0–10%), moderate (10–40%), and severe (>40%).

| District | Location | No. of Orchards Surveyed | Percent of Orchards (%) | | |
| | | | Infestation Severity | | |
			Mild	Moderate	Severe
Barmer	Location 1	25	12	60	28
	Location 2	25	0	84	16
Jodhpur	Location3	25	12	80	8
	Location 4	25	12	88	0
Total		100	9	78	13

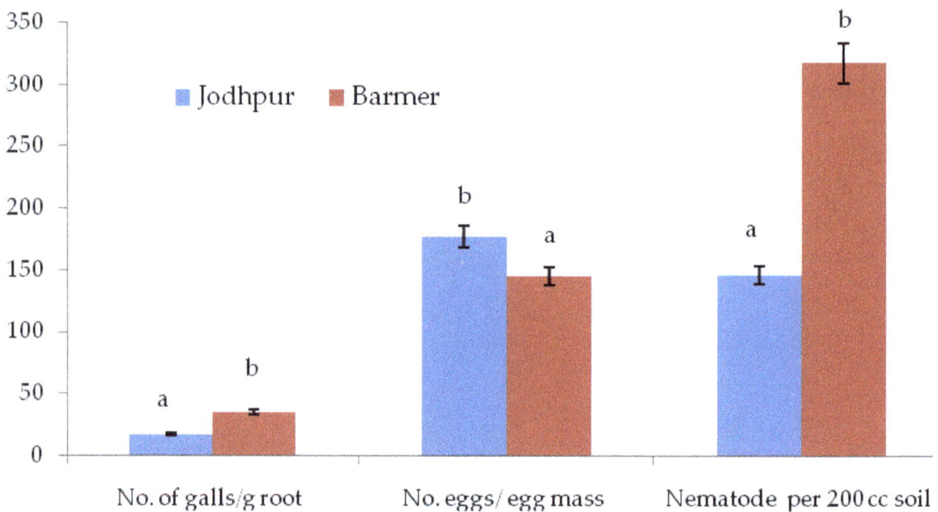

Figure 2. Root knot nematode population in pomegranate orchards located in two locations (Jodhpur and Barmer). Means of 100 samples across the surveyed orchards in four locations. Separately for each parameter, different letters indicate significant differences between locations assessed with the general linear model procedure followed by the least-squares means.

Plant height, canopy width, fruit yield, fruit weight, fruit size, juice content, TSS, and acidity in nematode-affected vis-à-vis healthy trees were recorded in all the surveyed orchards from all the locations. We found a significant reduction in plant growth, yield, and fruit quality in root knot nematode-affected orchard (Table 3). A significant yield reduction (40.2%) and a decrease in fruit size were measured in nematode-affected trees when the age of most of the orchards was only seven years. Juice content of fruits harvested from affected trees showed significantly lower juice compared to healthy plants, whereas TSS and acidity of juice were not very affected. Khan et al. [10] also reported 30 to 40% yield losses with poor quality fruits in current season in nematode-affected pomegranate orchards.

Table 3. Reduction in growth, yield, and fruit quality of pomegranate affected by root knot nematode compared to healthy plants.

Parameter	Healthy Trees		Nematode-Affected Trees		Reduction (%)
	Mean *	Range	Mean *	Range	
Plant height (cm)	222.0 b	180–280	167.0 a	100–230	24.6
Tree spread (cm)	223.0 b	165–250	169.4 a	85–222.5	24.2
Fruit yield (kg/tree)	20.9 b	17.8–31.3	12.5 a	2.8–16.6	40.2
Fruit weight (g/fruit)	227.3 b	176.0–280.3	167.5 a	130.2–220.8	26.3
Fruit length (cm)	8.0 a	7.15–9.27	6.2 a	5.8–7.9	22.0
Fruit breadth (cm)	7.8 b	6.9–8.6	5.6 a	5.4–7.3	28.6
Juice content (%)	30.4 a	25.0–36.2	24.2 a	22.5–28.6	20.4
TSS (%)	16.7 a	15.6–18.8	15.8 a	15.2–17.0	5.4
Acidity (%)	0.44 a	0.38–0.58	0.46 a	0.36–0.62	4.3

* Means of 25 healthy and 25 RKN-affected trees across the surveyed orchard. Within each row, different letters indicate significant differences in the measured parameter between healthy and affected trees according to ANOVA followed by the Tukey HSD test ($p \leq 0.05$).

3.2. Nematode Abundance and Identification of Species

Pattern of cuticular markings and the presence of high, squarish dorsal arch with smooth to wavy striation in the perineal area of the mature females confirmed the occurrence of only *M. incognita* in collected samples from all the locations in both the districts (Figure 3). In contrast to the present study, the association of more than one species of root knot nematode in pomegranate has been reported by different workers worldwide. Khan et al. [10] observed *Helicotylenchus digitus* and *M. incognita* to be the most frequently observed nematodes in pomegranate orchards. Similarly, Holland et al. [9] reported that *M. incognita* and *M. javanica* are the main species, whereas Darekar et al. [11] reported 10 different species of plant parasitic nematodes in pomegranate orchards with *M. incognita* being the most abundant. Singh et al. [6] summarized that *Meloidogyne* spp. are the most active at higher temperatures and optimum field capacity that favor their rapid multiplication, and this might be the reason of the dominance of *M. incognita* in arid regions where temperature remains high, and the orchards are planted under high density with regular water supply through drip irrigation systems.

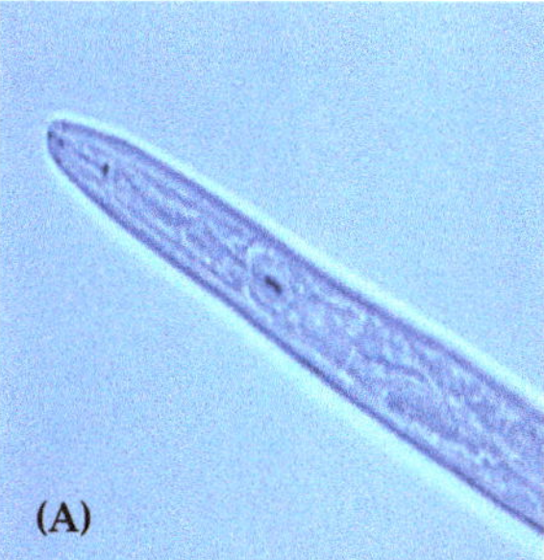

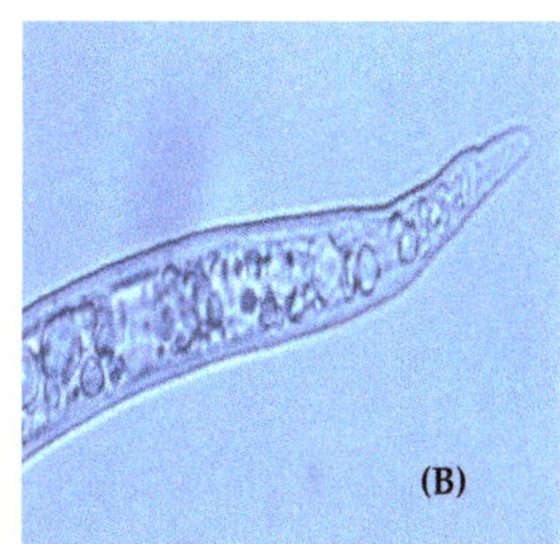

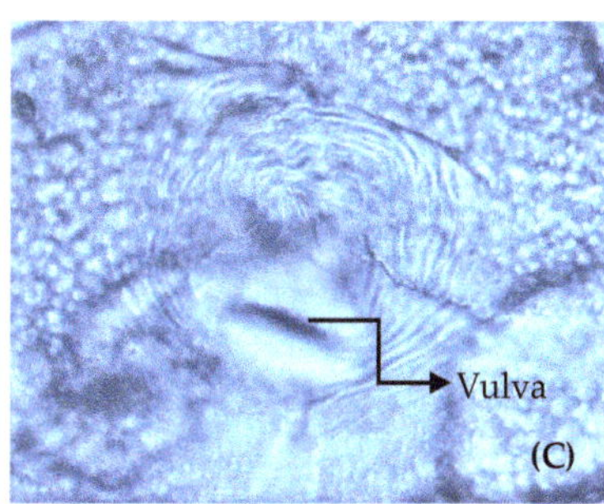

Figure 3. Morphology of female root knot nematode: (**A**) anterior extremity of the body; (**B**) posterior extremity of the body; (**C**) pattern of cuticular markings in the perineal area of mature female. Scale was approximately 10 µm in fluorescence microscopy with LEICA DM 3000; 1000× magnification.

3.3. Screening of Genotypes

One of the most important sustainable ways to manage root knot nematodes may be the adoption of resistant or tolerant genotypes against nematodes, either as rootstocks or breeding source to develop in the long term as improved resistant varieties. In order to have resistant or tolerant genotypes/cultivars against nematodes, we evaluated 27 different genotypes and cultivars of pomegranate collections. The results showed that the number of root knot galls g^{-1} root ranged from 28.5 to 56.5 at the third cycle of nematode growth (Table 4).

Therefore, the screening was unable to detect genotypes/varieties resistant or tolerant to *M. incognita*. Ahmadi et al. [22] also could not find resistant or tolerant cultivars to nematodes among 27 evaluated cultivars, whereas Shelke and Darekar [23] reported that, out of various genotypes, seven were moderately resistant to *M. incognita* race-2. Similarly, Ahire [15] reported that, out of 11 evaluated genotypes, 4 were moderately resistant, whereas the remainders were susceptible to root knot nematode. However, the present study clearly indicated that although all the genotypes were found susceptible to *M. incognita*, the relative abundance of infestation differed among the evaluated genotypes. It was also noted that variations in growth of nematode-inoculated vis-à-vis healthy seedlings of the genotypes IC318728, IC-318790, and IC-318723 were meager along with relatively lower number of root knot galls g^{-1} root.

In this part of the Indian arid region, commercial cultivation of pomegranate is relatively recent, but nematode infestation is becoming a major limiting factor for sustainable pomegranate production. Therefore, identifying tolerant/resistant genotypes or cultivars can be further used for the development of resistant rootstock or cultivars.

Table 4. Plant height, stem girth, and number of root knot galls (means $\pm$ standard errors) in healthy and nematode-inoculated trees of 27 genotypes.

Genotypes	Plant Height (cm) *			Stem Girth (mm) *			Root Knot Galls g^{-1} Root in Inoculated Plants **	Assigned Cluster
	Inoculated	Healthy	Variation (%)	Inoculated	Healthy	Variation (%)		
Amlidana	42.2 ± 1.2 de	48.9 ± 1.2 bc	15.8 ± 3.9 bc	6.7 ± 0.3 a	8.2 ± 0.3 a	22.8 ± 2.9 cd	40 ± 1 abc	4
Bassein seedless	43.7 ± 1.6 def	62.7 ± 1.6 fghi	43.4 ± 2.4 ijk	6.4 ± 0.5 a	9.7 ± 0.5 a	51.7 ± 2.0 lmn	46 ± 2 abc	1
Bhagwa	40.5 ± 2.1 cd	55.4 ± 2.2 cdef	36.7 ± 2.7 ghi	6.5 ± 0.7 a	8.8 ± 0.8 a	35.8 ± 2.6 efghi	51 ± ab	3
CAZRI Sel.	50.4 ± 2.1 efgh	70.3 ± 2.2 i	39.8 ± 2.6 hij	5.7 ± 0.9 a	8.3 ± 1.0 a	45.0 ± 2.3 jklm	32 ± 2 bcd	2
Co-white	51.7 ± 2.0 fgh	60.2 ± 2.4 efgh	16.4 ± 4.1 bc	7.3 ± 0.8 a	9.4 ± 0.9 a	28.8 ± 2.9 def	34 ± 2 bcd	2
Dholka	29.8 ± 2.7 a	45.5 ± 2.7 ab	34.5 ± 2.9 fgh	6.5 ± 0.8 a	9.5 ± 0.9 a	31.6 ± 2.8 defgh	35 ± 2 bc	2
G-137	50.2 ± 2.1 efgh	60.4 ± 2.4 efgh	20.3 ± 3.8 cd	6.7 ± 0.8 a	9.5 ± 0.9 a	41.9 ± 2.5 ijk	31 ± 3 cd	2
Ganesh	46.7 ± 2.2 defg	65.1 ± 2.4 ghi	39.4 ± 2.8 hij	6.7 ± 0.9 a	9.7 ± 0.9 a	44.7 ± 2.5 jkl	52 ± 2 ab	3
IC-318723	58.5 ± 2.0 h	65.3 ± 2.4 ghi	10.5 ± 5.4 ab	7.2 ± 0.8 a	9.0 ± 1.0 a	20.0 ± 3.7 bc	32 ± 3 bcd	2
IC318728	50.6 ± 2.1 efgh	52.4 ± 2.7 bcde	3.6 ± 9.3 a	7.0 ± 0.9 a	7.8 ± 1.1 a	11.4 ± 5.0 a	44 ± abc	1
IC-318754	42.9 ± 2.3 def	50.5 ± 2.8 bcd	15.0 ± 4.5 bc	8.1 ± 0.8 a	10.6 ± 1.0 a	23.6 ± 3.5 cd	38 ± 2 abc	4
IC-318790	45.2 ± 2.3 def	48.5 ± 2.9 bc	6.8 ± 6.7 a	7.4 ± 0.9 a	8.2 ± 1.1 a	9.7 ± 5.5 a	18 ± 4 d	5
Jalore seedless	41.6 ± 2.4 cde	62.5 ± 2.6 fghi	50.2 ± 2.4 kl	5.8 ± 1.0 a	8.9 ± 1.1 a	53.4 ± 2.3 mn	46 ± 2 abc	1
Jodhpur Red	25.7 ± 3.1 a	38.2 ± 3.4 a	48.6 ± 2.4 k	5.8 ± 1.0 a	8.8 ± 1.1 a	51.7 ± 2.3 lmn	42 ± 2 abc	1
Jodhpur Sel.	45.7 ± 2.4 def	58.5 ± 2.8 defg	28.0 ± 3.1 def	5.8 ± 1.0 a	8.3 ± 1.2 a	43.1 ± 2.4 ijkl	40 ± 3 abc	4
Jyoti	33.5 ± 2.9 abc	52.5 ± 3.1 bcde	56.7 ± 2.3 lm	6.4 ± 1.0 a	8.3 ± 1.2 a	29.8 ± 3.0 def	48 ± 2 abc	1
Kabuli yellow	30.2 ± 3.1 ab	39.5 ± 3.7 a	23.5 ± 3.2 cde	5.3 ± 1.2 a	7.5 ± 1.3 a	29.3 ± 3.1 def	42 ± 3 abc	1
Kasturi	33.3 ± 3.1 abc	44.6 ± 3.6 ab	25.3 ± 3.2 de	5.5 ± 1.2 a	8.8 ± 1.3 a	37.5 ± 2.8 fghij	38 ± 3 abc	4
KRS	45.6 ± 2.8 def	52.8 ± 3.4 bcde	15.8 ± 4.3 bc	5.6 ± 1.2 a	9.0 ± 1.3 a	60.7 ± 2.3 n	46 ± 3 abc	1
Mridula	44.8 ± 2.9 def	58.3 ± 3.4 defg	23.5 ± 3.6 cde	6.2 ± 1.2 a	10.2 ± 1.3 a	40.0 ± 2.4 gij	40 ± 3 abc	4
P-23	42.8 ± 3.0 def	62.5 ± 3.4 fghi	46.2 ± 2.7 jk	6.9 ± 1.2 a	9.8 ± 1.3 a	42.0 ± 2.3 ijk	47 ± 3 abc	1
P-26	45.5 ± 3.1 def	62.4 ± 3.5 fghi	37.1 ± 3.0 ghi	8.8 ± 1.1 a	11.2 ± 1.3 a	27.3 ± 2.8 cde	38 ± 3 abc	4
PhuleArakta	38.5 ± 3.5 bcd	56.0 ± 3.8 cdef	31.3 ± 3.3 efgh	7.0 ± 1.2 a	10.5 ± 1.3 a	31.5 ± 2.7 defg	45 ± 3 abc	1
Ruby	42.6 ± 3.5 def	55.2 ± 4.0 cdef	29.6 ± 3.6 efg	5.9 ± 1.4 a	8.8 ± 1.4 a	49.1 ± 2.2 klm	56 ± 3 a	3
Tabesta	46.3 ± 3.6 def	55.6 ± 4.2 cdef	20.3 ± 4.7 cd	7.8 ± 1.3 a	8.9 ± 1.4 a	14.1 ± 2.2 ab	42 ± 3 abc	1
Wonderful	42.7 ± 3.8 def	68.2 ± 3.9 hi	59.7 ± 3.0 m	6.5 ± 1.3 a	8.3 ± 1.4 a	27.6 ± 1.8 cde	44 ± 3 abc	1
Yercaud-1	55.2 ± 3.4 gh	58.4 ± 3.5 defg	5.8 ± 2.3 a	7.0 ± 1.2 a	7.9 ± 1.3 a	12.9 ± 1.5 ab	38 ± 3 abc	4

Within each column, different letters indicate significant differences between genotypes evaluated with (*) ANOVA followed by the Tukey HSD test ($p = 0.05$) or (**) generalized linear model followed by the least-squares means.

The hierarchical clustering suggests that the tested genotypes could be grouped in terms of susceptibility to the nematodes. Using the Ward linkage procedure, we built a dendrogram and classified these genotypes into five possible groups (Figure 4). The groups no. 1, 4, and 3 showed the highest number of root galls per gram of root, ranging from 38.5 to 56.5. In group 1, this parameter ranged from 42.3 to 48.3 and included 11 genotypes. In group 4, it ranged from 38.5 to 40.0 and included seven genotypes, whereas in group 3, it ranged from 51.5 to 56.5 and included three genotypes. The group no. 2 had a lower number of root galls per gram of root, ranging from 31.0 to 35.2, and included five genotypes. The group no. 5 showed a tolerant reaction against the nematode infestation, having the lowest number of galls per gram of root (18.5). Genotypes of group No. 2 and especially those of the group no. 5 showed resistance/tolerance reaction against root knot nematode; hence, they should be additionally studied to explore their suitability as a resistance source for the development of new resistant cultivar or as rootstocks for the effective long-term sustainable management of nematodes.

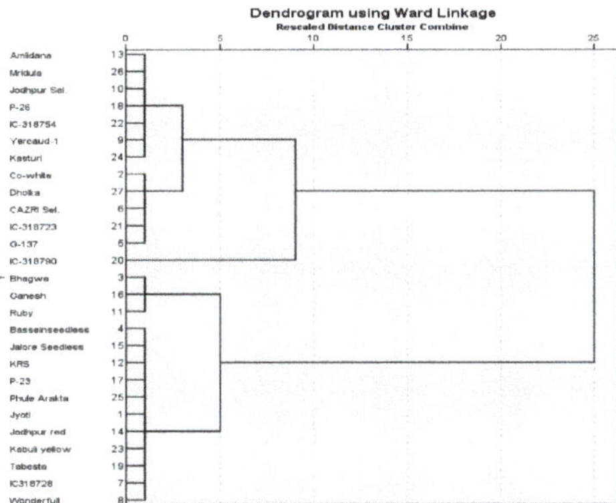

Figure 4. Dendrogram obtained with the hierarchical cluster analysis (Ward linkage method) using the data of root knot galls g^{-1} root. The numbers on the vertical axis represent a unique ID assigned to the genotypes.

3.4. Integrated Management of Root Knot Nematode

Table 5 clearly shows that all the treatments were significantly superior to untreated control in reducing the incidence of nematodes in terms of number of root galls g^{-1} root. However, the soil application of Carbofuran 20 g + Fluensulfone 20 g plant^{-1} was found to be the most effective compared to the other treatments in reducing the number of root galls (49.7%). This was closely followed by the treatment with Neemcake 500 g + *P. lilanicus* at 25 mL + Carbofuran at 20 g plant^{-1} + Fluensulfone at 20 g plant^{-1} (48.8%) and Neemcake 500 g + *P. lilanicus* at 25 mL + Fluensulfone 40 g plant^{-1} (46.4%). It was interesting to note that the soil application of Carbofuran at 40 g plant^{-1} or Fluensulfone 40 g plant^{-1} alone or in combination was not as much effective as its half dose applied in combination, i.e., Carbofuran at 20 g + Fluensulfone 20 g plant^{-1}. On the other hand, nematode population densities continued to increase (38.2%) in the rhizosphere of pomegranate plants in the absence of nematode management combinations (78.5 to 108.5 galls g^{-1} root). Different management treatments did not affect significantly plant growth in terms of plant height and canopy spread, but it was significantly superior to untreated control. Since pruning is a recommended practice in pomegranate cultivation, no significant variation in vegetative growth amongst different treatments was expected as all the plants were pruned uniformly. Selected trees in study were unable to produce any fruit at the time of initiation of experiments, but at the termination of the experiment, treated trees bore few fruits.

Since the fruit yield was negligible, data were not presented and discussed here. Nematicides reduced the populations of various *Meloidogyne* spp. in the soil [10], but once the symptoms have developed, they are incapable of completely eliminating those *Meloidogyne* species already in plant tissues [24]. Integrated approaches involving nematicides, soil amendments with organic sources, and different bioagents were proven to be effective for the control of root knot nematode in pomegranate and other fruit crops [22,23,25,26]. Similar to our study, the use of bioagents, viz., *Pacilomyces lilacinus* and *Trichoderma* with or without neem cakes was also reported to be effective by various researchers in different fruit crops [9,14,25,27]. Secondary metabolites produced by organic amendments were shown to cause a significant reduction of the reproduction of *M. incognita* on tomato in a field experiment [28]. Mechanisms involved in nematode control by endophytes such as *Paecilomyces* and *Trichoderma* with higher rate of organic materials may also trap and kill root knot nematode in the soil or root systems [25,26,29,30]. The other widely recognized mechanisms of bio-agents along with soil amendments include the production of toxins, enzymes, and other metabolic products, as well as the promotion of plant growth and induction of systemic resistance of host plants to pathogens [31].

Table 5. Effect of different treatments on number of root knot galls per gram of root, plant height, canopy spread, and fruit yield (means ± standard errors) of pomegranate trees.

Treatments	Nematode Galls g^{-1} Root *			Plant Height (cm) **			Canopy Spread (cm^2) **			Fruit Yield (kg tree^{-1}) **		
	Initial	Final	Reduction (%)	Initial	Final	Variation (%)	Initial	Final	Variation (%)	Initial	Final	Variation (%)
T1	74 ± 5 ab	52 ± 4 bcd	29.9 ± 1.5 c	200.0 ± 2.9 a	236.7 ± 3.7 cd	18.3 ± 3.1 bcd	172.5 ± 1.5 a	186.7 ± 4.3 a	8.23 ± 6.1 ab	4.8 ± 0.2 ab	5.2 ± 0.3 ab	8.3 ± 7.0 cd
T2	43 ± 4 d	30 ± 3 e	31 ± 0.9 cd	215.0 ± 1.7 bc	233.3 ± 2.7 bc	8.5 ± 5.2 ab	180.0 ± 2.6 b	197.8 ± 5.2 abc	9.8 ± 1.9 ab	5.3 ± 0.3 bcd	5.8 ± 0.2 b	9.4 ± 5.0 e
T3	91 ± 5 a	58 ± 4 bcd	36.1 ± 1.8 d	240.0 ± 4.6 e	243.3 ± 3.6 de	1.4 ± 3.1 a	188.9 ± 2.2 cd	195.0 ± 6.1 ab	3.2 ± 2.6 ab	5.4 ± 0.2 bd	5.9 ± 0.3 b	9.2 ± 4.0 de
T4	84 ± 5 ab	42 ± 4 cde	49.7 ± 2.1 e	210.0 ± 4.0 b	250.0 ± 5.5 efg	19.0 ± 2.3 cd	210.0 ± 1.9 f	221.7 ± 1.9 e	5.6 ± 2.0 ab	4.6 ± 0.2 a	5.2 ± 0.3 ab	13 ± 0.4 f
T5	79 ± 5 ab	52 ± 4 bcd	33.9 ± 4.9 cd	220.0 ± 2.7 cd	246.7 ± 3.4 ef	12.1 ± 1.7 bc	215.0 ± 2.0 g	228.9 ± 2.6 e	6.5 ± 1.8 ab	6.2 ± 0.2 e	7.0 ± 0.6 cde	12.9 ± 0.6 f
T6	73 ± 4 ab	62 ± 4 bc	15.2 ± 1.2 b	220.0 ± 1.7 cd	256.7 ± 3.1 g	16.7 ± 2.9 bcd	182.3 ± 1.8 b	186.7 ± 2.5 a	2.5 ± 1.0 ab	6.8 ± 0.1 fg	7.3 ± 0.3 e	7.4 ± 5.0 c
T7	76 ± 5 ab	41 ± 3 de	46.4 ± 2.3 e	210.0 ± 1.2 b	260.0 ± 5.2 g	23.8 ± 1.7 d	193.5 ± 1.0 de	219.4 ± 2.9 de	13.8 ± 2.2 b	5.5 ± 0.3 d	6.2 ± 0.2 bcd	12.7 ± 7.0 f
T8	66 ± 4 bc	47 ± 4 bcd	28.9 ± 2.7 c	225.0 ± 1.3 d	258.3 ± 3.1 g	14.8 ± 2.3 bcd	237.5 ± 0.9 h	245.0 ± 3.5 g	3.2 ± 1.8 ab	5.5 ± 0.2 d	5.8 ± 0.2 b	5.4 ± 7.0 b
T9	94 ± 5 a	64 ± b	32 ± 2.6 cd	210.0 ± 1.7 b	236.7 ± 2.3 cd	12.7 ± 2.7 bc	232.5 ± 1.3 h	242.2 ± 2.7 fg	4.2 ± 1.0 ab	5.3 ± 0.2 bcd	6.0 ± 0.4 bc	13.2 ± 6.0 fg
T10	51 ± 4 cd	46 ± 4 bcde	10.5 ± 1.4 b	200.0 ± 2.9 a	226.7 ± 3.0 b	13.4 ± 3.6 bc	195.2 ± 1.5 e	209.4 ± 5.2 cd	7.2 ± 0.9 ab	7.2 ± 0.1 g	7.8 ± 0.4 e	8.4 ± 6.0 cde
T11	79 ± 5 ab	40 ± de	48.8 ± 2.2 e	215.0 ± 1.7 bc	253.3 ± 1.2 fg	17.8 ± 5.5 bcd	187.5 ± 2.9 c	206.7 ± 3.9 bc	10.3 ± 1.3 ab	6.3 ± 0.2 ef	7.2 ± 0.4 de	14.2 ± 0.4 g
T12	71 ± 5 abc	47 ± 4 bcd	33.7 ± 2.0 cd	210 ± 2.3 b	236.7 ± 1.4 cd	12.7 ± 1.7 bc	215.0 ± 2.0 g	231.1 ± 2.7 ef	7.5 ± 2.7 ab	4.8 ± 0.1 abc	5.4 ± 0.2 b	12.5 ± 0.4 f
T13	78 ± 5 ab	108 ± 6 a	(+)38.2 ± 1.6 a	215 ± 1.3 bc	196.7 ± 2.5 a	(−)8.5 ± 4.6 a	237.5 ± 1.3 h	203.9 ± 3.1 bc	(−)16.5 ± 5.2 a	5.4 ± 0.3 bd	4.2 ± 0.6 a	(−)22.3 ± 0.4 a

Within each column, different letters indicate significant differences between genotypes evaluated with (*) generalized linear model followed by the least-squares means or (**) ANOVA followed by the Tukey HSD test (p = 0.05).

4. Conclusions

It is concluded from the present study that the majority of pomegranate orchards in the arid region of western Rajasthan are more or less infected by nematodes. Root knot nematode (*Meloidogyne incognita*) is the single dominant species found in all the pomegranate orchards surveyed during the study. Among the evaluated genotypes and varieties, all of them were found to be susceptible to root knot nematode, but the severity level of its infestation was variable; hence, a more detailed screening is needed on a larger population. Among the tested management approaches, there is no doubt that a combination of nematicides such as Carbofuran and Fluensulfan with their half dose was effective, but integrated approaches involving nematicides, bioagents, and organic amendments are more effective and sustainable, especially when nematode species are already in root tissue.

Author Contributions: Conceptualization, A.S. and R.K.K.; writing—original draft preparation, A.S.; writing—review and editing, A.S., P.S.K., P.K., Y.R., B.B., and K.S.J.; supervision, R.K.K.; project administration, A.S. All authors have read and agreed to the published version of the manuscript.

Funding: This research received no external funding.

Institutional Review Board Statement: Not applicable.

Informed Consent Statement: Not applicable.

Data Availability Statement: The datasets generated for this study are available on request to the principal/corresponding author.

Acknowledgments: The authors are grateful to the Director, ICAR-Central Arid Zone Research Institute, Jodhpur, for all necessary help required for conducting the research. The authors are also thankful to Vandita Kumari, Scientist (Agricultural Statistics), ICAR-Central Arid Zone Research Institute (CAZRI), Jodhpur, for statistical inferences.

Conflicts of Interest: The authors declare no conflict of interest.

References

1. Chandra, R.; Jadhav, V.T.; Sharma, J. Global scenario of pomegranate (*Punica granatum* L.) culture with special reference to India. *Fruits Veg. Cereals Sci. Biotechnol.* **2010**, *4*, 7–18.
2. Sarkhosh, A.; Yavari, A.M.; Zamani, Z. *The Pomegranate Botany, Production and Uses*; CAB International: Wallingford, UK, 2021; 559p.
3. Anonymous. Exports from India Pomegranate Fresh. 2020. Available online: https://agriexchange.apeda.gov.in (accessed on 5 November 2021).
4. Singh, A.; Meghwal, P.R.; Kumar, P. Quality pomegranate from thar desrt. *Indian Hortic.* **2019**, *1*, 29–32.
5. Chandra, R.; Suroshe, S.; Sharma, J.; Marathe, R.A.; Meshram, T.D. *Pomegranate Growing Manual*; Technical Bulletin; National Research Centre on Pomegranate: Solapur Maharastra, India, 2011; p. 58.
6. Singh, T.; Prajapati, A.; Maru, A.K.; Chaudhary, R.; Pate, D.J. Root-Knot Nematodes (*Meloidogyne* spp.) Infecting Pomegranate: A Review. *Agric. Rev.* **2019**, *40*, 309–313. [CrossRef]
7. Dasgupta, D.R.; Guar, H.S. The root-knot nematodes, *Meloidogyne* spp. in India. In *Plant Parasitic Nematodes of India: Problem and Progress*; Swarup, G., Dasgupta, D.R., Eds.; ICAR: New Delhi, India, 1986; pp. 139–178.
8. Sikora, R.; Coyne, D.; Hallmann, J.; Timper, P. *Plant Parasitic Nematodes in Subtropical and Tropical Agriculture*; CABI Publishing: Wallingford, UK, 2018.
9. Holland, D.; Hatib, K.; Bar-YaAkov, I. Pomegranate: Botany, Horticulture, Breeding. In *Horticultural Reviews*; Janick, J., Ed.; John Wiley & Sons, Inc.: Hoboken, NJ, USA, 2009; Volume 35, pp. 127–191.
10. Khan, M.R.; Jain, R.K.; Ghule, T.M.; Pal, S. Root knot Nematodes in India-a comprehensive monograph. In *All India Coordinated Research Project on Plant Parasitic Nematodes with Integrated Approach for Their Control*; Indian Agricultural Research Institute: New Delhi, India, 2014; 78p.
11. Darekar, K.S.; Shelka, K.S.; Mhare, N.L. Nematodes associated with fruit crops in Maharashtra state India. *Int. Nematol. Netw. Newsl.* **1990**, *7*, 11–12.
12. Hajihassani, A.; Davis, R.F.; Timper, P. Evaluation of selected nonfumigant nematicides on increasing inoculation densities of *Meloidogyne incognita* on cucumber. *Plant Dis.* **2019**, *103*, 3161–3165. [CrossRef] [PubMed]
13. Ahmadi, A.R.; Ansari-pour, B.; Almasi, H.; Hatami, H. *The Study on Reaction of Pomegranate Cultivars to Root-Knot Nematodes in Dastgert Station*; AGRIS, FAO: Rome, Italy, 2000; 15p.

14. Somasekhar, Y.M.; Ravichandra, N.G.; Jain, R.K. Bio-management of root knot (*Meloidogyne incognita*) infecting Pomegranate (*Punica granatum* L.) with combination of organic amendments. *Res. Crops* **2012**, *13*, 647–651.
15. Ahire, D.K.B. Screening of Different Rootstocks of Pomegranate (*Punica granatum* L.) against Root Knot Nematode. *Int. J. Curr. Microbiol. Appl. Sci.* **2021**, *10*, 288–294. [CrossRef]
16. Hartman, K.M.; Sasser, J.N. Identification of Meloidogyne species on the basis of differential host test and perineal pattern morphology. In *An Advanced Treatise on Meloidogyne*; Barker, K.R., Carter, C.C., Sasser, J.N., Eds.; North Carolina State University Graphics: Raleigh, NC, USA, 1985; Volume 2, pp. 69–77.
17. Jepson, S.B. *Identification of Root-Knot Nematodes (Meloidogyne Species)*; CAB International: Wallingford, UK, 1887.
18. Ranganna, S. *Handbook of Analysis and Quality Control for Fruit and Vegetable Products*, 2nd ed.; Tata McGraw-Hill Education Publication: New Delhi, India, 2017.
19. Cobb, N.A. *Estimating the Nematode Population of Soil*; United States, Department of Agriculture, Agricultural Technical Circular: Washington, DC, USA, 1918; 48p.
20. Van Bezooijen, J. *Methods and Techniques for Nematology (A Manual)*; Wageningen University: Wageningen, The Netherlands, 2006.
21. Chormule, A.J.; Mhase, N.L.; Gurve, S.S. Management of root- knot nematode, *Meloidogyne incognita* infesting grape under field conditions. *Ann. Plant Protec. Sci.* **2017**, *25*, 181–185.
22. Dawabah, A.A.; Al-Yahya, F.A.; Lafi, H.A. Integrated management of plant-parasitic nematodes on guava and fig trees under tropical field conditions. *Egypt. J. Biol. Pest Control.* **2019**, *29*, 2–9. [CrossRef]
23. Shelke, S.S.; Darekar, K.S. Reaction of pomegranate germplasm to root-knot nematode. *J. Maharashtra Agric. Univ.* **2000**, *25*, 308–310.
24. Sirias, H.C.I. Root-knot Nematodes and Coffee in Nicaragua: Management Systems, Species Identification and Genetic Diversity. PhD Thesis, Swedish University of Agricultural Sciences, Uppsala, Sweden, 2011; 57p.
25. Jagdev, G.H.; Mhase, N.L. Management of root- knot nematode, *Meloidogyne incognita* infesting fig under field conditions. *J. Entomol. Zool. Studies* **2019**, *7*, 193–197. [CrossRef]
26. Putten, W.H.V.D.; Cook, R.; Costa, S. Nematode interactions in nature: Models for sustainable control of nematode pests of crop plants. *Adv. Agron.* **2006**, *89*, 227–260.
27. Pawar, S.A.; Mhase, N.L.; Kadam, D.B.; Chandele, A.G. Management of Root-Knot Nematode, *Meloidogyne incognita*, Race-II infesting Pomegranate by using bioinoculants. *Indian J. Nemat.* **2013**, *43*, 92–94.
28. Radwan, M.A.; Farrag, S.A.A.; Abu-Elamayem, M.M.; Ahmed, N.S. Biological control of the root-knot nematode, *Meloidogyne incognita* on tomato using bioproducts of microbial origin. *Appl. Soil Ecol.* **2012**, *56*, 58–62. [CrossRef]
29. Yao, Y.R.; Tian, X.L.; Shen, B.M.; Mao, Z.C.; Chen, G.; Xie, B.Y. Transformation of the endophytic fungus *Acremonium implicatum* with GFP and evaluation of its biocontrol effect against *Meloidogyne incognita*. *World J. Microbiol. Biotechnol.* **2015**, *31*, 549–556. [CrossRef] [PubMed]
30. Schouten, A. Mechanisms involved in nematode control by endophytic fungi. *Annu. Rev. Phytopathol.* **2016**, *54*, 121–142. [CrossRef] [PubMed]
31. Nega, A. Review on concepts in biological control of plant pathogens. *J. Biol. Agric. Healthc.* **2014**, *4*, 33–54.

horticulturae

Article

Effect of Saline-Nutrient Solution on Yield, Quality, and Shelf-Life of Sea Fennel (*Crithmum maritimum* L.) Plants

Fabio Amoruso [1,2], Angelo Signore [2], Perla A. Gómez [3], María del Carmen Martínez-Ballesta [1,3], Almudena Giménez [1], José A. Franco [1,3], Juan A. Fernández [1,3] and Catalina Egea-Gilabert [1,3,*]

[1] Department of Agronomical Engineering, Universidad Politécnica de Cartagena, 30203 Cartagena, Murcia, Spain; f.amoruso17@uniba.it (F.A.); mcarmen.ballesta@upct.es (M.d.C.M.-B.); almudena.gimenez@upct.es (A.G.); josea.franco@upct.es (J.A.F.); juan.fernandez@upct.es (J.A.F.)
[2] Department of Agricultural and Environmental Science, University of Bari Aldo Moro, 70126 Bari, Italy; angelo.signore@uniba.it
[3] Institute of Plant Biotechnology, Universidad Politécnica de Cartagena, 30202 Cartagena, Murcia, Spain; perla.gomez@upct.es
*   Correspondence: catalina.egea@upct.es

Abstract: In this study, the effect of salinity (150 mM NaCl) compared to a control (9 mM NaCl) on growth, quality and shelf-life of fresh-cut sea fennel was evaluated. For that, sea fennel plants were cultivated in a hydroponic floating system and the sea fennel leaves were stored for 12 days at 5 °C. At harvest, leaves from plants grown in salinity had a lower content of NO_3^-, K^+ and Ca^{2+} and an increased Cl^- and Na^+ concentration when compared to the control. There was a positive effect in the aerial part with increased fresh weight due to salt stress, but a reduction in the root biomass. During storage, weight loss and colour changes were not significant while leaves' firmness was higher for control and increased during storage, probably due to lignification. Microbial growth (psychrophiles, yeast and moulds and enterobacteria) was higher at harvest for control and increased during storage, with no differences between treatments after 12 days at 5 °C. Sensory quality was similar for both treatments but leaves from NaCl treatment had a salty taste that was easily detected by panelists. These results show that saline-nutrient solution applied in hydroponics is a suitable system for sea fennel growth. It gives a slightly salty but high-quality product, acceptable as a "ready-to-eat" vegetable.

Keywords: salinity; microbial growth; sensory quality; floating system; ready-to-eat

Citation: Amoruso, F.; Signore, A.; Gómez, P.A.; Martínez-Ballesta, M.d.C.; Giménez, A.; Franco, J.A.; Fernández, J.A.; Egea-Gilabert, C. Effect of Saline-Nutrient Solution on Yield, Quality, and Shelf-Life of Sea Fennel (*Crithmum maritimum* L.) Plants. *Horticulturae* **2022**, *8*, 127. https://doi.org/10.3390/horticulturae8020127

Academic Editors: Rosario Paolo Mauro, Carlo Nicoletto and Leo Sabatino

Received: 30 December 2021
Accepted: 27 January 2022
Published: 30 January 2022

Publisher's Note: MDPI stays neutral with regard to jurisdictional claims in published maps and institutional affiliations.

1. Introduction

Sea fennel (*Crithmum maritimum* L.), also known as crest marine, marine fennel, samphire, and rock samphire [1], is a halophyte species, the sole one of the *Crithmum* genus [2,3], which belongs to the *Apiaceae* family.

This species is widespread in the Mediterranean coasts as well in the Canary Islands [2] and along the Atlantic coast of Portugal, England, Wales and Southern Ireland [4].

Being a perennial halophyte species, it is able to grow in sand hills or on rocky cliffs and is remarkably productive under saline conditions to exploit seawater, coastal lands, and other marginal areas otherwise useless [5], without requiring huge allocation and depletion of freshwater resources [6].

Its distinguishing sensory attributes in terms of taste, odour and colour has historically always found applications in culinary Mediterranean tradition and the food industry [1,7], and in some countries (e.g., Italy) its use is so long and rooted in time that such a product is included in the "List of Traditional Agri-Food Products" of the Italian Department for Agriculture [1]. Sea fennel importance is not limited only to the culinary uses (mainly as an appetizer), but also as carminative, diuretic or for treating obesity [8]. In addition, it is rich in several biologically active compounds (ascorbic acid, iodine, carotenoids, flavonoids, organic acids, phenolics, etc.) [9], exerting beneficial effects against oxidative or mutagenic mechanisms, and pathogenic bacteria [10], which is important for their healthy properties [11,12].

Apart from its use as a fresh product, Renna et al. [1] proposed sea fennel to be used in dried form with different techniques of drying, as this could lead to an "industrial production on a large scale and also to diversify local food through a micro-scale production".

Similar to other halophyte species, sea fennel has developed mechanisms to tolerate high salinity levels, particularly by accumulating Na^+ and Cl^- into the vacuoles [13]. Furthermore, Jiménez-Becker et al. [14] observed that sea fennel has the capacity, compared to other halophytes, to reduce the uptake of Cl^-, which results in a lower concentration within the leaves and to an increase in the concentration of soluble sugars and proline, in particular at high salinity levels (300 mM of NaCl).

Yet, even if products of halophytes species are produced more and more and sold in the markets worldwide, sea fennel may be still considered as a wild edible plant, since it has not undergone a structured programme for its genetic improvement and cultivation, even if it could be easily domesticated and engineered to exploit its beneficial elements content [15,16]. Recent knowledge suggests that sea fennel shows good potential as an emerging crop, despite studies on its cultivation techniques being limited [17]. A floating system seems to be particularly appropriate for baby leaf vegetable production since it allows precise control of plant nutrition and the maximisation of yield and quality of the product. Thus, Giménez et al. [18] demonstrated that the above system is a suitable method for growing *C. maritimum*. It is well known that cultivation conditions influence the quality of the raw material and therefore can modify its physiological behaviour and suitability for fresh-cut processing [19]. We hypothesise that any preharvest condition that stresses a plant, such as the salinisation of the nutrient solution, could affect the quality and shelf-life of the sea fennel, particularly increasing the phytochemicals of the plant.

In our vision, sea fennel has the potential for more extensive cultivation and for the ready-to-eat market as a baby-leaf vegetable, due not only to its organoleptic characteristics but also to its richness in terms of health-promoting compounds and its suitability for cultivation in saline conditions, an important aspect for the Mediterranean environment. This aspect would be crucial since soil salinity is currently the most important environmental stress limiting crop production in arid and semi-arid areas [20], and, in the near future this trend is expected to worsen [21]. For this purpose, we evaluated the effect of the salinity level of the nutrient solution in a floating system on the growth, quality, and shelf-life of *C. maritimum* during a storage period.

2. Materials and Methods

2.1. Plant Material and Growing Conditions

The experiment was conducted in an unheated greenhouse covered with thermal polyethylene located at the Experimental Agro-Food Station, Technical University of Cartagena (UPCT; lat. 3741′ N; long. 057′ W), using seeds provided by Semillas Cantueso, obtained in Dunas de Artola, (Málaga). Sowing was carried out manually into "styrofloat" trays of 60 cm × 41 cm containing peat. The trays were placed in a growth chamber at 20 °C for 5 days and then transferred to flotation beds, floating on fresh tap water with an electrical conductivity (EC) of 1.1 dS m^{-1} and a pH of 7.8. Aeration was provided using a blow pump connected to a pipe trellis positioned at the bottom of each flotation bed. Each level of treatment was carried out in 135 cm × 125 cm × 20 cm beds located at three places inside a greenhouse for all the experiments. A week after transferring to the floating beds, the plants were thinned, leaving a plant density around 400 plants m^{-2}, and the nutrient solution was added to the water and adjusted to EC 2.7 dS m^{-1} and the pH to 5.8 [22]. After 30 days, NaCl was added to the nutrient solution to half of the plants to reach a concentration of 150 mM, while the other half was set as control treatment (9 mM of NaCl). The EC and temperature of the nutrient solution were monitored during the growing cycle using sensors (CS547 Campbell Scientific Inc., Logan, UT, USA). Harvesting was carried out when plants had four–five pairs of leaves.

2.2. Analysis at Harvesting Time

Shoot fresh weight (FW), leaf area, specific leaf area (SLA) and root growth parameters were measured on 10 plants in each tray. Leaf area was measured using a leaf area meter (LICOR-3100 C; LICOR Biosciences Inc., Lincoln, NE, USA). Root length, area, and volume, and the number of branches were determined using a Winrhizo LA 1600 root counter (RegentInc., Quebec, QC, Canada) from pictures taken of each root system by a double-pass scanner incorporated in the counter. The dry weight (DW) of the shoot was determined by drying in an oven at 60 °C until constant weight.

At harvesting, the following biochemical parameters were measured in the sea fennel leaves: ions content, total phenolics and flavonoids content and total antioxidant capacity. The ions content was determined and quantified following the method described by Lara et al. [23] in the sea fennel leaves. Ions were extracted in triplicate per treatment. The extraction of 0.2 g of dry leaf samples of each treatment was carried out with 50 mL distilled water and continuous agitation in an orbital shaker (Stuart SSL1, Stone, UK) for 45 min at 110 rpm at 50 °C. Ion concentrations were determined by ion chromatography using a Metrosep A SUPP 5 column (Metrohm AG, Zofingen, Switzerland) at a flow rate of 0.7 mL min^{-1} for anions and a Metrosep C 2-250 column at a flow rate of 1.0 mL min^{-1} for cations, following the manufacturer's instructions.

The total phenolic content was determined using the Folin–Ciocalteu colorimetric method, modified by Everette et al. [24]. A 50 µL aliquot of the methanolic extract supernatant was mixed with 50 µL of Folin–Ciocalteu reagent and 750 µL of H_2O. The solution was incubated for 5 min and 150 µL of Na_2CO_3 was added. Then, it was incubated at room temperature for 2 h in darkness, after which the absorption at 765 nm was measured (HP 8453, Hewlett Packard). The measurement was expressed as mg gallic acid (GA) kg^{-1} FW. Each one of the three replicates was analysed in triplicate (instrumental replicate).

The total antioxidant capacity of the leaves was evaluated in terms of their ability to deactivate the DPPH radical according to Brand-Williams et al. [25], with the modifications described by Lopez-Marín et al. [26]. Briefly, a solution of 2,2- diphenyl-1 picryhydrazil (DPPH) in methanol was prepared. A 25 µL aliquot of the extract supernatant was mixed and 600 µL of DPPH stock solution added. The homogenate was shaken vigorously and kept in darkness for 15–20 min at room temperature. The absorbance at 517 nm was measured in a spectrophotometer (HP 8453, Hewlett Packard). The measurement was expressed as mg DPPH reduced kg^{-1} FW.

The total flavonoids content was evaluated according to Meda et al. [27]. The procedure consisted of mixing 50 µL of extract, 300 µL of methanol and 350 µL of a 2% $AlCl_3$ dilution in methanol. After a 15 min incubation in darkness at room temperature, the absorbance at 430 nm was measured. The measurement was expressed as mg rutin kg^{-1} FW.

2.3. Postharvest Product Handling and Analysis

Leaves free from defects were sanitised in a cold room (10 °C) by immersion in a solution containing 100 ppm NaClO and 0.2 g L^{-1} citric acid (2 min, 5 °C, pH 6.5). Then, they were rinsed with tap water (2 min, 5 °C) and finally excess of water was removed by a salad spinner (30 s). Twenty g of leaves were placed in polypropylene (PP) baskets (170 mm × 120 mm × 40 mm) and thermo-sealed on the top with a 25 µm thick film-oriented polypropylene (OPP). Three replicates for each irrigation treatment and storage time (processing day and after 6 and 12 days) were prepared and stored in darkness at 5 °C. Each sampling day, and before opening the baskets, atmosphere composition within the package was measured. For that, a 0.5 mL sample of the headspace was withdrawn with a gas-tight syringe and O_2 and CO_2 concentrations were determined by a gas chromatograph (7820A GC Agilent Technologies, Waldbroon, Germany). The gas chromatograph conditions were: oven at 80 °C, injector and detector at 250 °C, using H_2 and air as gas carriers at 35 mL min^{-1} and 350 mL min^{-1}, respectively. A stainless–steel column packed with PorapakQ (1/8″, 80/100 mesh size; Supelco Inc., Bellefonte, PA, USA) was used.

Microbial growth (mesophilic and psychrophilic aerobic bacteria, enterobacteria, and yeast and mould growth) was determined using standard enumeration methods. Samples of 1 g poured into a sterile stomacher bag (model 400 Bags 6141, London, UK) were homogenized with a 10 mL sterile peptone saline solution (pH 7; Scharlau Chemie SA, Barcelona, Spain) for 10 s in a masticator (Colwort Stomacher 400 Lab, Seward Medical, London, UK). For the enumeration of each microbial group, 10-fold dilution series were prepared in 9 mL of sterile peptone saline solution. Mesophilic, enterobacteria, and psychrotrophic were pour plated, and yeast and mould were spread plated. Media (Scharlau Chemie, Barcelona, Spain) and incubation conditions were as follows: plate count modified agar (PCA) for mesophilic and psychrotrophic aerobic bacteria (30 °C, 48 h and 5 °C for 7 days, respectively); violet red bile dextrose agar for enterobacteria (37 °C, 48 h); and rose Bengal agar for yeasts and moulds (3–5 days, 22 °C). All microbial counts were reported as log colony forming units per gram of product (log CFU g^{-1}). Each of the three replicates was analyzed by duplicate. The presence of *Listeria monocytogenes* was monitored according to the Regulation EC 1441/2007.

Weight loss was calculated as the difference between the initial weight of the samples at the beginning of storage and their final weight after 6 and 12 days. To normalize data, weight loss values were expressed as percentage of the initial value.

Firmness was measured at 22 °C using a texturometer (Brookfield, Canada). A compression test was carried out with a blade (1 mm width) at a force of 90 g and a speed of 10.0 mm s^{-1} to reach a leave deformation of 0.5 mm. Results were expressed in g.

Leaf colour was determined on three points of each replicate using a colorimeter (Minolta CR-400 Series, Ramsey, NJ, USA). Tristimulus parameters (L*, a*, b*) of the CIE Lab system were used to calculate the Hue angle = arctan (b*/a*) and chroma (C*) = $[(a*) 2 + (b*)^2]^{1/2}$.

2.4. Sensory Quality Panel

Sensory quality was analysed according to international standards (ASTM 1986) in a standardised room (UNE-EN ISO 8589 2007) equipped with ten testing boxes. Samples coded with three random digit numbers were served at room temperature. Still mineral water was used as palate cleanser. Evaluation was performed by 10 trained judges on day 0 and after 6 and 12 days of storage at 5 °C.

A 5-point scale was scored for colour, texture (crispness), flavour, aroma and global acceptance (5: excellent, 4: good, 3: fair, limit of usability, 2: poor; 1: extremely bad) and for defects as off-odours and mechanical damage (5: none; 4: slight; 3: moderate, limit of usability; 2: severe; 1: extreme) [28].

2.5. Statistical Analysis

A randomised complete block design with three replicates (beds) per both treatments, control and salinity, was used in the greenhouse. Each bed had three floating trays of 60 cm × 41 cm. Data were analysed using Statgraphics Plus. Analysis of variance (two-way ANOVA) was performed in which levels of salinity (9 and 150 mM), and storage time (0, 6 and 12 d) were included. When interactions were significant, they were included in the ANOVA, a least significant difference test was performed to compare level of salinity, and storage time. When the variables were measured at harvesting time, only salinity factor was included.

3. Results

3.1. Growth, Yield, and Quality Characteristic of C. maritimum at Harvesting

The salinity treatment did not affect the shoot FW and root parameters (Table 1). However, NaCl treatment reduced the leaf area and specific leaf area of *C. maritimum* plants, which indicates that the leaves were thicker when plants were grown with 150 mM NaCl.

Table 1. Influence of salinity treatment (control and 150 mM NaCl) on fresh weight, leaf area, specific leaf area (SLA), total root length, area root, diameter root and volume root of *C. maritimum* at harvest.

Treatments	Shoot Fresh Weight (g plant^{-1})	Shoot Dry Weight (g plant^{-1})	Leaf Area (cm^2 plant^{-1})	SLA (m^2 kg^{-1})	Total Root Length (cm)	Root Diameter (mm)	Root Volume (cm^3)
Control	2.10 ± 0.47 a	0.205 ± 0.005 a	3.43 ± 0.21 b	0.19 ± 0.012 b	112.24 ± 6.31 a	0.37 ± 0.01 a	0.59 ± 0.07 a
150 mM NaCl	2.45 ± 0.42 a	0.235 ± 0.005 a	2.37 ± 0.09 a	0.11 ± 0.005 a	109.92 ± 5.84 a	0.33 ± 0.02 a	0.46 ± 0.05 a

Values are the mean $\pm$ SE ($n = 6$). Values in the same column with different letters for each anion differ significantly according to LSD test ($p < 0.05$).

Some differences were observed regarding the contents of anions and cations in the leaf of *C. maritimum* at harvesting (Tables 2 and 3). With regard to anions, the content of nitrate was reduced by 17% in salinity conditions, while the content of chloride was increased by 3.7-fold. Furthermore, the addition of NaCl significantly reduced the content of bromide and sulphate, while the content of phosphate and oxalate was not affected by the salinity. Regarding cations, the content of sodium was found to increase by ca. 500%, whilst potassium, calcium and magnesium ions accumulated to a minor extent when sea fennel was grown in salinity conditions. These results agree with the hypothesis that sea fennel requires salt to grow, and it can tolerate high concentrations of salt. Finally, Cl$^-$ content was found to be systematically higher than Na$^+$, an imbalance that clearly indicates the existence in sea fennel plants of a regulatory mechanism to retain Na$^+$ far away from leaves since it could be a toxic element for the photosynthetic system.

Table 2. The content of anions (NO$_3^-$, Cl$^-$, Br$^-$, PO$_4^{3-}$, SO$_4^{2-}$, C$_2$O$_4^{2-}$) (mg kg^{-1} FW) in the leaf of *C. maritimum* under the different treatments (control and 150 mM NaCl) at harvesting.

Treatments	NO$_3^-$	Cl$^-$	Br$^-$	PO$_4^{3-}$	SO$_4^{2-}$	C$_2$O$_4^{2-}$
Control	1530.31 ± 586.47 b	1810.97 ± 120.56 a	152.00 ± 2.82 b	947.01 ± 94.14 a	1699.55 ± 32.28 b	88.19 ± 33.59 a
150 mM NaCl	1263.03 ± 19.79 a	6718.18 ± 1029.31 b	125.98 ± 3.46 a	1154.39 ± 142.73 a	671.25 ± 47.66 a	88.70 ± 28.87 a

Values are the mean $\pm$ SE ($n = 6$). Values in the same column with different letters for each anion differ significantly according to LSD test ($p < 0.05$).

Table 3. The content of cations (Na$^+$, K$^+$, Ca^{2+}, Mg^{2+}) (mg kg^{-1} FW) in the leaf of *C. maritimum* under the different treatments (control and 150 mM NaCl) at harvesting.

Treatments	Na$^+$	K$^+$	Ca^{2+}	Mg^{2+}
Control	777.57 ± 54.46 a	3642.49 ± 109.48 b	1108.29 ± 20.91 b	379.10 ± 12.46 b
150 mM NaCl	4639.45 ± 703.02 b	1070.45 ± 345.86 a	532.05 ± 143.54 a	203.42 ± 10.47 a

Values are the mean $\pm$ SE ($n = 6$). Values in the same column with different letters for each cation differ significantly according to LSD test ($p < 0.05$).

Salinity significantly reduced the phenolics content in sea fennel leaves at harvesting by 6% but increased the flavonoids content by 10% (Table 4) However, no significant differences were found between treatments with respect to antioxidant capacity.

Table 4. Total phenolics content, total flavonoids, and total antioxidant capacity in leaves of *C. maritimum* under the different treatments (control and 150 mM NaCl).

Treatments	Total Phenolics (mg GA kg^{-1} FW)	Total Flavonoids (mg Rutin kg^{-1} FW)	Total Antioxidant Capacity (mg DPPH$_{reduced}$ kg^{-1} FW)
Control	887.43 ± 11.95 b	1966.89 ± 45.17 a	112.24 ± 6.31 a
150 mM NaCl	833.53 ± 9.42 a	2167.24 ± 22.09 b	109.92 ± 5.84 a

Values are the mean ± SE (n = 6). Values in the same column with different letters differ significantly according to LSD test ($p < 0.05$).

3.2. Postharvest Quality

Sea fennel leaves lost water moderately in both treatments during the storage time (data not shown). Particularly, after 12 days of storage, sea fennel leaves obtained from plants grown under salinity presented a higher weight loss (1.24%) than those grown in the control (0.32%).

Leaves' firmness (Table 5) was higher for control than for those grown under salty conditions and increased slightly during storage in both treatments, most likely related to lignification. The leaves of plants grown with NaCl treatment had less turgor due to higher dehydration, presenting lower firmness than the control.

Table 5. Effect of 150 mM NaCl addition in the nutrient solution on *C. maritimum* leaf firmness (mm) at 0, 6 and 12 days of storage at 5 °C.

	Sea Fennel Firmness (g)
Salinity Treatment (A)	
Control	596.27 ± 20.43 b [x]
150 mM NaCl	457.94 ± 28.27 a
Storage (B)	
0 days	494.48 ± 35.26 a [y]
6 days	515.05 ± 34.01 a
12 days	571.80 ± 28.37 a
Significant Differences	
A	***
B	ns
A × B	ns

Asterisks indicates significance at *** $p < 0.001$; ns: non-significant. Different letters in the same column indicate significant differences. Values are the mean ± SE ([x] n = 45, [y] n = 30).

A passive modified atmosphere was generated inside the packages, which was related to the respiration rate of the produce. After 6 days of storage at 5 °C, no differences in O_2 and CO_2 concentration in the atmosphere within the packages were observed between treatments (Figure 1). However, after 12 days of storage, CO_2 concentration was slightly higher and O_2 concentration was moderately lower within the baskets of sea fennel leaves grown with 150 mM NaCl. It could indicate a higher respiration rate for these leaves, probably induced by the pre-harvest stressing conditions of salinity. However, the trend seems to be that both treatments were close to reaching the steady-state and, probably, at that moment differences between treatments would be minimal.

As regards the colour of the sea fennel leaves at harvest, the plants treated with NaCl, presented a luminosity (L* parameter) about 6% higher than the control (Table 6). Due to that, hue values were lower for those leaves than for the control. Salinity slightly affected leaf colouration towards lighter colours. However, it was almost undetected by the sensory panel. The colour parameters did not change significantly over the 12 days of monitoring. Therefore, salinity did not adversely affect the colour, keeping marketability at values that resemble those of the control.

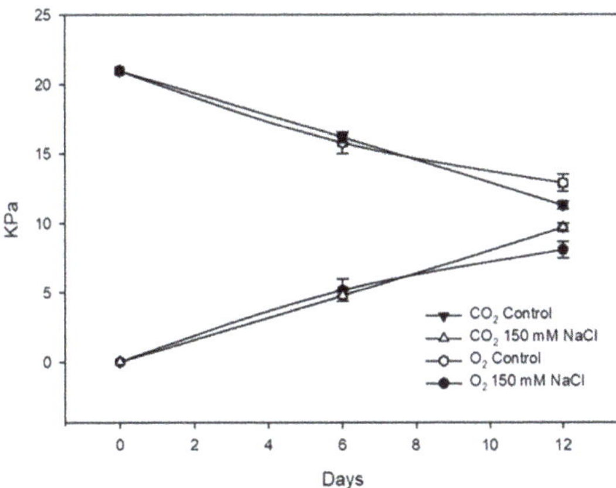

Figure 1. Effect of the salinity of the growing nutrient solution on the atmosphere composition within packages of fresh-cut sea fennel stored for 12 d at 5 °C. Values are the mean $\pm$ SE ($n = 3$).

Table 6. Effect of 150 mM NaCl addition in the nutrient solution on *C. maritimum* leaf colour parameters at 0, 6 and 12 days of storage at 5 °C.

	L	a	b	HUE	Chroma
Salinity Treatment (A)					
Control	37.20 ± 0.66 a ^{x}	-11.31 ± 0.39 a	18.32 ± 1.01 a	121.98 ± 1.08 b	22.26 ± 0.76 a
150 mM NaCl	39.14 ± 0.40 b	-10.94 ± 0.27 a	20.62 ± 0.55 a	118.01 ± 0.70 a	23.35 ± 0.54 a
Storage (B)					
0 days	38.51 ± 0.62 a ^{y}	11.29 ± 0.42 a	20.25 ± 1.12 a	119.43 ± 1.96 a	21.09 ± 0.94 a
6 days	37.95 ± 1.07 a	10.84 ± 0.36 a	18.98 ± 1.28 a	120.04 ± 1.14 a	23.58 ± 0.56 a
12 days	38.04 ± 0.66 a	11.26 ± 0.49 a	19.18 ± 0.94 a	120.51 ± 1.04 a	22.71 ± 0.98 a
Significant Differences					
A	*	ns	ns	*	ns
B	ns	ns	ns	ns	ns
A $\times$ B	ns	ns	ns	ns	ns

Asterisk indicates significance at * $p < 0.05$, ns: non-significant. Different letters in the same column indicate significant differences. Values are the mean $\pm$ SE (^{x} $n = 9$, ^{y} $n = 6$).

Microbial load (psychrophiles, yeast and moulds and enterobacteria) was higher at harvest for control leaves and increased during storage at 5 °C. *Listeria* was not detected in any treatment. The results in Table 7 show that there was a significant interaction between salinity treatments and storage for psychrophilic bacteria, enterobacteria and yeast and mould counts.

Psychrophilic bacteria counts were significantly higher in control leaves at harvest, but after 6 days of storage, there were no significant differences between treatments (Figure 2). A similar trend was found for Enterobacteria.

The sensory quality, even when decreasing, was acceptable for both treatments at the end of storage (Table 8). The most important changes were observed in texture and freshness, mainly related to a lower crispness associated with the water loss. The leaves obtained from salinity had a salty taste which was not observed in the control. However, this hint of salt was not unpleasant. The samples did not present strange smells in any case.

Table 7. Psychrophilic bacteria, mesophilic bacteria, enterobacteria and yeast and moulds counts (log CFU g^{-1}) of *C. maritimum* leaves after different salinity treatment (control and 150 mM NaCl) and storage at 5 °C for 0, 6 and 12 days.

	Psychrophilic Bacteria (log UFC g^{-1})	Mesophilic Bacteria (log UFC g^{-1})	Enterobacteria (log UFC g^{-1})	Yeast and Moulds (log UFC g^{-1})
Salinity Treatment (A)				
Control	5.81 ± 0.28 b ^{x}	5.40 ± 0.31 a	5.15 ± 0.38 b	3.89 ± 0.18 b
150 mM NaCl	5.25 ± 0.45 a	5.24 ± 0.36 a	3.90 ± 0.98 a	3.34 ± 0.12 a
Storage (B)				
0 days	4.23 ± 0.44 a ^{y}	4.05 ± 0.11 a	1.89 ± 0.83 a	3.09 ± 0.16 a
6 days	5.99 ± 0.08 b	5.61 ± 0.09 b	5.61 ± 0.13 b	3.82 ± 0.10 b
12 days	6.38 ± 0.01 b	6.30 ± 0.01 c	6.21 ± 0.07 c	3.84 ± 0.21 b
Significant Differences				
A	*	ns	***	**
B	***	***	***	**
A × B	*	ns	***	ns

Asterisk indicates significances at * $p < 0.05$, ** $p < 0.01$, *** $p < 0.001$; ns: non-significant. Different letters in the same column indicate significant differences. Values are the mean ± SE (^{x} $n = 9$, ^{y} $n = 6$).

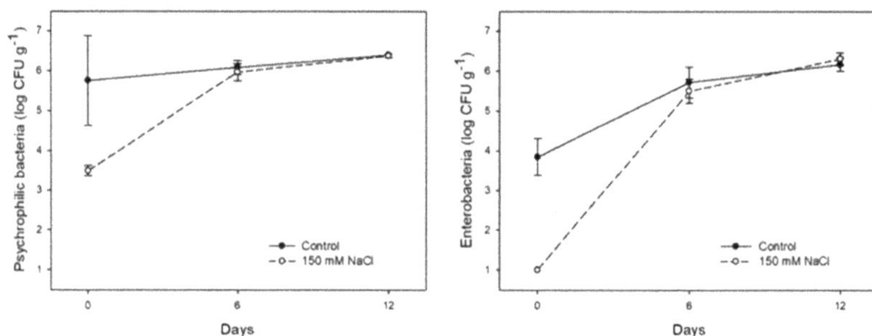

Figure 2. Psychrophilic bacteria and enterobacteria (log CFU g^{-1}) of *C. maritimum* leaves with different salinity treatment (control and 150 mM NaCl) after storage at 5 °C for 0, 6 and 12 days. Values are the mean ± SE ($n = 3$).

Table 8. Effect of 150 mM NaCl addition in the nutrient solution on sensory quality of *C. maritimum* leaves at 0, 6 and 12 days of storage at 5 °C.

	Sensorial Quality					
	Day 0		Day 6		Day 12	
	Control	150 mM NaCl	Control	150 mM NaCl	Control	150 mM NaCl
Acceptance						
Visual appearance	5	4	4.5	4	4	3.5
Colour	4	4	4	4	4	4
Texture (Crispness)	5	5	4	3.5	3	3
Flavour (Freshness)	5	4	4	4	3	3
Aroma	5	5	5	5	5	5
Global acceptance	5	4.5	4	3.5	3.5	3
Alterations						
Off-odours	5	5	5	5	5	5
Mechanical damage	4	4	4	4	3.5	3.5

4. Discussion

In this study the impact of salinity (150 mM NaCl) on plant growth, quality and self-life of *C. maritimum* was analysed. The salinity had no effect on the biomass and growth traits measured, indicating that sea fennel is a facultative halophyte with moderate tolerance to salinity, which does not require salt for maximal growth [29,30]. Similarly, Jiménez-Becker et al. [14] did not find differences in this species with respect to biomass growth when NaCl concentrations of 100 mM, 200 mM, and 300 mM were used. Nevertheless, the response to the salinity of *C. maritimum* is population-dependent, being this trait often correlated with the growth of the plants in their natural habitat [29]. The leaf area was reduced under 150 mM NaCl, in agreement with previous results of Hamed et al. [29], provoking a reduction of the specific leaf area and consequently an increase in leaf thickness and succulence (measured as leaf FW:leaf area ratio [31], the latter being one of the major factors involved in plant salt tolerance, the main quality trait for stimulated growth in halophytes [32]. However, Jiménez-Becker et al. [14] detected a decrease in leaves FW:DW ratio (leaves succulence) together with a decrease in plant water content in the plants grown with 300 mM with respect to that of those grown with 100 mM NaCl; therefore, the salt concentration in the nutrient solution is one of the main factors to consider in leaf succulence of halophytes plants. It is important to bear in mind that succulence together with firmness and juiciness procure leaf texture, which is an important sensory attribute for determining the post-harvest quality and consumers' acceptance [33]. Consequently, acquiring leaf succulence in the crop cycle and keeping it during post-harvest through adequate technology could be a useful strategy for guaranteeing the quality and shelf-life of sea fennel.

Sea fennel can be also a good source of daily minerals required in a healthy diet. The increase in Na^+ and Cl^- as the result of NaCl salinity was a common and expected response that was previously reported in sea fennel [32,34] since these elements are compartmentalized in vacuoles to avoid causing cytotoxicity [35]. However, differences between Na^+ and Cl^- accumulation in the aerial part were observed. It was postulated that differences in ion charge are responsible for the more expensive energetically sequestration of Na^+ compared to the sequestration of Cl^-, as the potential inside the vacuole is positive relative to the cytoplasm [36]. This would explain that the Cl^- content in the aerial part was found to be systematically higher than Na^+, under control and saline conditions. In this study, the K^+/Na^+ ratio dropped dramatically with salinity treatment as it was reported in Tunisian [29,37] and Argelian *C. maritimum* populations [34], when increasing salinity concentrations were applied, although the degree of resilience was population dependent. Maintaining a high K^+/Na^+ ratio is likely to be important to avoid the effects of ion toxicity under salt stress [38]. In our study, the accumulation of Na^+ in the control plant leaves was lower than in other halophytes [39]. In addition, K^+ was accumulated 4.68-fold higher than Na^+, which could mean that sea fennel grown in a floating system could be suitable to cover part of the amount of K^+ required daily. However, due to the high Na^+ concentration, it would be better to use it as a meal accompaniment or as a condiment [7,39], instead of as a main fresh vegetable dish. In control plants, Ca^{2+} concentration was 2-fold than the Ca^{2+} accumulation found by Sánchez-Faure et al. [40] in sea fennel plants grown in their natural habitat. Despite saline treatment reducing the available Ca^{2+}, its content in sea fennel leaves remained relatively high (532.05 mg kg^{-1} FW), with the potential benefit of preventing salt-induced oxidative damages, due to the protecting function of Ca^{2+} when plants face extreme heat, dry, or saline conditions [41,42]. The above-mentioned Ca^{2+} reduction with the salinity treatment could be due not only to the Na^+ accumulation but also to its reduce mobility and transport to the shoot under salinity stress [43,44]. An adequate Ca^{2+} intake for adults of 750 mg per day was marked by EFSA. Therefore, 100 g of fresh sea fennel grown in our conditions may represent 15.8% (control plants) and 7.6% (plant grown with 150 mM NaCl) of the daily recommended doses.

On the other hand, nitrate, bromide, and sulphate were reduced in the leaves of plants treated with NaCl, confirming a reduction in the absorption capacity of nutrients by the roots under salt stress [45]. The difference in nitrate accumulation in response to salinity is generally linked with the inhibition of NO_3^- uptake by Cl^- [46], which could happen by the interaction between these ions at the site of entry and for ion transport [47,48]. The nitrate content in the plants studied was generally quite low, and lower than the maximum legislated in the EU (Commission Regulation (EC) No 1258/2011) for other leafy vegetables such as spinach, lettuce, or rocket plant (2000–7000 mg kg^{-1}).

Salinity increased the content of total flavonoids but decreased phenolic content, while total antioxidant capacity was unaffected. Plants vary widely in their phenolic composition and content also accordingly to genetics and environmental conditions [49]. Our results agree with those of Labiad et al. [50], who demonstrated an increase in flavonoids content in NaCl treated sea fennel plants. Similarly, Yuan et al. [51] demonstrated on radish sprouts that moderate concentration of NaCl (100–150 mM) reduced total phenolic content while total antioxidant capacity remained unchanged. More recently, Emami Bistgani et al. [52] observed an increase in total phenolic content by around 20% after saline irrigation (60 mM NaCl) was applied to *Thymus vulgaris* and *Thymus daenensis*, compared with control plants. Additionally, an increase in leaf flavonoid content by 38.6% and 36.6% was observed in plants grown under salt stress conditions after the application of 60 and 90 mM NaCl. Plants cope with salinity-induced stress by altering metabolic processes and stimulating antioxidant activity to scavenge free radicals and ions chelators. Therefore, salt tolerance seems to be favoured by increased antioxidative compounds against oxidative stress induced by a toxic ion action [53]. Flavonoids are frequently induced by abiotic stress and promote roles in plant protection [54] as happened in our study. Hence, based on previous evidence and current data, it is possible to affirm that salt stress (150 mM NaCl) could be a feasible approach to keep, or even increase, the content of health-promoting compounds in *C. maritimum*.

Few studies have examined the storage conditions for edible halophytes leaves, with a clear lack of knowledge on *C. maritimum* shelf-life. The storage period of halophytes is usually limited to around a week, so high-tech storage and shipping conditions are required for longer periods [55]. The results presented here show that crop cultivation in controlled soilless conditions, even when salty, can yield production of high quality and good storability. The leaves kept their marketability until 12 days at 5 °C. Concerning the colour, NaCl produced clearer leaves, probably due to the presence of salt crystals. To corroborate this, a detailed microscopy study would be needed. These results are in agreement with D'Imperio et al. [56] who found similar colour parameters on wild sea fennel collected along sea shoreline, which is the natural habitat of this species. Colour is among the first quality parameters catching the attention of consumers with a strong influence on consumers' choice and opinion about the food quality [1,7]. Changes of colour observed in our study were subtle and undetected by the trained sensory panel.

Modified atmosphere packaging is commonly used for fresh produce quality maintenance, prolonging shelf-life, and decreasing the microbial growth on perishable commodities [57]. The atmosphere reached in our experiments seems to be adequate, since no off-odours related to anaerobic metabolism were detected. The high relative humidity inside the packages made the weight loss almost negligible, indicating that the modified atmosphere is convenient for retaining succulence and firmness. The relatively lower firmness of leaves obtained from salinity did not affect the shelf-life.

The reduced microbial load at harvest for leaves grown with 150 mM NaCl would be related to the fact that they had less aerial biomass, so microorganisms appeared later and/or in fewer number than in the control samples. However, at the end of storage, that difference was negligible. Abadias et al. [58] obtained similar values (10^6 to 10^7 CFU g^{-1}) of yeast and moulds in fresh-cut lettuce grown under salty conditions. Enterobacteriaceae, a common species in raw vegetables, even when reduced with NaCl, was still present, being an indicator of contamination, that should be carefully avoided in a floating system. The absence of Enterobacteriaceae is an ideal starting point prior to storage and commercialisation.

5. Conclusions

A saline-nutrient solution may be used successfully in hydroponic-grown sea fennel plants to enhance raw product quality and post-harvest shelf-life. The product presents a high concentration of flavonoids, a good sensory quality, and a reduction of microbial load. Consequently, it could be said that the saline treatment can be useful for the hydroponic culture of sea fennel obtaining a product with good marketability as an emerging crop for fresh consumption.

Nonetheless, variability in yields and chemical composition with geographical origins, harvesting and post-harvesting conditions needs to be explored and better understood prior to large-scale commercialisation for both farmers and consumers.

Author Contributions: Conceptualization, M.d.C.M.-B., C.E.-G. and J.A.F. (Juan A. Fernández); methodology, A.G., F.A. and J.A.F. (José A. Franco); formal analysis, A.S.; investigation, A.G., F.A. and P.A.G.; resources, A.S. and J.A.F. (Juan A. Fernández); data curation, A.G. and P.A.G.; writing—original draft preparation, F.A., A.G., P.A.G., M.d.C.M.-B and J.A.F. (Juan A. Fernández); writing—review and editing C.E.-G., J.A.F. (José A. Franco), P.A.G., A.S., M.d.C.M.-B. and J.A.F. (Juan A. Fernández); supervision, J.A.F. (José A. Franco), C.E.-G., P.A.G. and M.d.C.M.-B.; funding acquisition, J.A.F. (Juan A. Fernández). All authors have read and agreed to the published version of the manuscript.

Funding: This research received no external funding.

Institutional Review Board Statement: Not applicable.

Informed Consent Statement: Not applicable.

Data Availability Statement: Not applicable.

Acknowledgments: The elaboration of the manuscript was supported by a grant from the University of Bari Aldo Moro (reference Premio di Studio GLOBAL THESIS) awarded to Fabio Amoruso.

Conflicts of Interest: The authors declare no conflict of interest.

References

1. Renna, M.; Gonnella, M.; Caretto, S.; Mita, G.; Serio, F. Sea fennel (*Crithmum maritimum* L.): From underutilized crop to new dried product for food use. *Genet. Resour. Crop Evol.* **2017**, *64*, 205–216. [CrossRef]
2. United States Department of Agriculture-Agriculture Research Service (USDA). U.S. National Plant Germplasm System [WWW Document]. Available online: https://npgsweb.ars-grin.gov/gringlobal/taxon/taxonomydetail?id=402237 (accessed on 15 November 2021).
3. Sánchez-Hernández, E.; Buzón-Durán, L.; Andrés-Juan, C.; Lorenzo-Vidal, B.; Martín-Gil, J.; Martín-Ramos, P. Physicochemical Characterization of *Crithmum maritimum* L. and *Daucus carota* subsp. gummifer (Syme) Hook. fil. and Their Antimicrobial Activity against Apple Tree and Grapevine Phytopathogens. *Agronomy* **2021**, *11*, 886. [CrossRef]
4. Atia, A.; Barhoumi, Z.; Mokded, R.; Abdelly, C.; Smaoui, A. Environmental eco-physiology and economical potential of the halophyte *Crithmum maritimum* L. (Apiaceae). *J. Med. Plants Res.* **2011**, *5*, 3564–3571.
5. Perez-Vizcaino, F.; Duarte, J.; Jiménez, R.; Santos-Buelga, C.; Osuna, A. Antihypertensive effects of the flavonoid quercetin. *Pharmacol. Rep.* **2009**, *61*, 67–75. [CrossRef]
6. Atzori, G.; de Vos, A.C.; van Rijsselberghe, M.; Vignolini, P.; Rozema, J.; Mancuso, S.; van Bodegom, P.M. Effects of increased seawater salinity irrigation on growth and quality of the edible halophyte *Mesembryanthemum crystallinum* L. under field conditions. *Agric. Water Manag.* **2017**, *187*, 37–46. [CrossRef]
7. Renna, M.; Gonnella, M. The use of the sea fennel as a new spice-colorant in culinary preparations. *Int. J. Gastron. Food Sci.* **2012**, *1*, 111–115. [CrossRef]
8. Siracusa, L.; Kulisic-Bilusic, T.; Politeo, O.; Krause, I.; Dejanovic, B.; Ruberto, G. Phenolic composition and antioxidant activity of aqueous infusions from *Capparis spinosa* L. and *Crithmum maritimum* L. before and after submission to a two-step in vitro digestion model. *J. Agric. Food Chem.* **2011**, *59*, 12453–12459. [CrossRef] [PubMed]
9. Giungato, P.; Renna, M.; Rana, R.; Licen, S.; Barbieri, P. Characterization of dried and freeze-dried sea fennel (*Crithmum maritimum* L.) samples with headspace gas-chromatography/mass spectrometry and evaluation of an electronic nose discrimination potential. *Food Res. Int.* **2019**, *115*, 65–72. [CrossRef]
10. Souid, A.; Della Croce, C.M.; Frassinetti, S.; Gabriele, M.; Pozzo, L.; Ciardi, M.; Abdelly, C.; Hamed, K.B.; Magné, C.; Longo, V. Nutraceutical potential of leaf hydro-ethanolic extract of the edible halophyte *Crithmum maritimum* L. *Molecules* **2021**, *26*, 5380. [CrossRef]

11. Pistrick, K. Notes on neglected and underutilized crops Current taxonomical overview of cultivated plants in the families Umbelliferae and Labiatae. *Genet. Resour. Crop Evol.* **2002**, *49*, 211–221. [CrossRef]

12. Ozcan, B.; Tzeremes, P.G.; Tzeremes, N.G. Energy consumption, economic growth and environmental degradation in OECD countries. *Econ. Model.* **2020**, *84*, 203–213. [CrossRef]

13. Tran, D.Q.; Konishi, A.; Cushman, J.C.; Morokuma, M.; Toyota, M.; Agarie, S. Ion accumulation and expression of ion homeostasis-related genes associated with halophilism, NaCl-promoted growth in a halophyte *Mesembryanthemum crystallinum* L. *Plant Prod. Sci.* **2020**, *23*, 91–102. [CrossRef]

14. Jiménez-Becker, S.; Ramírez, M.; Plaza, B.M. The influence of salinity on the vegetative growth, osmolytes and chloride concentration of four halophytic species. *J. Plant Nutr.* **2019**, *42*, 1838–1849. [CrossRef]

15. Petropoulos, S.A.; Karkanis, A.; Martins, N.; Ferreira, I.C.F.R. Edible halophytes of the Mediterranean basin: Potential candidates for novel food products. *Trends Food Sci. Technol.* **2018**, *74*, 69–84. [CrossRef]

16. Renna, M.; Cocozza, C.; Gonnella, M.; Abdelrahman, H.; Santamaria, P. Elemental characterization of wild edible plants from countryside and urban areas. *Food Chem.* **2015**, *177*, 29–36. [CrossRef] [PubMed]

17. Renna, M. Reviewing the prospects of sea fennel (*Crithmum maritimum* L.) as emerging vegetable crop. *Plants* **2018**, *7*, 92. [CrossRef]

18. Giménez, A.; Martínez-Ballesta, M.D.; Egea-Gilabert, C.; Gómez, P.A.; Artés-Hernández, F.; Pennisi, G.; Orsini, F.; Crepaldi, A.; Fernández, J.A. Combined Effect of Salinity and LED Lights on the Yield and Quality of Purslane (*Portulaca oleracea* L.) Microgreens. *Horticulturae* **2021**, *7*, 180. [CrossRef]

19. Nicola, S.; Fontana, E. Fresh-cut produce quality: Implications for a systems approach. In *Postharvest Handling: A System Approach*; Florkowski, W.J., Shewfelt, R., Breuckner, B., Prussia, S.E., Eds.; Academic Press: San Diego, CA, USA; Elsevier: Amsterdam, The Netherlands, 2014; pp. 217–273.

20. Ali, A.Y.A.; Ibrahim, M.E.H.; Zhou, G.; Nimir, N.E.A.; Elsiddig, A.M.I.; Jiao, X.; Zhu, G.; Salih, E.G.I.; Suliman, M.S.E.S.; Elradi, S.B.M. Gibberellic acid and nitrogen efficiently protect early seedlings growth stage from salt stress damage in Sorghum. *Sci. Rep.* **2021**, *11*, 6672. [CrossRef]

21. Machado, R.; Serralheiro, R. Soil salinity: Effect on vegetable crop growth. Management practices to prevent and mitigate soil salinization. *Horticulturae* **2017**, *3*, 30. [CrossRef]

22. Egea-Gilabert, C.; Fernández, J.A.; Migliaro, D.; Martínez-Sánchez, J.J.; Vicente, M.J. Genetic variability in wild vs. cultivated *Eruca vesicaria* populations as assessed by morphological, agronomical and molecular analyses. *Sci. Hortic.* **2009**, *121*, 260–266. [CrossRef]

23. Lara, L.J.; Egea-Gilabert, C.; Niñirola, D.; Conesa, E.; Fernández, J.A. Effect of aeration of the nutrient solution on the growth and quality of purslane (*Portulaca oleracea*). *J. Hortic. Sci. Biotechnol.* **2011**, *86*, 603–610. [CrossRef]

24. Everette, J.D.; Bryant, Q.M.; Green, A.M.; Abbey, Y.A.; Wangila, G.W.; Walker, R.B. Thorough study of reactivity of various compound classes toward the folin-Ciocalteu reagent. *J. Agric. Food Chem.* **2010**, *58*, 8139–8144. [CrossRef] [PubMed]

25. Brand-Williams, W.; Cuvelier, M.-E.; Berset, C. Use of a free radical method to evaluate antioxidant activity. *LWT-Food Sci. Technol.* **1995**, *28*, 25–30. [CrossRef]

26. López-Marín, J.; Gálvez, A.; del Amor, F.M.; Albacete, A.; Fernández, J.A.; Egea-Gilabert, C.; Pérez-Alfocea, F. Selecting vegetative/generative/dwarfing rootstocks for improvingfruit yield and quality in water stressed sweet peppers. *Sci. Hortic.* **2017**, *214*, 9–17. [CrossRef]

27. Meda, A.; Lamien, C.E.; Romito, M.; Millogo, J.; Nacoulma, O.G. Determination of the total phenolic, flavonoid and proline contents in Burkina Fasan honey, as well as their radical scavenging activity. *Food Chem.* **2005**, *91*, 571–577. [CrossRef]

28. Tomás-Callejas, A.; López-Velasco, G.; Camacho, A.B.; Artés, F.; Artés-Hernández, F.; Suslow, T.V. Survival and distribution of *Escherichia coli* on diverse fresh-cut baby leafy greens under preharvest through postharvest conditions. *Int. J. Food Microbiol.* **2011**, *151*, 216–222. [CrossRef]

29. Hamed, K.B.; Debez, A.; Chibani, F.; Abdelly, C. Salt response of *Crithmum maritimum*, an oleagineous halophyte. *Trop. Ecol.* **2004**, *45*, 151–159.

30. Hamed, K.B.; Castagna, A.; Salem, E.; Ranieri, A.; Abdelly, C. Sea fennel (*Crithmum maritimum* L.) under salinity conditions: A comparison of leaf and root antioxidant responses. *Plant Growth Regul.* **2007**, *53*, 185–194. [CrossRef]

31. Amanullah; Khan, I.; Jan, A.; Jan, M.T.; Khalil, S.K.; Shah, Z.; Afzal, M. Compost and nitrogen management influence productivity of spring maize (*Zea mays* L.) under deep and conventional tillage systems in Semi-arid regions. *Commun. Soil Sci. Plant Anal.* **2015**, *46*, 1566–1578. [CrossRef]

32. Yepes, L.; Chelbi, N.; Vivo, J.-M.; Franco, M.; Agudelo, A.; Carvajal, M.; del Carmen Martinez-Ballesta, M. Analysis of physiological traits in the response of Chenopodiaceae, Amaranthaceae, and Brassicaceae plants to salinity stress. *Plant Physiol. Biochem.* **2018**, *132*, 145–155. [CrossRef]

33. Damerum, A.; Chapman, M.A.; Taylor, G. Innovative breeding technologies in lettuce for improved post-harvest quality. *Postharvest Biol. Technol.* **2020**, *168*, 111266. [CrossRef] [PubMed]

34. Hamdani, F.; Derridj, A.; Rogers, H.J. Diverse salinity responses in *Crithmum maritimum* tissues at different salinities over time. *J. Soil Sci. Plant Nutr.* **2017**, *17*, 716–734. [CrossRef]

35. Ksouri, R.; Ksouri, W.M.; Jallali, I.; Debez, A.; Magné, C.; Hiroko, I.; Abdelly, C. Medicinal halophytes: Potent source of health promoting biomolecules with medical, nutraceutical and food applications. *Crit. Rev. Biotechnol.* **2012**, *32*, 289–326. [CrossRef] [PubMed]
36. Rea, P.A.; Poole, R.J. Vacuolar H^+-translocating pyrophosphatase. *Ann. Rev. Plant Physiol. Mol. Biol.* **1993**, *44*, 157–180. [CrossRef]
37. Ben Hamed, K.; Ben Youssef, N.; Ranieri, A.; Zarrouk, M.; Abdelly, C. Changes in content and fatty acid profiles of total lipids and sulfolipids in the halophyte *Crithmum maritimum* under salt stress. *J. Plant Physiol.* **2005**, *162*, 599–602. [CrossRef] [PubMed]
38. Flowers, T.J.; Munns, R.; Colmer, T.D. Sodium chloride toxicity and the cellular basis of salt tolerance in halophytes. *Ann. Bot.* **2015**, *115*, 419–431. [CrossRef] [PubMed]
39. Castañeda-Loaiza, V.; Oliveira, M.; Santos, T.; Schüler, L.; Lima, A.R.; Gama, F.; Salazar, M.; Neng, N.R.; Nogueira, J.M.F.; Varela, J. Wild vs cultivated halophytes: Nutritional and functional differences. *Food Chem.* **2020**, *333*, 127536. [CrossRef] [PubMed]
40. Sánchez-Faure, A.; Calvo, M.M.; Pérez-Jiménez, J.; Martín-Diana, A.B.; Rico, D.; Montero, M.P.; del Carmen Gómez-Guillén, M.; López-Caballero, M.E.; Martínez-Alvarez, O. Exploring the potential of common iceplant, seaside arrowgrass and sea fennel as edible halophytic plants. *Food Res. Int.* **2020**, *137*, 109613. [CrossRef]
41. Jiang, Y.; Huang, B. Effects of calcium on antioxidant activities and water relations associated with heat tolerance in two cool-season grasses. *J. Exp. Bot.* **2001**, *52*, 341–349. [CrossRef]
42. Hernández, J.A.; Aguilar, A.B.; Portillo, B.; López-Gómez, E.; Beneyto, J.M.; García-Legaz, M.F. The effect of calcium on the antioxidant enzymes from salt-treated loquat and anger plants. *Funct. Plant Biol.* **2003**, *30*, 1127–1137. [CrossRef]
43. Coskun, D.; Britto, D.T.; Jean, Y.-K.; Kabir, I.; Tolay, I.; Torun, A.A.; Kronzucker, H.J. K+ efflux and retention in response to NaCl stress do not predict salt tolerance in contrasting genotypes of rice (*Oryza sativa* L.). *PLoS ONE* **2013**, *8*, e57767. [CrossRef] [PubMed]
44. Yuriko, O.; Osakabe, K.; Shinozaki, K.; Tran, L. SP Response of plants to water stress. *Front. Plant Sci.* **2014**, *5*, 86.
45. Chrysargyris, A.; Tzionis, A.; Xylia, P.; Tzortzakis, N. Effects of salinity on tagetes growth, physiology, and shelf life of edible flowers stored in passive modified atmosphere packaging or treated with ethanol. *Front. Plant Sci.* **2018**, *9*, 1765. [CrossRef] [PubMed]
46. Aslam, M.; Huffaker, R.C.; Rains, D.W. Early effects of salinity on nitrate assimilation in barley seedlings. *Plant Physiol.* **1984**, *76*, 321–325. [CrossRef]
47. Varma, S.; Rogers, D.M.; Pratt, L.R.; Rempe, S.B. Design principles for K+ selectivity in membrane transport. *J. Gen. Physiol.* **2011**, *137*, 479–488. [CrossRef]
48. Rubinigg, M.; Posthumus, F.; Ferschke, M.; Elzenga, J.T.M.; Stulen, I. Effects of NaCl salinity on 15 N-nitrate fluxes and specific root length in the halophyte *Plantago maritima* L. *Plant Soil* **2003**, *250*, 201–213. [CrossRef]
49. De Abreu, I.N.; Mazzafera, P. Effect of water and temperature stress on the content of active constituents of *Hypericum brasiliense* Choisy. *Plant Physiol. Biochem.* **2005**, *43*, 241–248. [CrossRef]
50. Labiad, M.H.; Giménez, A.; Varol, H.; Tüzel, Y.; Egea-Gilabert, C.; Fernández, J.A.; Martínez-Ballesta, M.-C. Effect of exogenously applied methyl jasmonate on yield and quality of salt-stressed hydroponically grown sea fennel (*Crithmum maritimum* L.). *Agronomy* **2021**, *11*, 1083. [CrossRef]
51. Yuan, G.; Wang, X.; Guo, R.; Wang, Q. Effect of salt stress on phenolic compounds, glucosinolates, myrosinase and antioxidant activity in radish sprouts. *Food Chem.* **2010**, *121*, 1014–1019. [CrossRef]
52. Emami Bistgani, Z.; Hashemi, M.; DaCosta, M.; Craker, L.; Filippo, M.; Morshedloo, M.R. Effect of salinity stress on the physiological characteristics, phenolic compounds and antioxidant activity of *Thymus vulgaris* L. and *Thymus daenensis* Celak. *Ind. Crops Prod.* **2019**, *135*, 311–320. [CrossRef]
53. Rakhmankulova, Z.F.; Shuyskaya, E.V.; Shcherbakov, A.V.; Fedyaev, V.; Biktimerova, G.Y.; Khafisova, R.R.; Usmanov, I.Y. Content of proline and flavonoids in the shoots of halophytes inhabiting the South Urals. *Russ. J. Plant Physiol.* **2021**, *62*, 71–79. [CrossRef]
54. Grace, S.G.; Logan, B.A. Energy dissipation and radical scavenging by the plant phenylpropanoid pathway. *Philos. Trans. R. Soc. Lond. Ser. B Biol. Sci.* **2000**, *355*, 1499–1510. [CrossRef] [PubMed]
55. Gago, C.; Sousa, A.R.; Juliao, M.; Miguel, G.; Antunes, D.C.; Panagopoulos, T. Sustainable use of energy in the storage of halophytes used for food. *Int. J. Energy Environ.* **2011**, *4*, 592–599.
56. D'Imperio, M.; Renna, M.; Cardinali, A.; Buttaro, D.; Serio, F.; Santamaria, P. Calcium biofortification and bioaccessibility in soilless "baby leaf" vegetable production. *Food Chem.* **2016**, *213*, 149–156. [CrossRef] [PubMed]
57. Lopes, M.; Sanches-Silva, A.; Castilho, M.; Cavaleiro, C.; Ramos, F. Halophytes as source of bioactive phenolic compounds and their potential applications. *Crit. Rev. Food Sci. Nutr.* **2021**, *2*, 1–24. [CrossRef]
58. Abadias, M.; Usall, J.; Anguera, M.; Solsona, C.; Viñas, I. Microbiological quality of fresh, minimally-processed fruit and vegetables, and sprouts from retail establishments. *Int. J. Food Microbiol.* **2008**, *123*, 121–129. [CrossRef] [PubMed]

horticulturae

Article

Analysis of Free Sugars, Organic Acids, and Fatty Acids of Wood Apple (*Limonia acidissima* L.) Fruit Pulp

Shrinivas Lamani [1], Konerira Aiyappa Anu-Appaiah [2], Hosakatte Niranjana Murthy [1,*], Yaser Hassan Dewir [3,*] and Jesamine J. Rikisahedew [4]

[1]   Department of Botany, Karnatak University, Dharwad 580003, Karnataka, India; shrinivasgl92@gmail.com
[2]   Department of Microbiology and Fermentation Technology, CSIR-CFTRI, Mysuru 570020, Karnataka, India; anuapps@yahoo.com
[3]   Plant Production Department, College of Food & Agriculture Sciences, King Saud University, Riyadh 11451, Saudi Arabia
[4]   Institute of Agricultural and Environmental Sciences, Estonian University of Life Sciences, Fr. R. Kreutzwaldi 1, 51006 Tartu, Estonia; jesamine@emu.ee
*   Correspondence: hnmurthy60@gmail.com (H.N.M.); ydewir@ksu.edu.sa (Y.H.D.)

Abstract: Wood apple (*Limonia acidissima* L.) is an underutilized, fruit-yielding tree that is native to India and Sri Lanka. Wood apple trees are also cultivated in India, Sri Lanka, Bangladesh, Myanmar, Thailand, Malaysia, Vietnam, Kampuchea, Laos, and Indonesia for delicious fruits and medicinal purposes. The major objective of the present work was the analysis of the nutritional status of wood apple fruit pulp.The fruits are rich in total carbohydrates (24.74 ± 0.19%), total proteins (9.30 ± 0.16%), oil (0.99 ± 0.01%), fiber (3.32 ± 0.02%), and ash (2.73 ± 0.12%). Further analysis and quantification of free sugars, organic acids, and fatty acid methyl esters were carried out by using high-performance liquid chromatography (HPLC) and gas chromatographic (GC) methods. In total, five sugars and nine organic acids were detected and quantified. The predominant sugars were fructose (16.40 ± 0.23%) and glucose (14.23 ± 0.10%), whereas the predominant organic acids were D-tartaric (4.01 ± 0.03%), ascorbic (4.51 ± 0.05%), and citric acid (4.27 ± 0.04%). The oil content of fruit pulp was 0.99 ± 0.01% and GC-MS analysis revealed that, it comprise of 16 fatty acid methyl esters. The percentage of saturated fatty acids were 32.17 ± 0.35%, that includes palmitic (18.52 ± 0.12%) and stearic acids (9.02 ± 0.08%), whereas the unsaturated fatty acids were 51.98 ± 0.94%, including oleic acid (23.89 ± 0.06%), α-linolenic acid (16.55 ± 0.26%), linoleic acid (10.02 ± 0.43%), and vaccenic acid (1.78 ± 0.23%).

Keywords: fatty acids; free sugars; GC-MS; HPLC; organic acids; UPLC; wood apple

Citation: Lamani, S.; Anu-Appaiah, K.A.; Murthy, H.N.; Dewir, Y.H.; Rikisahedew, J.J. Analysis of Free Sugars, Organic Acids, and Fatty Acids of Wood Apple (*Limonia acidissima* L.) Fruit Pulp. *Horticulturae* **2022**, *8*, 67. https://doi.org/10.3390/horticulturae8010067

Academic Editors: Rosario Paolo Mauro, Carlo Nicoletto, Leo Sabatino and Luigi De Bellis

Received: 30 November 2021
Accepted: 2 January 2022
Published: 11 January 2022

Publisher's Note: MDPI stays neutral with regard to jurisdictional claims in published maps and institutional affiliations.

1. Introduction

Underutilized fruits are considered as a vital source of essential amino acids, vitamins, mineral elements, and dietary fiber. In addition to these constituents, fruits are also rich in proteins, fats, carbohydrates, and reducing and non-reducing sugars [1–6]. Underutilized fruits are not only a primary source of food, but also have immense therapeutical potential due to the presence of bioactive compounds, including antioxidants such as unsaturated fats, organic acids, and sugars [1–4,7,8]. Nutraceutical compounds, which are essential for the human diet and major constituents for the storage of energy and the structural and functional composition of cells, were achieved from oils and fats [9]. Fatty acids are related to the prevention of coronary heart diseases, diabetes, cancer, and depression [10]. Fatty acids are also considered as the major components of cell membranes and have biological activities that act to influence cell and tissue metabolism, function, and responsiveness to hormones and other signals [11]. Free sugars and organic acids contribute to the taste of fruit pulp. Sugars are considered as common source of energy in living organisms and are the major constituents present in nature [12]. In recent years, polysaccharides have attracted much attention from researchers because they play a major role with respect to structural material, respiratory substrate, and synthesis of macromolecules and also in the mechanisms of plant tissue resistance to desiccation [13,14]. Aside from their traditional uses, free sugars are also used as a potential source to improve the quality of foods [15].

Organic acids in fruits are rapidly oxidized in the metabolism; hence, they do not have adverse effects on the body. Organic acids, including citric and ascorbic acids, lower the pH and inhibit/kill the growth of spoilage-causing organisms in food products; that is why they are commonly used as a preservative in many food processing industries [16]. One of the nutritionally and medicinally important, underutilized, fruit-yielding tree species is *Limonia acidissima* L., belongs to Rutaceae family and it is native to India and Sri Lanka. It is commonly called as wood apple, elephant apple, or monkey fruit. It is found throughout the plains of India, particularly in drylands up to an elevation of 450 m, and is a deciduous, slow-growing, erect tree that reaches a height of up to approximately 12 to 15 m with a few upward-reaching branches bending outwards near the summit where they are subdivided into slender branchlets drooping at the tips. The spines are axillary, short, straight, and 2–3 cm long on some of the zigzag twigs. The leaves are deciduous, alternate, dark green, and 3–5 inches long with oil glands and a slight lemon scent when crushed. The flowers are normally bisexual, small, numerous, dull red or greenish yellow in color (Figure 1) [17]. It has also been widely introduced and naturalized in Myanmar, Thailand, Malaysia, Vietnam, Kampuchea, Laos, and Indonesia. Wood apple is a seasonal fruit; in India, the fruit matures in the month of October and is available until January. The maximum yield per plant varies from 40.50 to 70.00 kg, as does that of *Aegle marmelos* [18].

Figure 1. Wood apple (*Limonia acidissima* L.)—Tree habit (**A**), fruits (**B**), pulp (**C**), and seeds (**D**).

Wood apple is used by the tribal and rural population of the developing world, and also usage of such underutilized fruits facilitates an extra economic benefit to rural folk [2]. The pulp of the wood apple is eaten raw or blended with coconut milk and sugar syrup into juice. The pulp can be also be processed into a beverage or frozen like ice cream. The pulp is also made into chutney, preserves, jelly, marmalade, jam, jelly, and toffee [19]. Wood apple as a whole and its parts, such as leaves, roots, bark, and unripe and ripened fruits, have been used as medicine for the treatment of various ailments. Ethnopharmacological properties have been documented in various studies. The decoction of the leaves is used in the treatment of constipation and vomiting [20]. Bark and stem are used in treating liver diseases and hemorrhoids, respectively [21]. Fruits were used for the treatment of blood impurities, leucorrhoea, and as a diuretic [22]. Fruits are also used as a stomachic, stimulant, astringent, aphrodisiac, diuretic, cardiotonic, tonic to liver and lungs, a cure for coughs, hiccups, asthma, leucorrhoea, diarrhea, and also for wound healing activity [23]. Wood apple fruits are considered as a potential source of natural antioxidants and seed oil [24,25]. In earlier communication, we reported the fatty acid profile, and tocopherol content of the seed oil, and nutritional analysis of the seed cake of wood apple [25]. However, the composition of fatty acids, free sugars, and organic acids of wood apple fruit pulp is not yet reported. Therefore, the present work aimed to investigate the fatty acid composition in the pulp oil, free sugars, and organic acids, and the nutritional aspects of wood apple fruit pulp using advanced instruments such as GC-MS, UPLC-ELSD, and HPLC-PDA.

2. Materials and Methods

2.1. Sample Collection and Separation

Matured wood apple (*Limonia acidissima* L.) fruits were collected from Jogimatti reserve forest, Chitradurga District, Karnataka, India ($14°11'13.0''$ N, $76°23'36.8''$ E). Twenty matured plants were randomly selected from the site and 20 fruits were collected from each plant and pooled together for further study. A minimum distance of 20 m was maintained between the selected plants. Fruits were collected in their ripened stage, indicated by brown color during January. Pulp was collected from fresh fruits completely and stored at $-20\ °C$ for future utilization. The separated pulp was dried by keeping in hot air oven (Serwell Instrument Incorporation, Bangalore, India) at $38 \pm 2\ °C$ until it reached constant weight and was stored for future use.

2.2. Chemicals and Standards

n-Hexane, methanol, potassium hydroxide, hydrochloric acid, 14% methanolic boron trifluoride (w/v), benzene, helium, diethyl ether, ethanol, acetonitrile, triethylamine, sulphuric acid, milli-Q water, acetone, phenolphthalein, chloroform, hanus iodine solution, potassium iodide, sodium thiosulfate, sodium hydroxide, PBS solution, bovine serum albumin, copper sulphate, folin–ciocalteu reagent, anthrone reagent, and sodium carbonate were of analytical grade from Merck, Bangalore (India). Fructose, glucose, sucrose, rhamnose, maltose, D-galacturonic acid, oxalic acid, D-tartaric acid, malic acid, ascorbic acid, acetic acid, citric acid, succinic acid, pyruvic acid, and gallic acid were purchased from Sigma-Aldrich, Bangalore, India.

2.3. Extraction of Oil from Wood Apple Fruit Pulp

Moisture-free pulp powder was used to extract oil with n-hexane using the cold-pressed extraction method. The extract was concentrated using a rotary evaporator at 40 °C. The extracted oil was used for the analysis of fatty acid composition and the defatted fruit pulp was used for nutritional and nutraceutical analysis.

2.4. Morphological and Proximate Characterization of Wood Apple Fruits

The morphological characters of fruits were assayed from three different lots of fruits; each group consisted of 10 randomly collected matured fruits. Fruit length, width, and circumferance were measured using a Vernier caliper and the average weight of wood apple fruits was estimated gravimetrically. Separated wood apple fruit pulp was used to evaluate the proximate composition, including moisture, total ash, and crude fiber, and were determined according to the standard methods of AOAC [26]. Protein, carbohydrate, and crude fat contents were quantified by the method of Sadasivam and Manickam [27]. Titratable acidity, total soluble solids, and pH were assayed according to AOAC [28] methods. Briefly, 10 g of defatted fruit pulp was thoroughly mixed with 50 mL of milli-Q water at room temperature and used for analysis. Titratable acidity was determined as a percentage of citric acid. A total soluble solid was calculated directly using a refractometer (Erma handheld refractometer, Japan) and expressed in °Brix. The pH was measured by homogenization of 1 mg dried pulp sample in 10 mL of milli-Q water and then by immersing the electrode directly into a mixture of pulp using a pH meter (CD Instrumental Pvt. Ltd., Bangalore, India) [29]. The ash content in the sample was assessed according to the AOAC method [28].

2.5. Extraction and Quantification of Total Phenolic Content

The method of Singleton et al. [30] was used to assess the total phenolic content. Defatted pulp powder (2 g) was extracted with 20 mL of aqueous methanol (80%, *v/v*). Aliquots of filtrate (0.5 mL) were taken into tubes and 3 mL of milli-Q water and 0.5 mL of Folin–Ciocalteu reagent (1:1) were added and homogenized thoroughly. After that, 2 mL of 20% (*w/v*) sodium carbonate was added and incubated for 30 min at room temperature in the dark. Absorbance was recorded at 760 nm. The amount of phenolics was expressed as a percentage of gallic acid equivalent.

2.6. Fatty Acid Composition
Sample Preparation and Analysis of Fatty Acid Methyl Esters (FAMEs)

FAMEs were prepared according to Morison and Smith's method [31]. 1 mL of 0.7 N methanolic KOH (*w/v*) was used to saponify 10 mg of oil and incubated for 1 h at 60 °C in a water bath. It was cooled and 2 mL of hexane was added to discard the organic phase. The reaction was neutralized by adding 1 mL of 0.7 N methanolic HCL (*v/v*) and 2 mL of hexane. The hexane pool was collected and evaporated to dryness under nitrogen air atmosphere. The evaporated tubes were methylated with 0.7 mL of 14% methanolic boron trifluoride and 0.3 mL of benzene. 2 mL of hexane was added and the FAMEs were collected and water washed. The hexane pool was collected again and evaporated and used for GC-MS analysis. Agilent 7890 B GC5977 A MSD GC-MS (Agilent Technologies, Santa Clara, CA, USA) system was used for the analysis of FAMEs. The separation was achieved with an Agilent CP8824 column (60 m length, 0.25 μm, i.d. 250 μm). The GC column temperature initially set at 80 °C and held for 1 min, then increased to 20 °C/min to 130 °C, then 8.5 °C/min to 160 °C, then 2.75 °C/min to 200 °C and held for 3 min. Helium, with a flow rate of 1 mL per min, was used as a carrier gas. The injection volume was 1 μL and the split ratio was 20:1. MS data were collected in an electron ionization mode at 70 eV scanning from m/z 40 to 500. Fatty acid methyl ester in each sample was identified with their mass spectrum data and results were confirmed by mass spectral library search (NIST 2.0 g).

2.7. Sugars and Organic Acids
Sample Preparation Technique

Defatted fruit pulp (5 g) was extracted with 30 mL of aqueous methanol (70%, *v/v*) in a round bottom flask at 40 °C with constant stirring. The same was repeated twice and the supernatant was collected and evaporated to dryness using a rotary evaporator (IKA-RV 10 digital rotary evaporator, IKA, Königswinter, Germany), and the concentrated sample was diluted to 10 mL with milli-Q water. For organic acid and sugar analysis, 10 μL of samples were was injected into the HPLC system [16,32].

2.7.1. LC Conditions for Sugars and Organic Acids
2.7.1.1. Investigation of Sugar Profile by UPLC-ELSD

Sugar compositions of the sample were achieved with a ultra-high performance liquid chromatography system (Waters AQUITY UPLC H-Class) coupled with evaporative light scattering detector (ELSD) described by Koh et al. [33] with minor modifications. Separation was achieved with a UPLC® BEH amide column (2.1 × 100 mm i.d., 1.7 μm particle size), and the temperature was maintained at 35 °C. The condition of ELSD was: 200, gain; 40 psi, gas pressure; 40 °C, temperature of drift tube; cooling, nebulizer. Gradient elution program with acetonitrile: milli-Q water (80% and 30%) used as eluents A and B, respectively, with 0.2% triethylamine (TEA). The eluent flow rate was set at 0.12 mL/min and the volume of injection was 2 µL. The gradient program followed as 60% A for 0–10 min, 60–100% A for 10–10.01 min, and equilibrated at 18 min. All samples were analyzed in triplicate and the analytes were identified by comparison (retention time and characteristic absorption) with authentic high purity standards (Sigma) and quantified through external standardization.

2.7.1.2. Investigation of Organic Acids Profile by HPLC-PDA

Organic acid profile was investigated with high-performance liquid chromatography (Shimadzu UFLC) equipped with SPD 20A detector and LC 20 AD system controller (photodiode array detector). The chromatographic separation was carried out with reverse-phase Phenomenex C18 column (Luna® 5 µm C18 (2) 100 Å, LC Column 250 × 4.6 mm) using an injection volume of 5 µL and detected at 210 nm. The mobile phase employed was 0.005 N sulphuric acid at a flow rate of 1 mL per minute in an isocratic mode of elution [34]. All samples were analyzed in triplicate and the analytes were identified by comparison (retention time and characteristic absorption) with authentic high purity standards (Sigma) and quantified through external standardization.

2.7.2. Statistical Analysis

All the determined values were estimated in triplicate (n = 3). Results are presented as mean $\pm$ standard error (SE). Statistical analysis was carried out by using SPSS software.

3. Results and Discussion

3.1. Morphological and Proximate Characterization of Wood Apple Fruits

The collected wood apple fruits were round in shape and the average size of fruits, including length, width, and circumference, showed 7.42 ± 0.19, 7.43 ± 0.10, and 23.42 ± 0.29 cm, respectively, and the average weight of fruits was 179.45 ± 1.42 g (Table 1). The studied morphological characters of wood apple fruits strongly agreed with the results of Singh et al. [35]. The nutritional estimation of fruit pulp is shown in Table 1 and it strongly agreed with the reports available [36,37]. Moisture content in the fruits is a considerably great parameter to assess the quality as it influences the texture, taste, shelf life, and growth of the microbes. Results showed 58.89 ± 1.21% of moisture, which is comparatively less than *Annona squamosa* and *Aegle mormelos* (78.54% and 64.71%, respectively) [38,39]. Crude fat constitutes true fat and also the phospholipids, sterols, essential oils, and fat-soluble pigments. The crude fat value (0.99 ± 0.01%) strongly agreed with the reports of Cheema et al. [39] regarding *Aegle mormelos* (0.49%) and *Ziziphus nummularia* (0.93%).

The parameters which decide the quality, including total soluble solids, titratable acidity, and pH, shown in the fruit pulp were 19.52 ± 0.17 °Brix, 4.61 ± 0.13%, and 3.61 ± 0.09%, respectively. To assess the internal quality of fruits, these parameters were considered as principal factors. The studied fruit pulp showed an approximately higher value of TSS compared to *Annona squamosa*, 23.10 °Brix [38], and lower than Golden Delicious apple cultivar, 13.40% [40]. With respect to the titratable acidity, the studied fruit pulp showed 4.61 ± 0.13% which is higher than *Aegle mormelos*, 1.2% [41]. The pH value reported in the fruit pulp was 3.61 ± 0.09, which is comparatively similar to the Golden Delicious apple cultivar, 3.79% [40], and less than *Phoenix dactylifera*, 6.20, and sel-42 papaya cultivar, 4.03% [42,43].

Total ash content in the current study was 2.73 ± 0.12% and it is comparable to the reported values of kumquat fruit [44]. The total ash content of the pulp was directly proportional to the presence of mineral elements. The crude fiber value obtained from the current study showed wood apple is a good source of dietary fiber, which is 3.32 ± 0.02%. Results revealed that the studied fruit pulp has a lower content of dietary fiber than kumquat fruit (5.31%) [44]. It has been proposed that consumption of fruits with high dietary fiber has various health benefits including being a carrier of antioxidants in the gastrointestinal tract to reduce risk factors related to cardiovascular and intestinal diseases [45]. Total protein content was 25.24 ± 0.07% and it is comparable with African mangosteen (31.76%) and higher than kumquat fruits (7.38%) [6,44]. A total of 24.74 ± 0.19% of carbohydrate content was observed in the studied wood apple fruit pulp. Total carbohydrate content was comparatively higher than *Annona squamosa* (10.80%) and kumquat fruits (5.23%) [38,44]. The total phenolics content was 0.50 ± 0.01% and the phenolic values are comparable with Mauritian citrus fruits [46]. Nutritional analysis results reflect that wood apple fruit pulp is an important source to meet nutritional requirements because it plays a major role in human nutrition and health.

Table 1. Proximate and total phenolic composition of wood apple fruit pulp [1].

Sl. No.	Parameters	Composition
1	Fruit size (cm)	
	Length	7.42 ± 0.19
	Width	7.43 ± 0.10
	Circumference	23.42 ± 0.29
2	Fruit weight (g)	179.45 ± 1.42
3	Moisture (% FW)	58.89 ± 1.21
4	Oil (% FW)	0.99 ± 0.01
5	TSS (°Brix)	19.52 ± 0.17
6	pH	3.61 ± 0.09
7	Titratable acidity (% FW)	4.61 ± 0.13
8	Ash (% FW)	2.73 ± 0.12
9	Crude fiber (% FW)	3.32 ± 0.02
10	Total protein (% FW)	9.30 ± 0.16
11	Total carbohydrate (% FW)	24.74 ± 0.19
12	Total phenolics (% DW) [2]	0.50 ± 0.01

[1] Mean ± standard error of triplicate determinations.. TSS—Total soluble solids; pH—Hydrogen ion concentration; [2] Phenolic content was estimated as gallic acid equivalents.

3.2. Fatty Acid Composition

There are no previous reports on the wood apple fruit pulp fatty acid profile. The crude fat content in wood apple fruit pulp was 0.99 ± 0.01% and the fatty acid profile represents the higher amount of unsaturated fatty acids. The fruits are rich in contents of unsaturated fatty acids are valuable sources for treating various ailments. Oils are essential for a healthy life because they are important natural sources of essential fatty acids [47,48]. Unsaturated fatty acids act as an important flavor precursor which plays a role in the storage and release of aroma components and directly influences aroma development [49].

The results showed (Table 2 and Figure 2) that the sum of saturated fatty acids (SFAs) was 32.17 ± 0.35%, whereas the monounsaturated fatty acids (MUFAs) were 26.20 ± 0.33%, and the polyunsaturated fatty acids (PUFAs) were 25.78 ± 0.61% in the pulp oil of wood apple fruit. The abundance of unsaturated fatty acids was observed earlier in jujube fruits of 'Choutrana', 'Mahres', 'Mahdia', and 'Sfax' cultivars (68.5% to 72.4%) [50]. The most abundant PUFA detected in wood apple pulp oil was α-linolenic acid (16.55 ± 0.26%), followed by 10.02 ± 0.43% of linoleic acid. The α-linolenic acid and linoleic acid contents of the wood apple fruit pulp were higher than the observed results in the varieties of *Ziziphus jujuba* [48]. The results showed that wood apple pulp oil is a good source of essential fatty acids. Oleic acid (23.89 ± 0.06%) was the most predominant monounsaturated fatty acid observed and this is higher as compared to the observed results of genotypes of *Ziziphus jujuba* [47]. MUFAs are very well known for many health benefits and are a good source to decrease the success of low-density lipoprotein (LDL) cholesterol levels which bring down the heart-related risks [51].

Table 2. Fatty acid methyl ester compositions of wood apple (*Limonia acidissima* L.) fruit pulp oil [1].

Peak	t_R (min)	Common Name	Fatty Acid Methyl Esters	Value (%)
1	7.93		Cyclooctasiloxane, hexadecamethyl-	0.45 ± 0.05
2	9.52		Cyclononasiloxane, octadecamethyl-	0.38 ± 0.05
3	9.66	Acetophenone	Acetophenone	2.34 ± 0.06
4	9.95	Lauric acid	Dodecanoic acid, methyl ester (C12:0)	1.62 ± 0.02
5	12.03	Myristic acid	Methyl tetradecanoate (C14:0)	1.74 ± 0.04
6	13.21	Pentadecylic acid	Pentadecanoic acid, methyl ester (C15:0)	0.44 ± 0.02
7	14.5	Palmitic acid	Hexadecanoic acid, methyl ester (C16:0)	18.52 ± 0.12
8	15.09		Phenol, 2,4-bis(1,1-dimethylethyl)-	7.70 ± 0.06
9	17.42	Stearic acid	Methyl stearate (C18:0)	9.02 ± 0.08
10	18.03	Oleic acid	9-Octadecenoic acid (Z)-, methyl ester (C18:1n9c)	23.89 ± 0.06
11	18.16	Vaccenic acid	11-Octadecenoic acid, methyl ester (C18:1n7)	1.78 ± 0.23
12	19.11	Linoleic acid (ω- 6)	9,12-Octadecadienoic acid (Z,Z)-, methyl ester (C18:2n6)	9.23 ± 0.35
13	20.46	α-Linolenic acid (ω-3)	9,12,15-Octadecatrienoic acid, methyl ester, (Z,Z,Z)-(C18:3n3)	16.55 ± 0.26
14	20.67	Arachidic acid	Eicosanoic acid, methyl ester (C20:1)	0.81 ± 0.06
15	21.38	Paullinic acid (ω-7)	cis-13-Eicosenoic acid, methyl ester (C20:1n7)	0.52 ± 0.03
16	25.60		2,5-di-tert-Butyl-1,4-dimethoxybenzene	2.77 ± 0.07
		$\sum$ Saturated fatty acids		32.17 ± 0.35
		$\sum$ Monounsaturated fatty acids		26.20 ± 0.33
		$\sum$ Polyunsaturated fatty acids		25.78 ± 0.61

[1] Means ± standard error of triplicate determinations.

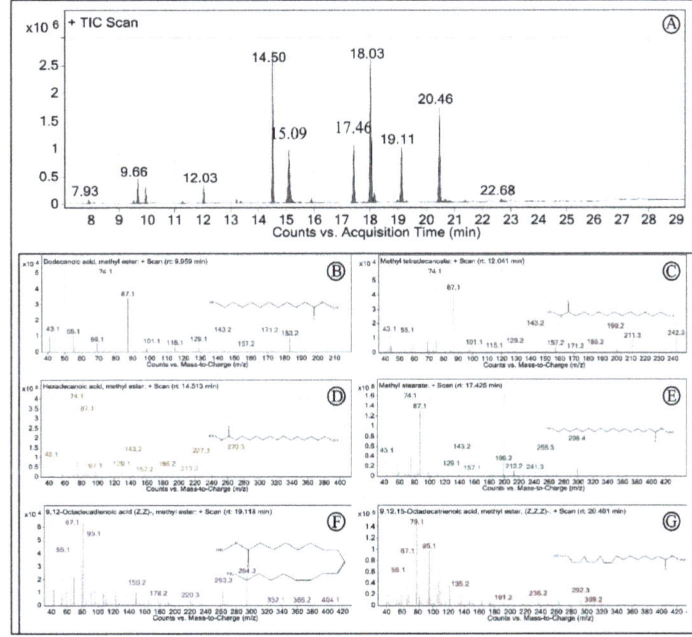

Figure 2. GC-MS chromatograms of wood apple (*Limonia acidissima* L.) fruit pulp oil fatty acid methyl esters (FAMEs) (**A**) and MS spectra of major FAMEs of analyzed fatty acids. Dodecanoic acid (**B**), methyl tetradecanoate (**C**), methyl palmitate (**D**), methyl stearate (**E**), methyl linoleate (**F**), methyl linolenate (**G**).

The monounsaturated, fatty acid-rich oil is advisable in terms of nutrition and provides enough stability to be used for baking purposes [52]. The most abundant saturated fatty acid present in oil was palmitic acid (18.52 ± 0.12%), whereas the stearic acid, myristic acid, lauric acid, arachidic acid, and pentadecylic acids were 9.02 ± 0.08%, 1.74 ± 0.04%, 1.62 ± 0.02%, 0.81 ± 0.06%, and 0.44 ± 0.02%, respectively. Saturated fatty acids observed in wood apple fruit pulp were higher than the genotypes of *Ziziphus jujuba* [47]. A higher content of saturated fatty acids in oils represents that they have more oxidative stability, as well as that they are well accepted for cooking, including frying and baking, and are considered as important constituents for cell membranes, secretary, and transport lipids with crucial roles in protein palmitoylation and palmitoylated signal molecules [11,53]. Wood apple pulp oil showed a higher percentage of PUFAs, which have the important natural sources of essential fatty acids, omega-3 and omega-6, and fruits with good fatty acid profiles have great health potential for both fresh consumption and/or industrial processing.

3.3. Investigation of Sugar Profile by UPLC-ELSD

The profiling of free sugars was achieved with the UPLC-ELSD and the results are shown in Figure 3 and Table 3. Sugars are much needed to decide the quality and maturity characters of fruits, whereas fructose and glucose are the vital components [54]. The total free sugar content in defatted fruit pulp was 31.59 ± 0.17% DW, whereas the abundant sugars present in wood apple fruit pulp were fructose (16.40 ± 0.23%) and glucose (14.23 ± 0.10%), and the minor contents of sugars were rhamnose (0.24 ± 0.01%), sucrose (0.13 ± 0.01%), and maltose (0.57 ± 0.03%). During the period of total fruit maturation, the sugar content is increased and, at the time when the matured fruit detaches from the tree, it reaches the maximum level and the increased sugar content in wood apple is because of starch decomposition [55]. The fructose and glucose abundance in repined fruits makes the pulp soft and sweeter, which influence the quality of consumer acceptance [54], whereas the unripe fruits are very sour and astringent [56],

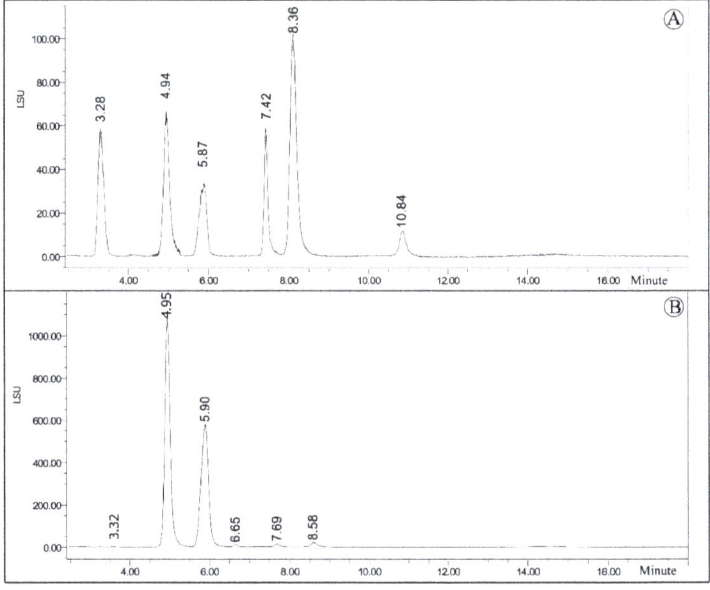

Figure 3. UPLC-ELSD chromatograms of free sugar standards (**A**) and wood apple (*Limonia acidissima* L.) fruit pulp (**B**).

Table 3. Free sugar composition (%) of wood apple defatted fruit pulp [1].

Sl. No.	Retention Time	Free Sugars	Composition [2]
1	3.32	Rhamnose	0.24 ± 0.01
2	4.93	Fructose	16.40 ± 0.23
3	5.89	Glucose	14.23 ± 0.10
4	7.71	Sucrose	0.13 ± 0.01
5	8.33	Maltose	0.57 ± 0.03
	Total sugars		31.59 ± 0.17

[1] DW—Dry weight. [2] Means $\pm$ standard error of triplicate determinations.

3.4. Investigation of Organic Acid Profile by HPLC-PDA

The present study showed nine organic acids, including galacturonic acid, oxalic acid, tartaric acid, malic acid, ascorbic acid, acetic acid, citric acid, succinic, and pyruvic acid. Fruit flavor may be contributed to by the presence of different types of organic acids in fruits. The results of the organic acid analysis is shown in Table 4 and Figure 4. The total amount of organic acid present in wood apple pulp was $17.46 \pm 0.23\%$ DW. The major organic acids were ascorbic acid ($4.51 \pm 0.05\%$), citric acid ($4.27 \pm 0.04\%$), and tartaric acid ($4.01 \pm 0.03\%$), whereas, the amount of succinic acid ($1.83 \pm 0.03\%$), galacturonic acid ($0.93 \pm 0.02\%$), acetic acid ($0.81 \pm 0.06\%$), pyruvic acid ($0.79 \pm 023\%$), malic acid ($0.23 \pm 0.01\%$), and oxalic acid ($0.05 \pm 0.01\%$) were in minor concentrations. The citric acid content in passion fruit (4.35%) was approximately equal, whereas in *Ficus carica* it was 0.69% and *Ziziphus jujuba* 0.19% [57,58]. Malic and succinic acids in *Ziziphus jujuba* were 0.22% and 0.01% respectively [57]. Many studies revealed that the fruit's of citrus family is majorly constituted of citric acid, malic acid, quininic acid, and oxalic acid. Together, these organic acids influence the citrus fruit acidity. Among these organic acids, citric acid is the largest contributor to the acidity taste and accounted for 79%, 71%, and 45% of total organic acids in acidic lemon, lime, and orange, respectively [59].

Table 4. Organic acid compositions of wood apple defatted fruit pulp (% DW) [1].

Sl. No.	t_R (min)	Organic Acids	Composition [2]
1	3.06	D-Galacturonic acid	0.93 ± 0.02
2	3.37	Oxalic acid	0.05 ± 0.01
3	3.55	D-Tartaric acid	4.01 ± 0.03
4	4.48	Malic acid	0.23 ± 0.01
5	4.74	Ascorbic acid	4.51 ± 0.05
6	5.59	Acetic acid	0.81 ± 0.06
7	8.74	Citric acid	4.27 ± 0.04
8	9.59	Succinic acid	1.83 ± 0.03
9	13.31	Pyruvic acid	0.79 ± 0.01
	Total organic acids		17.46 ± 0.23

[1] DW—Dry weight. [2] Means $\pm$ standard error of triplicate determinations.

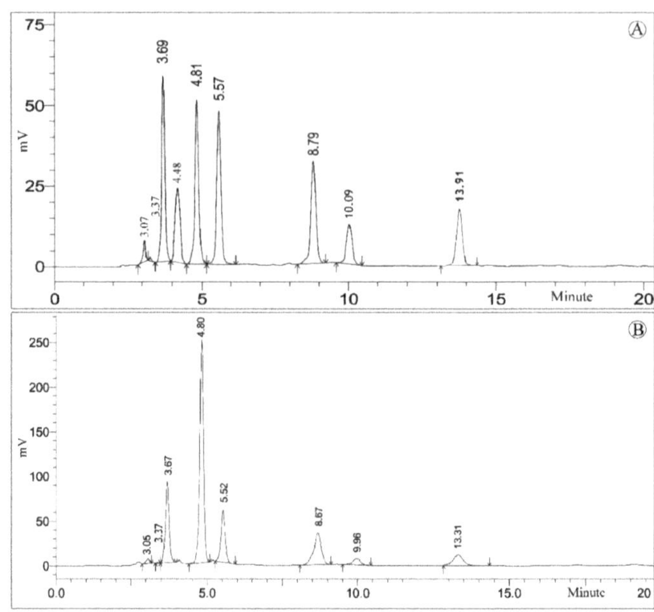

Figure 4. HPLC-PDA chromatograms of organic acid standards (**A**) and wood apple (*Limonia acidissima* L.) fruit pulp (**B**).

4. Conclusions

The characterization of wood apple fruit pulp showed a higher percentage of carbohydrate (24.74 ± 0.19%), protein (9.30 ± 0.16%), ash (2.73 ± 0.12%), and fiber (3.32 ± 0.02%), which is comparable to other citrus fruits such as bael (*Aegle mermelos*) and kumquat (*Fortunella marginata*). Wood apple fruits are also rich in organic acids, such as D-tartaric, ascorbic, and citric acid, which are accountable for the taste and quality of the fruits. The total free sugar content in defatted wood apple fruit pulp was 31.59 ± 0.17% DW. Fructose (16.40 ± 0.23%) and glucose (14.23 ± 0.10%) were the major sugars of the fruit pulp. Fatty acid profiling revealed that wood apple possesses a higher percentage of unsaturated fatty acids including oleic, α-linolenic, linoleic, and vaccenic acids. The results of the present study showed that the wood apple fruits are rich in nutritional components.

Author Contributions: Conceptualization, H.N.M. and K.A.A.-A.; methodology, H.N.M., K.A.A.-A., and Y.H.D.; Formal analysis, H.N.M., K.A.A.-A., Y.H.D., and J.J.R.; investigation and data curation, S.L. and H.N.M.; validation, S.L., H.N.M., K.A.A.-A., Y.H.D., and J.J.R.; writing—original draft preparation, S.L., H.N.M., K.A.A.-A., and Y.H.D.; writing—review and editing, S.L., H.N.M., K.A.A.-A., Y.H.D., and J.J.R. All authors have read and agreed to the published version of the manuscript.

Funding: Researchers Supporting Project number (RSP-2021/375), King Saud University, Riyadh, Saudi Arabia.

Institutional Review Board Statement: Not applicable.

Informed Consent Statement: Not applicable.

Data Availability Statement: All data are presented within the article.

Acknowledgments: The authors acknowledge Researchers Supporting Project number (RSP-2021/375), King Saud University, Riyadh, Saudi Arabia.

Conflicts of Interest: The authors declare no conflict of interest.

References

1. Murthy, H.N.; Bapat, V.A. Importance of underutilized fruits and nuts. In *Bioactive Compounds in Underutilized Fruits and Nuts, Reference Series in Phytochemistry*; Murthy, H.N., Bapat, V.A., Eds.; Springer Nature: Cham, Switzerland, 2020; pp. 3–19.
2. Murthy, H.N.; Dalawai, D. Bioactive compounds of wood apple (*Limonia acidissima* L.). In *Bioactive Compounds in Underutilized Fruits and Nuts, Reference Series in Phytochemistry*; Murthy, H.N., Bapat, V.A., Eds.; Springer Nature: Cham, Switzerland, 2020; pp. 543–569.
3. Murthy, H.N.; Dalawai, D.; Dewir, Y.H.; Ibrahim, A. Phytochemicals and biological activities of *Garcinia morella* (Gaertn.) Desr.: A review. *Molecules* **2020**, *25*, 5690. [CrossRef]
4. Murthy, H.N.; Yadav, G.G.; Dewir, Y.H.; Ibrahim, A. Phytochemicals and biological activity of desert date (*Belanites aegyptica* (L.) Delile). *Plants* **2020**, *10*, 32. [CrossRef]
5. Manohar, S.H.; Naik, P.M.; Patil, L.M.; Karikatti, S.I.; Murthy, H.N. Chemical composition of *Garcinia xanthochymus* seeds, seed oil, and evaluation of its antimicrobial and antioxidant activity. *J. Herbs Spices Med. Plants* **2014**, *20*, 148–155. [CrossRef]
6. Joseph, K.S.; Bolla, S.; Joshi, K.; Bhat, M.; Naik, K.; Patil, S.; Bendre, S.; Gangappa, B.; Haibatti, V.; Payamalle, S.; et al. Determination of chemical composition and nutritive value with fatty acid compositions of African Mangosteen (*Garcinia livingstonei*). *Erwerbs-Obstbau* **2016**, *59*, 195–202. [CrossRef]
7. Murthy, H.N.; Dalawai, D.; Mamatha, U.; Angadi, N.B.; Dewir, Y.H.; Al-Suhaibani, N.A.; El-Hendawy, S.; Al-Ali, A.M. Bioactive constituents and nutritional composition of *Bridelia stipularis* L. Blume fruits. *Int. J. Food Prop.* **2021**, *24*, 796–805. [CrossRef]
8. Murthy, H.N.; Dandin, V.S.; Dalawai, D.; Park, S.Y.; Paek, K.Y. Bioactive compounds from *Garcinia* fruits of high economic value for food and health. In *Bioactive Molecules in Food, Reference Series in Phytochemistry*; Merllion, J.M., Ramawat, K.G., Eds.; Springer Nature: Cham, Switzerland, 2018; pp. 1–26.
9. Yoshida, H.; Hirakawa, Y.; Abe, S. Roasting influences on molecular species of triacylglycerols in sunflower seeds (*Helianthus annuus* L.). *Food Res. Int.* **2001**, *34*, 613–619. [CrossRef]
10. Timilsena, Y.P.; Vongsvivut, J.; Adhikari, R.; Adhikari, B. Physicochemical and thermal characteristics of Australian chia seed oil. *Food Chem.* **2017**, *228*, 394–402. [CrossRef]
11. German, J.B. Dietary lipids from an evolutionary perspective: Sources, structures and functions. *Matern. Child Nutr.* **2011**, *7*, 2–16. [CrossRef]
12. Westman, E.C. Is dietary carbohydrate essential for human nutrition? *Am. J. Clin. Nutr.* **2002**, *75*, 951–953. [CrossRef]
13. Yu, S.M. Cellular and genetic responses of plants to sugar starvation. *Plant Physiol.* **1999**, *121*, 687–693. [CrossRef] [PubMed]
14. Finkle, B.J.; Zavala, M.E.; Ulrich, J.M. Cryoprotective compounds in the viable freezing of plant tissues. In *Cryopreservation of Plant Cells and Organs*; Kartha, K.K., Ed.; CRC Press Inc.: Boca Raton, FL, USA, 1985; pp. 75–115.
15. Bekers, M.; Marauska, M.; Grube, M.; Karklina, D.; Duma, M. New prebiotics for functional food. *Acta Aliment.* **2004**, *33*, 31–37. [CrossRef]
16. Rai, A.K.; Prakash, M.; Anu Appaiah, K.A. Production of *Garcinia* wine: Changes in biochemical parameters, organic acids and free sugars during fermentation of Garcinia must. *Int. J. Food Sci. Technol.* **2010**, *45*, 1330–1336. [CrossRef]
17. Orwa, C.; Mutua, A.; Kindt, R.; Jamnadass, R.; Anthony, S. *Agroforestry Database: A Tree Reference and Selection Guide Version 4.0*; World Agroforestry Centre: Nairobi, Kenya, 2009.
18. Singh, A.K.; Singh, S.; Singh, R.S.; Joshi, H.K.; Sharma, S.K. Characterization of bael (*Aegle marmelos*) varieties under rainfed hot semi-arid environment of western India. *Indian J. Agric. Sci.* **2014**, *84*, 1236–1242.
19. Lim, T.K. *Limonia acidissima* . In *Edible Medicinal and Non-Medicinal Plants: Fruits*; Lim, T.K., Ed.; Springer Science & Business Media: Berlin/Heidelberg, Germany, 2012; Volume 4, pp. 884–889.
20. Chatterjee, T.K. *Herbal Options*, 3rd ed.; Books and Allied (P) Ltd.: Calcutta, India, 2003; pp. 203–256.
21. Joshi, R.K.; Badakar, V.M.; Kholkute, S.D.; Khatib, N. Chemical composition and antimicrobial activity of the essential oil of the leaves of *Feronia elephantum* (Rutaceae) from North West Karnataka. *Nat. Prod. Commun.* **2011**, *6*, 141–143. [CrossRef] [PubMed]
22. Gupta, R.; Johri, S.; Saxena, A.M. Effect of ethanolic extract of *Feronia elephantum* Correa fruits on blood glucose levels in normal and streptozotocin induced diabetic rats. *NPR* **2009**, *8*, 32–36.
23. Ilango, K.; Chitra, V. Wound healing and antioxidant activities of the fruit pulp of *Limonia acidissima* Linn (Rutaceae) in rats. *Trop. J. Pharm. Res.* **2010**, *9*, 223–230. [CrossRef]
24. Ilaiyaraja, N.; Likhith, K.R.; Sharath Babu, G.R.; Khanum, F. Optimization of extraction of bioactive compounds from *Feronia limonia* (wood apple) fruit using response surface methodology (RSM). *Food Chem.* **2015**, *173*, 348–354. [CrossRef] [PubMed]
25. Lamani, S.; Anu-Appaiah, K.A.; Murthy, H.N.; Dewir, Y.H.; Rihan, H.Z. Fatty acid profile, tocopherol content of seed oil, and nutritional analysis of seed cake of wood-apple (*Limonia acidissima* L.), an underutilized fruit-yielding tree species. *Horticulturae* **2021**, *7*, 275. [CrossRef]
26. AOAC. *Official Methods of Analysis of AOAC International*; AOAC: Gaithersburg, MD, USA, 2005.
27. Sadasivam, S.; Manickam, A. *Biochemical Methods*, 3rd ed.; New Age International (P) Limited, Publishers: New Delhi, India, 2008.
28. AOAC. *Official Methods of Analysis of Association of Analytical Chemists*, 15th ed.; AOAC: Washington, DC, USA, 1998.
29. Anonymous. *Manual of Methods of Analysis of Foods: Fruit and Vegetable Products. Lab. Manual 5*; Food Safety and Standards Authority of India; Ministry of Health and Family Welfare: New Delhi, India, 2015; Volume 40, p. 11.

30. Singleton, V.L.; Orthofer, R.; Lamuela-Raventos, R.S. Analysis of total phenols and other oxidation substrates and antioxidants by means of Folin-ciocalteau Reagent. *Methods Enzymol.* **1999**, *299*, 152–178.
31. Morison, W.R.; Smith, L.M. Preparation of fatty acid methyl esters and dimethylacetals from lipids with boron fluoride-methanol. *J. Lipid Res.* **1964**, *5*, 600–608. [CrossRef]
32. Ghfar, A.A.; Wabaidur, S.M.; Ahmed, A.Y.B.H.; Alothman, Z.A.; Khan, M.R.; Al-Shaalan, N.H. Simultaneous determination of monosaccharides and oligosaccharides in dates using liquid chromatography—Electrospray ionization mass spectrometry. *Food Chem.* **2015**, *176*, 487–492. [CrossRef] [PubMed]
33. Koh, D.; Park, J.; Lim, J.; Yea, M.; Bang, D. A rapid method for simultaneous quantification of 13 sugars and sugar alcohols in food products by UPLC-ELSD. *Food Chem.* **2018**, *240*, 694–700. [CrossRef] [PubMed]
34. Jayaprakasha, G.K.; Sakariah, K.K. Determination of organic acids in *Garcinia cambogia* (Desr.) by high-performance liquid chromatography. *J. Chromatogr. A* **1998**, *806*, 337–339. [CrossRef]
35. Singh, A.K.; Singh, S.; Yadav, V.; Sharma, B.D. Genetic variability in wood apple (*Feronia limonia*) from Gujarat. *Indian J. Agric. Sci.* **2016**, *86*, 1504–1508.
36. Poongodi Vijayakumar, T.; Punitha, K.; Banupriya, L. Drying characteristics and quality evaluation of wood apple (*Limonia acidissima* L.) fruit pulp powder. *Int. J. Curr. Trends Res.* **2013**, *2*, 147–150.
37. Sonawane, S.K.; Bagul, M.B.; LeBlanc, J.G.; Arya, S.S. Nutritional, functional, thermal and structural characteristics of *Citrullus lanatus* and *Limonia acidissima* seed flours. *J. Food Meas. Charact.* **2016**, *10*, 72–79. [CrossRef]
38. Abdualrahman, M.A.Y.; Ma, H.; Zhou, C.; Yagoub, A.E.A.; Ali, A.O.; Tahir, H.E.; Wali, A. Postharvest physicochemical properties of the pulp and seed oil from *Annona squamosa* L. (Gishta) fruit grown in Darfur region, Sudan. *Arab. J. Chem.* **2019**, *12*, 4514–4521. [CrossRef]
39. Cheema, J.; Yadav, K.; Sharma, N.; Saini, I.; Aggarwal, A. Nutritional quality characteristics of different wild and underutilized fruits of Terai region, Uttarakhand (India). *Int. J. Fruit Sci.* **2016**, *17*, 72–81. [CrossRef]
40. Wu, J.; Gao, H.; Zhao, L.; Liao, X.; Chen, F.; Wang, Z.; Hu, X. Chemical compositional characterization of some apple cultivars. *Food Chem.* **2007**, *103*, 88–93. [CrossRef]
41. Sagar, V.R.; Kumar, R. Effect of drying treatments and storage stability on quality characteristics of bael powder. *J. Food Sci. Technol.* **2014**, *51*, 2162–2168. [CrossRef] [PubMed]
42. Siddeeg, A.; Zeng, X.; Ammar, A.; Han, Z. Sugar profile, volatile compounds, composition and antioxidant activity of sukkari date palm fruit. *J. Food Sci. Technol.* **2019**, *56*, 754–762. [CrossRef]
43. Kelebek, H.; Selli, S.; Gubbuk, H.; Gunes, E. Comparative evaluation of volatiles, phenolics, sugars, organic acids and antioxidant properties of Sel-42 and Tainung papaya varieties. *Food Chem.* **2015**, *173*, 912–919. [CrossRef] [PubMed]
44. e Souza, C.S.; Anunciacao, P.C.; Della Lucia, C.M.; Rodrigues das Dores, R.G.; de Miranda Milagres, R.C.R.; Pinheiro Sant'Ana, H.M. Kumquat (*Fortunella marginata*): A good alternative for ingestion of nutrients and bioactive compounds. *Proceedings* **2021**, *70*, 105. [CrossRef]
45. Saura-Calixto, F. Dietary fiber as a carrier of dietary antioxidants: An essential physiological function. *J. Agric. Food. Chem.* **2011**, *59*, 43–49. [CrossRef]
46. Ramful, D.; Bahorun, T.; Bourdon, E.; Tarnus, E.; Aruoma, O.I. Biactive phenolics and antioxidant propensity of flavedo extracts of Mauritian citrus fruits: Potential prophylactic ingredients for functional food application. *Toxicology* **2010**, *278*, 75–87. [CrossRef]
47. San, B.; Yildirim, A.N. Phenolic, alpha-tocopherol, beta-carotene and fatty acid composition of four promising jujube (*Ziziphus jujuba* Miller) selections. *J. Food Compos. Anal.* **2010**, *23*, 706–710. [CrossRef]
48. Reche, J.; Almansa, M.S.; Hernandez, F.; Carbonell-Barrachina, A.A.; Legua, P.; Amoros, A. Fatty acid profile of peel and pulp of Spanish jujube (*Ziziphus jujuba* Mill.) fruit. *Food Chem.* **2019**, *295*, 247–253. [CrossRef]
49. Sanchez-Salcedo, E.M.; Sendra, E.; Carbonell-Barrachina, A.A.; Martínez, J.J.; Hernandez, F. Fatty acids composition of Spanish black (*Morus nigra* L.) and white (*Morus alba* L.) mulberries. *Food Chem.* **2016**, *190*, 566–571. [CrossRef]
50. Elaloui, M.; Laamouri, A.; Albouchi, A.; Cerny, M.; Mathieu, C.; Vilarem, G.; Hasnaoui, B. Chemical compositions of the tunisian *Ziziphus jujuba* oil. *Emir. J. Food Agric.* **2014**, *26*, 602–608. [CrossRef]
51. FAO/WHO. Fats and fatty acids in human nutrition. In *Report of an Expert Consultation*; FAO/WHO: Geneva, Switzerland, 2010.
52. Petukhov, I.; Malcolmson, L.J.; Przybylski, R.; Armstrong, L. Frying performance of genetically modified canola oils. *JAOCS* **1999**, *76*, 627–632. [CrossRef]
53. Matthaus, B. Use of palm oil for frying in comparison with other high-stability oils. *Eur. J. Lipid Sci. Technol.* **2007**, *109*, 400–409. [CrossRef]
54. Hu, W.; Sun, D.W.; Pu, H.; Pan, T. Recent developments in methods and techniques for rapid monitoring of sugar metabolism in fruits. *Compr. Rev. Food Sci. Food Saf.* **2016**, *15*, 1067–1079. [CrossRef]
55. Murrinie, E.D.; Yudono, P.; Purwantoro, A.; Sulistyaningsih, E. Morphological and physiological changes during growth and development of wood apple (*Feronia limonia* (L.) Swingle) fruit. *Int. J. Bot.* **2017**, *13*, 75–81. [CrossRef]
56. Seymour, G.B.; Taylor, J.E.; Tucker, C.A. *Biochemistry of Fruit Ripening*; Chapman and Hall: London, UK, 1993.
57. Gao, Q.H.; Wu, C.S.; Wang, M.; Xu, B.N.; Du, L.J. Effect of drying of jujubes (*Ziziphus jujuba* Mill.) on the contents of sugars, organic acids, α-tocopherol, β-carotene, and phenolic compounds. *J. Agric. Food Chem.* **2012**, *60*, 9642–9648. [CrossRef] [PubMed]

58. Slatnar, A.; Klancar, U.; Stampar, F.; Veberic, R. Effect of Drying of Figs (*Ficus carica* L.) on the contents of sugars, organic acids and phenolic compounds. *J. Agric. Food Chem.* **2011**, *59*, 11696–11702. [CrossRef] [PubMed]
59. Albertini, M.; Carcouet, E.; Pailly, C.O.; Gambotti, C.; Luro, F.; Berti, L. Changes in organic acids and sugars during early stages of development of acidic and acidless citrus fruit. *J. Agric. Food Chem.* **2006**, *54*, 8335–8339. [CrossRef]

horticulturae

Article

Seaweed Extract Improves *Lagenaria siceraria* Young Shoot Production, Mineral Profile and Functional Quality

Beppe Benedetto Consentino [1], Leo Sabatino [1,*], Rosario Paolo Mauro [2], Carlo Nicoletto [3], Claudio De Pasquale [1], Giovanni Iapichino [1] and Salvatore La Bella [1]

[1] Department of Agricultural, Food and Forest Sciences (SAAF), University of Palermo, 90128 Palermo, Italy; beppebenedetto.consentino@unipa.it (B.B.C.); claudio.depasquale@unipa.it (C.D.P.); giovanni.iapichino@unipa.it (G.I.); salvatore.labella@unipa.it (S.L.B.)

[2] Department of Agriculture, Food and Environment (Di3A), University of Catania, Via Valdisavoia, 5, 95123 Catania, Italy; rosario.mauro@unict.it

[3] Department of Agronomy, Food, Natural Resources, Animals and Environment (DAFNAE), University of Padova, 35020 Legnaro, Italy; carlo.nicoletto@unipd.it

* Correspondence: leo.sabatino@unipa.it; Tel.: +39-09123862252

Citation: Consentino, B.B.; Sabatino, L.; Mauro, R.P.; Nicoletto, C.; De Pasquale, C.; Iapichino, G.; La Bella, S. Seaweed Extract Improves *Lagenaria siceraria* Young Shoot Production, Mineral Profile and Functional Quality. *Horticulturae* **2021**, *7*, 549. https://doi.org/10.3390/horticulturae7120549

Academic Editor: Douglas D. Archbold

Received: 15 November 2021
Accepted: 30 November 2021
Published: 3 December 2021

Abstract: Vegetable landraces represent the main source of biodiversity in Sicily. *Lagenaria siceraria* is appreciated by Southern Mediterranean consumers for its immature fruits and young shoots. Plant-based biostimulants supply, such as seaweed extract (SwE), is a contemporary and green agricultural practice applied to ameliorate the yield and quality of vegetables. However, there are no studies concerning the effects of SwE on *L. siceraria*. The current study evaluated the effects of SwE foliar application (0 or 3 mL L^{-1}) on five *L. siceraria* landraces (G1, G2, G3, G4 and G5) grown in greenhouses. Growth traits, first female flower emission, fruit yield, young shoot yield, fruit firmness, young shoot nitrogen use efficiency (NUE$_{ys}$) and specific young shoot quality parameters, such as soluble solids content (SSC), mineral profile, ascorbic acid, and polyphenols, were appraised. Plant height and number of leaves at 10, 20 and 30 days after transplant (DAT) were significantly higher in plants treated with SwE as compared with untreated plants. Treating plants with SwE increased marketable fruit yield, fruit mean mass, young shoot yield and number of young shoots by 14.4%, 15.0%, 22.2%, 32.4%, and 32.0%, respectively as compared with untreated plants. Relevant increments were also recorded for NUE$_{ys}$, P, K, Ca, Mg, ascorbic acid and polyphenols concentration. SwE application did not significantly affect total yield and SSC. Furthermore, SwE treated plants produced a lower number of marketable fruits than non-treated plants. The present study showed that SwE at 3 mL L^{-1} can fruitfully enhance crop performance, young shoot yield and quality of *L. siceraria*.

Keywords: plant-based biostimulants; foliar application; bottle gourd landraces; greenhouse cultivation; crop production; NUE

1. Introduction

Lagenaria siceraria (Mol.) Stand., generally known as the bottle gourd or white-flowered, is a climbing annual monoecius plant belonging to the cucurbitaceous family native to Africa (Zimbabwe). *L. siceraria* species comprises two different subspecies: *L. siceraria* ssp. *siceraria* and *L. siceraria* ssp. *asiatica*. Tribal groups sited in the Northern Telangana area use its dry fruit shells as bottles, pots, music instruments or as fishing tools [1]. Bottle gourd is also used in traditional Indian medicine as cardiotonic, aphrodisiac, hepatoprotective, analgesic, anti-inflammatory and diuretic [2–4]. Nowadays, the bottle gourd is grown in India and the Mediterranean area—mostly in Sicily—for its immature fruits, young leaves and shoots, these last being consumed as green leafy vegetables.

In Sicily, over an area of 26,000 km^2, Raimondo et al. [5] estimated 3252 taxa. Consequently, Sicily is an essential centre of origin and differentiation of a number of vegetables [6–11] cultivated both in the open field and in a protected environment. Although

Sicily is not the area of origin for *L. siceraria*, bottle gourd landraces cultivated in Sicily show significant diversity [8]. Herbaceous grafting is considered a toolbox to face biotic and abiotic plant distresses related to monocropping in intensive protected vegetable cultivation systems [12–14]. In this regard, *L. siceraria* is also used as a rootstock for watermelon to improve growth, yield, fruit quality, biotic and/or abiotic stresses tolerance [15,16].

Bottle gourd yield and quality depend on diverse factors such as climatic conditions, soil fertility, agronomical practices and diseases [17]. Currently, to enhance crop production, modern agriculture usually adopts high quantities of fertilizers which, however, have a deleterious environmental impact [18]. Thus, there are considerable research efforts to find new green cultivation technologies to boost the yield and quality of vegetables. In regard to these considerations, biostimulant applications is a valuable and eco-friendly technology to improve vegetable quality traits [18–24]. Among different classes of biostimulants, seaweed extracts (SwEs) are very appreciated. They are composed of different types of seaweeds, although the most used in agriculture are brown algae (e.g., *Ecklonia maxima* and *Ascophillum nodosum*). These algae are appreciated for their content of polysaccharides, betaines, micro- and macronutrients and hormones, which improve plant production and overall quality [22,25,26]. Their positive effects on plants under optimal, sub-optimal or unfavourable conditions are related to several biochemical and physiological mechanisms such as the elicitation of enzymes involved in carbon and nitrogen metabolic paths, the stimulation of phytohormones synthesis and the improvement in mineral uptake and accumulation through the increase of the root system size [27–29]. However, the application of SwEs on *Lagenaria siceraria* has not been examined yet. The SwEs supply might affect immature fruits, young shoot yield and quality.

Taking into account all the abovesaid and considering that: (i) bottle gourd is an underutilised species; (ii) immature fruits, young leaves and shoots of bottle gourd are, however, very appreciated by Mediterranean consumers [30]; (iii) seaweed extracts may boost plant performance of vegetables, the purpose of the current work was to appraise the influence of seaweed extract on yield and quality of fruits and young shoots of five local landraces of *L. siceraria* grown in greenhouses.

2. Materials and Methods

2.1. Experimental Field and Treatments

The study was performed in Marsala, during the winter-spring period of 2019, in an experimental field of the Department of Agricultural, Food, and Forestry Sciences of the University of Palermo (SAAF) (latitude 12°26′ N, longitude 37°47′ E, altitude 37 m). Seeds of five *L. siceraria* landraces (coded G1, G2, G3, G4 and G5) [8]—from self-pollinated flowers—were sown on 10 December 2018 in plug trays (66 cells) filled with a peat moss-based substrate (FAP, Padova, Italy). Plug plants were transplanted on 15 January 2019 in an unheated greenhouse, 2 m between rows and 1 m intra-row, obtaining a plant density of 0.5 plant m^{-2}. The soil hosting the experiment was composed of sand (<78%) at a pH of 8.3 and high activity limestone at 9.0%. The exchangeable K_2O (655 mg kg^{-1}), P (70 mg kg^{-1}), total N (2.2%), and organic matter (8 t·ha^{-1}) were also determined [31–33].

Plants were fertigated through a drip irrigation system, with 80, 50 and 80 kg ha^{-1} of N, in form of ammonium nitrate (Yara Italia S.p.A., Milan, Italy), P_2O_5, in form of superphosphate (Siriac, Ragusa, Italy) and K_2O, in form of potassium sulphate (Fertilsud s.r.l., Barletta, Italy), respectively. During the whole experiment, the conventional bottle gourd cultivation technique was followed, and plant needs were satisfied as recommended [34]. Genotypes were separated by an insect-proof net. At the floral anthesis stage, all-female flowers were manually pollinated, and a clip insulator was applied to prevent cross-pollination among landraces.

The seaweed extract application was performed with an extract of *Ecklonia maxima* (Kelpstar®, Mugavero fertilizers, Palermo, Italy). This seaweed extract was produced via a cold micronisation process to not alter the seaweed components. This product was composed of 1% of organic nitrogen, 10% of organic carbon, phytohormones (11 mg L^{-1}

of auxins and 0.03 mg L^{-1} of cytokinins) and 30% of organic components characterised by a nominal molecular weight < 50 kDa. Treatments were administered weekly by foliar spray starting seven days after transplant. One L m^{-2} of the SwE-based solution was supplied for each application.

Two doses of seaweed extract (0 and 3 mL L^{-1}) were combined with five *L. siceraria* landraces (G1, G2, G3, G4 and G5) in a randomised blocks design. All treatments were replicated 3 times (15 plants per replication) obtaining 30 experimental plots (2 seaweed extract doses × 5 genotypes × 3 replicates), resulting in a total of 450 plants.

Maximum and minimum temperatures inside the greenhouse were collected by a data logger (Figure 1).

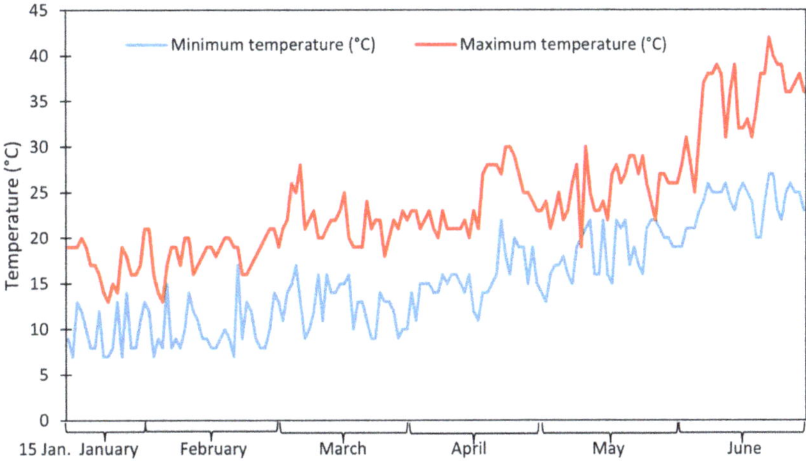

Figure 1. Daily maximum and minimum temperature recorded from 15 January 2019 to 30 June 2019 inside the experimental greenhouse.

2.2. Plant Growth, Fruit Yield and Firmness

Plant growth features, fruit yield and yield-related traits were recorded on all plants. Fruit firmness was collected on 15 randomly selected fruits per replicate. Plant height and number of leaves at 10, 20 and 30 days after transplant (DAT) were recorded. First female flower emission was recorded and expressed as DAT. Immediately after harvest, total yield (kg plant^{-1}), marketable yield (kg plant^{-1}), number of marketable fruits (No.) and fruit mean mass (kg) were collected. Fruit firmness was measured via a digital penetrometer (FR-5120, Lutron electronic enterprise Co., Ltd., Taiwan) and the values were expressed as Newton (N).

2.3. Young Shoot Yield, Nutritional and Functional Components and NUE$_{ys}$

All the shoots used for yield, nutritional and functional assessments were 30 cm in length.

After harvest, young shoot yield (kg plant^{-1}) and number of young shoots per plant (No. plant^{-1}) were recorded on all young shoots produced.

Five young shoots per plant, randomly selected from each replicate and collected from the 2nd and 3rd harvest, were washed with distilled water and used to determine nutritional and functional compounds. To appraise soluble solid content (SSC), 100 g of young shoot sample was juiced and clarified. Subsequently, SSC was appraised via a refractometer (MTD-045 nD, Three-In-One Enterprises Co., Ltd., New Taipei, Taiwan) and was expressed as °Brix. The ascorbic acid concentration was evaluated by a reflectometer (Merck RQflex10 Reflectoquant®, Sigma-Aldrich, Saint Louis, MO, USA) and ascorbic acid strips (Merck, Darmstadt, Germany) and the value was expressed as mg 100 g^{-1} fresh weight (fw). Polyphenols' concentration was measured following the Folin–Ciocâlteau method [35] (absorbance at 750 nm). Polyphenols value was presented as gallic acid equivalent (GAE) 100 g^{-1} dry weight (dw). Calcium (Ca), potassium (K) and magnesium (Mg) concentrations were assessed using the procedure reported by Morand and Gullo [36]. Phosphorous (P) concentration in shoots was appraised following the Fogg and Wilkinson method [37]. Young shoot nitrogen (N) concentration was measured using the Kjeldahl procedure. All the mineral concentrations were presented as g kg^{-1} dw.

Nitrogen use efficiency (NUE$_{ys}$) was calculated as follow: young shoot yield (t)/N application rate (kg).

2.4. Statistics

All data were analysed via the SPSS software v.20 package (StatSoft, Inc., Chicago, IL, USA) using a two-way Analysis of Variance (ANOVA). Tuckey's HSD test ($p \leq 0.05$) was used for multiple comparisons of means. Data reported as percentage were subjected to an arcsin transformation prior to ANOVA as follow: $\varnothing = \arcsin(p/100)^{1/2}$. A heat map summarising all *L. siceraria* traits was performed via the online program clustvis (https://biit.cs.ut.ee/clustvis/, accessed on 14 September 2021) with a Euclidean distance as the similarity measure and hierarchical clustering with complete linkage.

3. Results

3.1. Plant Growth Traits, Production Features, NUE and Fruit Firmness

ANOVA for plant height at 10, 20, 30 DAT and for number of leaves at 10 and 20 DAT showed no significant interaction between SwE doses and genotype (Table 1).

Table 1. Effect of the seaweed extract treatments (SwE) and genotypes (G) on plant height at 10, 20 and 30 DAT and on the number of leaves at 10 and 20 DAT of *L. siceraria* plants.

Treatments	Plant Height 10 DAT (cm)		Plant Height 20 DAT (cm)		Plant Height 30 DAT (cm)		Number of Leaves 10 DAT (No.)		Number of Leaves 20 DAT (No.)	
Seaweed extract dose (mL L^{-1})										
0	20.7	b	30.2	b	48.5	b	6.8	b	9.4	b
3	25.3	a	35.7	a	74.3	a	9.0	a	13.2	a
Genotype										
G1	22.9	b	31.1	b	60.4	b	7.5	b	10.8	b
G2	23.2	b	30.5	b	61.9	b	8.3	ab	9.8	b
G3	16.0	c	25.9	c	48.5	c	4.5	c	8.3	c
G4	26.1	a	38.0	a	69.0	a	9.7	a	13.3	a
G5	26.6	a	39.4	a	67.0	a	9.7	a	14.2	a
Significance										
SwE	***		***		***		***		***	
G	***		***		***		***		***	
SwE × G	NS		NS		NS		NS		NS	

Values present in a column and followed by different letters are significantly dissimilar at $p \leq 0.05$. NS, *** non-significant or significant at 0.001, respectively.

Regardless of the genotype, SwE meaningfully enhanced the aforesaid plant growth parameters. On the other hand, irrespective of the SwE application, G4 and G5 genotypes revealed the highest plant height at 10, 20 and 30 DAT and the highest number of leaves at 10 and 20 DAT, whereas, G3 landrace had the lowest values (Table 1).

Statistic on the number of leaves at 30 DAT displayed a significant interaction between SwE and genotype (Figure 2).

Overall, plants treated with SwE revealed a higher number of leaves at 30 DAT compared with the untreated ones. G4 and G5 landraces treated with SwE showed the highest values, followed by G1 and G2 landraces treated at 3 mL L^{-1} SwE. G3 untreated plants had the lowest value (Figure 2).

ANOVA for first female flower emission did not have a significant interaction between SwE and genotype (Figure 3).

Regardless of the genotype, SwE application delayed the first female flower emission (Figure 3). Disregarding the SwE treatment, G3 landrace had the earliest female flower emission, followed by G2 landrace. G4 and G5 landraces revealed the latest female flower emission (Figure 3).

ANOVA for yield and yield-related traits did not reveal a significant influence of the interaction SwE × G (Table 2).

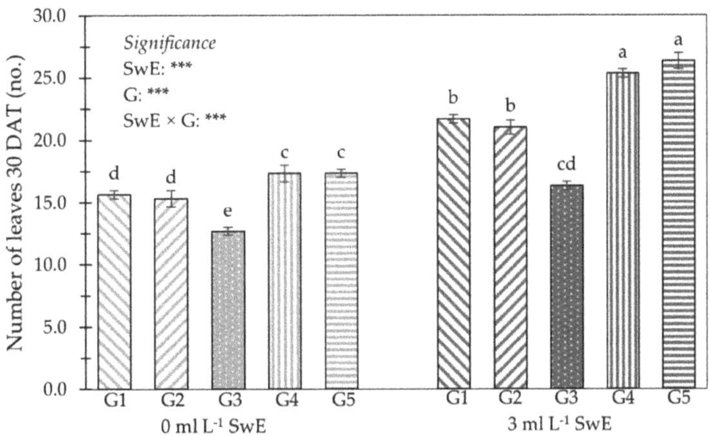

Figure 2. Effect of the seaweed extract treatments (SwE) and genotypes (G) on the number of leaves at 30 DAT of *L. siceraria*. Values with different letters indicate significant differences at $p \leq 0.05$. *** significant at 0.001. Bars represent the standard error.

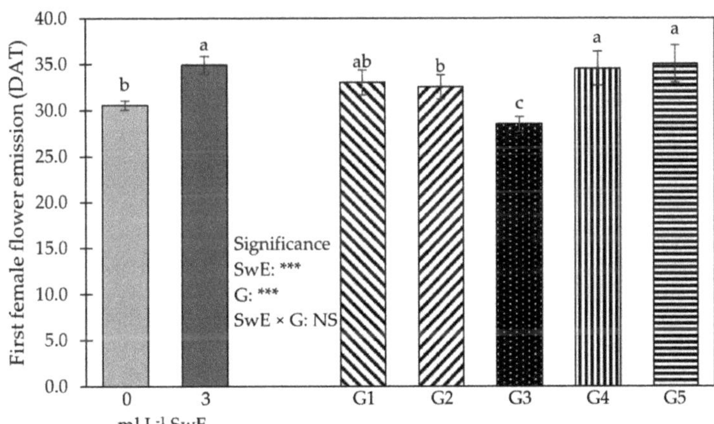

Figure 3. Effect of the seaweed extract treatments (SwE) and genotypes (G) on first flower emission of *L. siceraria*. Values with different letters indicate significant differences at $p \leq 0.05$. NS, *** non−significant or significant at 0.001, respectively. Bars represent the standard error.

Table 2. Effect of the seaweed extract treatments (SwE) and genotypes (G) on fruits total yield, fruits marketable yield, No. of marketable fruits, fruit mean mass, young shoot yield and number of *L. siceraria*.

Treatments	Fruits								Young Shoots			
	Total Yield (kg plant^{-1})		Marketable Yield (kg plant^{-1})		Marketable Fruits (No. plant^{-1})		Mean Mass (kg)		Yield (kg plant^{-1})		Number (No. plant^{-1})	
Seaweed extract dose (mL L^{-1})												
0	4.86	a	4.0	b	4.5	a	0.9	b	2.04	b	51.0	b
3	4.89	a	4.6	a	4.0	b	1.1	a	2.70	a	67.3	a
Genotype												
G1	6.80	a	6.0	a	4.4	b	1.4	a	2.35	c	58.8	b
G2	5.30	c	4.6	c	4.9	ab	0.9	b	2.45	bc	61.0	b
G3	5.67	b	5.0	b	5.6	a	0.9	b	1.60	d	40.2	c
G4	3.12	e	2.8	e	3.0	c	1.0	b	2.73	a	68.3	a
G5	3.50	d	3.1	d	3.2	c	1.0	b	2.70	ab	67.3	b
Significance												
SwE	NS		***		**		***		***		***	
G	***		***		***		***		***		***	
SwE × G	NS		NS		NS		NS		NS		NS	

Values present in a column and followed by different letters are significantly dissimilar at $p \leq 0.05$. NS, **, *** non−significant or significant at 0.01 or 0.001, respectively.

SwE application did not affect total fruit yield (Table 2). Conversely, notwithstanding the biostimulant treatment, G1 had the highest total yield, followed by G3, which in turn had a higher total fruit yield than the G2 landrace. The lowest total fruit yield value was recorded in the G4 landrace.

When averaged over genotype, fruit marketable yield was increased by SwE treatment (Table 2). Regardless of the SwE application, data collected on fruit marketable yield followed the trend recognised for total fruit yield.

SwE non-treated plants showed a greater number of marketable fruits compared with the treated ones (Table 2). Averaged over the SwE application, G3 landrace showed the highest number of marketable fruits, followed by G1 landrace. Whereas G2 plants did not meaningfully diverge neither from G3 plants nor from G1 plants. G4 and G5 landraces had the lowest number of marketable fruits.

SwE treatment significantly increased fruit mean mass compared with the control (Table 2). Regardless of the SwE application, the G1 genotype gave the highest fruit mean mass compared with the other genotypes.

Young shoot yield in SwE treated plants was higher by 22.2% compared to untreated control (Table 2). Averaged over SwE application, the G4 genotype showed the highest young shoot yield, whereas the G3 landrace revealed the lowest one.

SwE application boosted the number of young shoots (Table 2). G4 landrace gave the highest number of young shoots, followed by G1, G2 and G5. The genotype G3 gave the lowest value.

Statistic for NUE$_{ys}$ underlined no significant interaction SwE × G (Figure 4).

Regardless of the genotype, SwE treated plants displayed the highest NUE$_{ys}$ value (Figure 4). Averaged over SwE treatment, G4 and G5 landraces gave the highest NUE$_{ys}$ value. However, the G5 landrace did not significantly differ neither from the G4 landrace nor from the G2 landrace. The lowest values were observed in the G3 landrace (Figure 4).

Statistic on fruit firmness revealed no significant interaction SwE × G (Figure 5).

Fruits from plants treated with SwE revealed a higher firmness than fruits from untreated plants. When averaged over SwE treatment, G2 and G3 landraces displayed the highest firmness, followed by G1 landrace. G4 and G5 landraces had the lowest values (Figure 5).

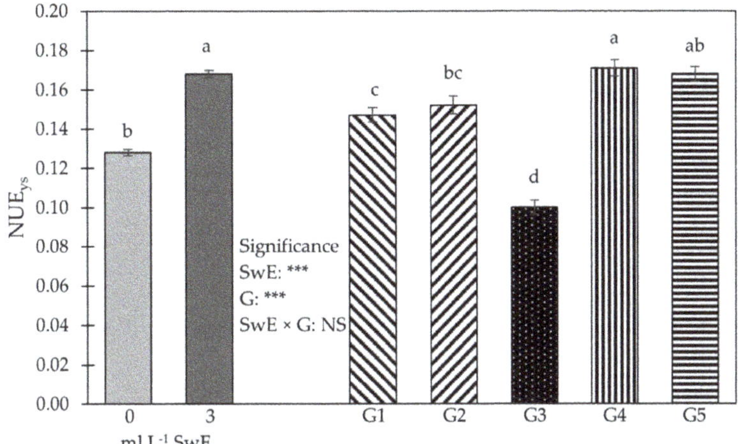

Figure 4. Effect of the seaweed extract treatments (SwE) and genotypes (G) on shoot nitrogen use efficiency of *L. siceraria*. Values with different letters indicate significant differences at $p \leq 0.05$. NS, *** non−significant or significant at 0.001, respectively. Bars represent the standard error.

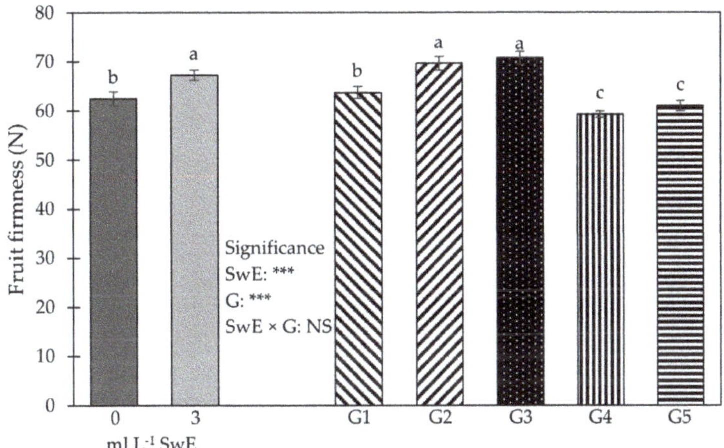

Figure 5. Effect of the seaweed extract treatments (SwE) and genotypes (G) firmness of *L. siceraria* fruits. Values with different letters indicate significant differences at $p \leq 0.05$. NS, *** non−significant or significant at 0.001, respectively. Bars represent the standard error.

3.2. Young Shoot Nutritional Properties, Mineral Profile and Functional Components

Statistical analysis for SSC did not show a significant interaction between SwE and G (Figure 6).

Averaged over genotype, SwE treatment did not affect young shoot SSC (Figure 6). Contrariwise, when averaged over SwE application, G2, G4 and G5 landraces had the highest SSC, followed by G1 genotype which in turn had a higher SSC value than G3 landrace (Figure 6).

The mineral profile was mainly influenced by SwE application and genotype. However, ANOVA for N, P, K, Ca, and Mg concentrations highlighted no significant interaction SwE × G (Table 3).

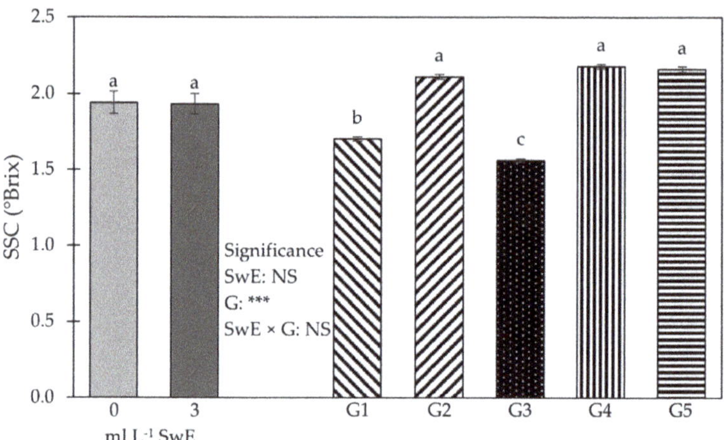

Figure 6. Effect of the seaweed extract treatments (SwE) and genotypes (G) on young shoot soluble solid content of *L. siceraria*. Values with different letters indicate significant differences at $p \leq 0.05$. NS, *** non−significant or significant at 0.001, respectively. Bars represent the standard error.

Table 3. Effect of the seaweed extract treatments (SwE) and genotypes (G) on nitrogen (N), phosphorous (P), potassium (K), calcium (Ca) and magnesium (Mg) concentrations of *Lagenaria siceraria* young shoots.

Treatments	N (g kg^{-1} DW)		P (g kg^{-1} DW)		K (g kg^{-1} DW)		Ca (g kg^{-1} DW)		Mg (g kg^{-1} DW)	
Seaweed extract dose (mL L^{-1})										
0	1.86	a	5.14	b	19.30	b	25.77	b	9.54	b
3	1.74	b	5.30	a	19.75	a	26.00	a	10.00	a
Genotype										
G1	1.80	c	5.30	b	19.80	a	26.10	a	9.62	b
G2	1.70	d	5.11	c	19.08	b	25.60	b	9.60	b
G3	1.60	e	4.99	d	18.72	b	25.00	c	9.44	b
G4	1.94	b	5.38	a	20.13	a	26.40	a	10.11	a
G5	2.00	a	5.33	ab	19.82	a	26.30	a	10.02	a
Significance										
SwE	***		***		***		***		***	
G	***		***		***		*		***	
SwE × G	NS		NS		NS		NS		NS	

Values present in a column and followed by different letters are significantly dissimilar at $p \leq 0.05$. NS, *, *** non−significant or significant at 0.05 or 0.001, respectively.

Averaged over genotype, the highest N concentration was observed in young shoots from untreated plants (Table 3). Disregarding the SwE treatment, genotype G5 revealed the highest N concentration, followed by G4 landrace which in turn displayed a higher N concentration than the G1 landrace. The lowest value was observed in young shoots from the G3 genotype.

P, K, Ca, and Mg contents in plants exposed to 3 mL L^{-1} of SwE was higher by 3.1%, 2.3%, 0.9% and 4.8%, respectively compared with untreated plants (Table 3). Averaged over SwE application, G4 and G5 landraces revealed the highest P concentration, followed by G1 landrace. However, young shoots from the G5 landrace did not show a significant difference in terms of P concentration. The lowest P concentration was recorded in young shoots from the G3 genotype. The highest K concentration was found in young shoots from G1, G4 and G5 landraces (19.80, 20.13 and 19.82 g kg^{-1} dw, respectively), whereas, G2 and G3 genotypes had the lowest values. Data on Ca concentration followed a similar trend to that described for K concentration. Regardless of SwE application, G4 and G5 landraces had a higher Mg concentration compared with the other genotypes.

ANOVA for shoot ascorbic acid concentration showed a significant interaction between SwE and G (Figure 7).

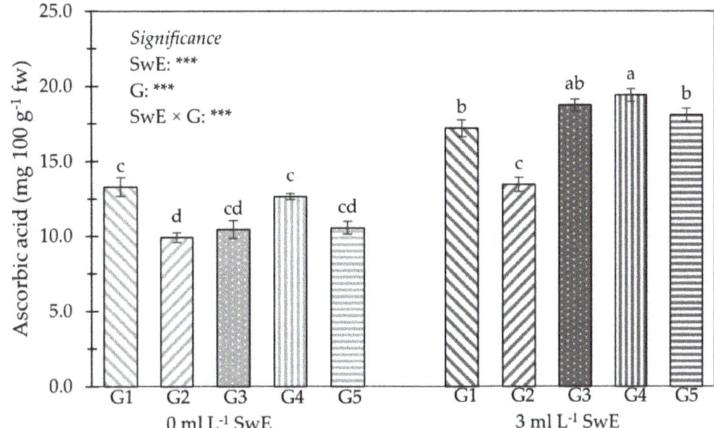

Figure 7. Effect of the seaweed extract treatments (SwE) and genotypes (G) on the ascorbic acid concentration of *L. siceraria* young shoots. Values with different letters indicate significant differences at $p \leq 0.05$. *** significant at 0.001. Bars represent the standard error.

Overall, SwE treatment enhanced ascorbic acid concentration in G1, G2, G3, G4 and G5 young shoots by 29.5%, 35.9%, 79.6%, 53.4% and 71.6%, respectively (Figure 7). The combinations 3 mL L^{-1} SwE $\times$ G3 and 3 mL L^{-1} SwE $\times$ G4 gave the highest ascorbic acid content. However, when G3 landraces were exposed to SwE did not significantly differ neither from G4 nor from G5 in terms of ascorbic acid concentration. The lowest ascorbic acid concentration was detected in young shoots from the 0 mL L^{-1} SwE $\times$ G2 combination.

ANOVA for polyphenols concentration showed a significant interaction SwE $\times$ G (Figure 8).

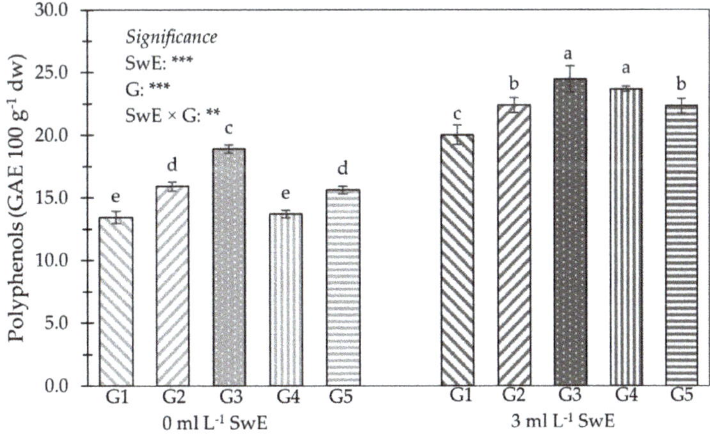

Figure 8. Effect of the seaweed extract treatments (SwE) and genotypes (G) on polyphenols concentration of *L. siceraria* young shoots. Values with different letters indicate significant differences at $p \leq 0.05$. ***, ** significant at 0.01 or 0.001, respectively. Bars represent the standard error.

As for ascorbic acid concentration in all tested genotypes, polyphenols were significantly enhanced by SwE treatment. The highest values were recorded in G3 and G4 landraces supplied with 3 mL L^{-1} of SwE, followed by G2 and G5 landraces (Figure 7). The lowest values were observed in young shoots harvested from G1 and G4 untreated plants.

3.3. Heat Map Analysis of the Whole Data Set

A data heat-map analysis of all assessed features (agronomic, nutritional and functional) was performed to realise a graphical appraisal of the influences determined by the experimental factors on *L. siceraria* (Figure 9).

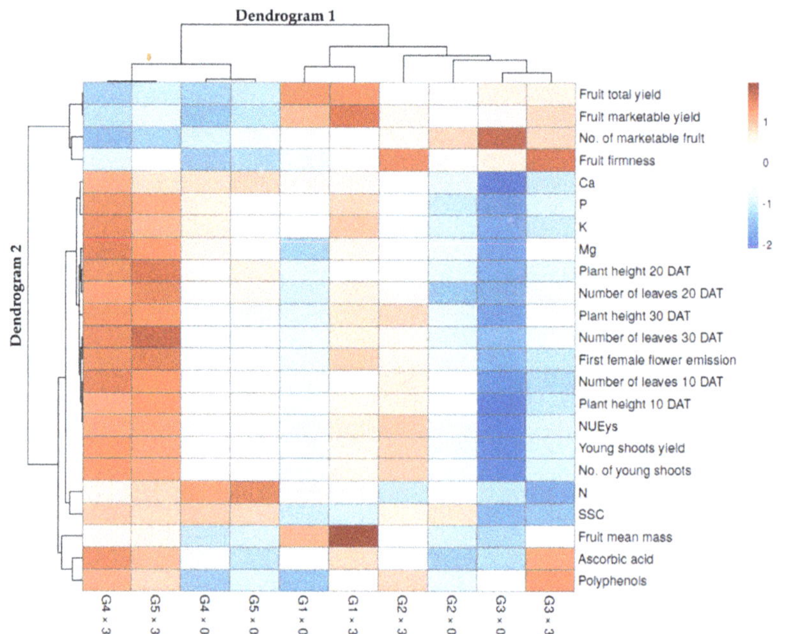

Figure 9. Heat map assessment including all *L. siceraria* plant traits in response to seaweed extract treatments (0 or 3 mL L^{-1}) and genotypes (G1, G2, G3, G4 or G5). The heat map picture was produced via https://biit.cs.ut.ee/clustvis/ (accessed on 14 September 2021).

The heat-map output consisted of two dendrograms, Dendrogram 1 sited on the top containing all the combinations of SwE doses and genotypes and Dendrogram 2, located on the left side, comprising all traits that influenced this distribution. Dendrogram 1 presented two main clusters, the first on the left included the G4 and G5 landraces both treated and untreated with SwE. The other site on the right side contained G1, G2 and G3 landraces both treated and untreated with SwE. Expressly, in the left cluster of Dendrogram 1, the G4 and G5 landraces treated with SwE were parted from the other G4 and G5 untreated controls due to the higher P, K, Mg, plant height (at 10, 20 and 30 DAT), number of leaves (at 10, 20 and 30 DAT), first female flower emission, NUE$_{ys}$, young shoot yield, number of young shoots, fruit mean mass, ascorbic acid and polyphenols. The group on the left included G4 and G5 landraces treated with SwE. Within this group, the G4 × 3 combination was clearly separated by lower total yield, marketable yield, number of marketable fruits, plant height at 10 and 20 DAT, number of leaves at 20 and 30 DAT and N. While, the group on the right side included the G4 and G5 landraces untreated plants. Inside this cluster, the G4 × 0 combination was separated by higher P, K, Mg and ascorbic acid and lower total yield, marketable yield, fruit firmness, N, fruit mean mass and polyphenols.

On the right of the Dendrogram 1, two main groups were documented, the one on the left included the combinations G1 × 0 and G1 × 3, separated from the G2 and G3 landraces treated with 0 or 3 mL L^{-1} of SwE, that had, in particular, lower fruit total yield, fruit marketable yield, Ca, P, K, N, Mg, plant height at 20 DAT and fruit mean mass, but the higher number of marketable fruits, firmness and polyphenols.

The group on the left side comprised G1 untreated control and G1 SwE treated plants; the G1 × 0 combination was separated from G1 × 3 by lower fruit marketable yield, fruit firmness, P, K, Mg, plant height (at 10, 20 and 30 DAT), number of leaves (at 10, 20 and 30 DAT), first female flower emission, NUE$_{ys}$, young shoot yield, number of young shoots, fruit mean mass, ascorbic acid and polyphenols.

The cluster on the right comprised G2 and G3 untreated and SwE treated combinations. In this group, the G2 × 3 combination was evidently parted from G2 × 0, G3 × 0 and G3 × 3 combinations by higher Ca, P, K, plant height (at 10, 20 and 30 DAT), number of leaves (at 10, 20 and 30 DAT), first female flower emission, NUE$_{ys}$, young shoot yield and number of young shoots. The right side of this cluster included G2 × 0, G3 × 0 and G3 × 3 treatments; the G2 landrace untreated control was parted by higher Ca, P, K, Mg, plant height (at 10, 20 and 30 DAT), number of leaves (at 10 and 30 DAT), first female flower emission, NUE$_{ys}$, young shoot yield, number of young shoots, N and SSC. The right part of this group comprised G3 landraces SwE treated and untreated control. G3 × 0 combination was separated from G3 × 3 combination by lower fruit marketable yield, fruit firmness, Ca, P, K, Mg, plant height (at 10, 20 and 30 DAT), number of leaves (at 10, 20 and 30 DAT), first female flower emission, NUE$_{ys}$, young shoot yield, number of young shoots, fruit mean mass, ascorbic acid and polyphenols.

4. Discussion

In the present work, we investigated the effect of SwE application on yields and young shoot quality features of five *L. siceraria* landraces. Irrespective of the genotype, SwE supply improved plant height and the number of leaves. These results are coherent with those of Rouphael et al. [38] who, investigating the influence of *Ecklonia maxima* SwE on production, quality and physiological traits of zucchini squash cultivated under saline conditions, found that, regardless of the salinity treatments, SwE enhances plant aerial weight. Findings are also in line with those of La Bella et al. [23] who, examining the influence of the SwE of *E. maxima* and molybdenum enrichment on yield, quality and NUE in spinach plants, highlighted that SwE application boosts growth plant features. Without regard to the SwE treatments, the G4 and G5 landraces performed better than the other tested genotypes, while the G3 landrace had the lowest plant growth features. As reported by Weiner [39], this was probably because plants may grow efficiently until they achieve the threshold size for reproduction. Once they accomplish this size, a certain fraction of sources is assigned to reproduction. Indeed, the G3 landrace revealed the earliest female flower emission. On the other hand, when averaged over genotype, SwE application delayed first female flower emission, in accordance with the aforesaid vegetative vs. reproductive competitive activity.

There are researches on the favourable effect of diverse plant biostimulants on the marketable yield of various vegetables [22,28]. In this respect, the results of the present study are in agreement with those reported by Ali et al. [40], who underlined that *Ascophyllum nodosum*-based SwE improved tomato yield in a soilless system by 54% compared with the control. These results were related to the *A. nodosum* SwE polysaccharides content which in turn enhances yield promoting endogenous hormone homeostasis [41]. Outcomes agree with those of Colla et al. [28] who, studying the effect of different classes of biostimulants on yield and fruit quality of tomato cultivated under greenhouse, found that SwE 'Kelpak' increase marketable yield compared with the control. Furthermore, the results are sustained by Hussain et al. [42] who found that SwE of *Durvillaea potatorum* and *A. nodosum* augment tomato marketable yield. Data are also consistent with those of Hassan et al. [43] and La Bella et al. [23]. In this study, the marketable fruit increase prompted by the SwE application was due to a higher fruit mean mass rather than to the higher number of marketable fruits. These results are in contrast with those of Colla et al. [28], who reported that SwE treatment improves the number of fruits per plant but did not affect fruit mean mass. Thus, we may hypothesise that the plant yield response to SwE application is genotype-dependent. G4 and G5 landraces revealed the lowest fruit yields. Averaged over genotype, SwE supply elicited young shoot production, both in terms of yield and number. In this respect, there are reports that brown seaweed extracts comprise phytohormones (IAA, cytokinin, GA, polyamines and ABA) [25,44,45]. Consequently, we may assume that the growth eliciting effects of SwE are linked to their effect. Wally et al. [46] stated that the SwE phytohormone-like action might also be triggered by chemical compounds included in the extract rather than by the phytohormones themselves. On the other hand, irrespective of the SwE application, G4 and G5 landraces had the highest young shoot yield traits. Thus, it seems that young shoot production is a genotype-associated trait and is negatively related to fruit production. Results showed that SwE treatment increased NUE_{ys}. These findings are coherent with those of Di Mola et al. [46] who investigated the effect of plant-based biostimulants on nitrogen use and uptake efficiency, yield and quality of leafy vegetables cultivated under different nitrogen regimes, found that treated plants had a higher NUE than untreated ones. Moreover, the results are in agreement with previous studies concerning the influence of *E. maxima* SwE application and molybdenum supply on yield, quality and NUE of spinach [23]. Outcomes showed that the SwE application enhanced fruit firmness. Overall, this is consistent with previous research on the influence of plant-based biostimulants on different tomato cultivars [47]. As reported by Basile et al. [47] and Cozzolino et al. [48] the SwE effect on fruit firmness is related to the higher Ca uptake and accumulation of SwE treated plants compared with the control. Indeed, as pointed out by Hocking et al. [49], calcium-pectin cross-links play an imperative function in defining the resistance of cell walls and, thus, the characteristics of the physical and structural fruit. Furthermore, since it is assumed that auxins partake in Ca transport and fruit uptake [49], the present study suggested that the SwE may have an auxin-like action in *L. siceraria*, improving calcium nutrition. Regardless of the SwE application, statistics showed a significant influence of the genotype on fruit firmness. Findings showed that SwE did not significantly affect young shoot SSC. In this respect, the results concur with those stated by Colla et al. [28] and La Bella et al. [23] who found no significant effect of the SwE on SSC on tomato and spinach, respectively. ANOVA displayed a significant effect of the genotype on young shoot SSC values. A similar response was previously reported for zucchini squash by Rouphael et al. [50] and Rouphael et al. [51], and it was related to a reduction in water accumulation in the fruit without influence on the biosynthesis and accumulation of organic solutes. Thus, since G2, G4 and G5 landraces gave the highest SSC young shoot yield and considering that G2, G4 and G5 were more productive than the other genotypes in terms of young shoots, but less performing in terms of fruit yields, it seems that the aforesaid landraces had a resource translocation mainly toward to the shoots rather than to the fruits compared to the G1 and G3 landraces.

The results displayed that the SwE application reduced N concentration in young shoots. In this respect, there are contrasting reports. Krouk et al. [52] and Castaings et al. [53] state that different *A. nodosum*-based SwE upregulated the expression of a nitrate transporter gene NRT1.1. which enhance nitrogen sensing and auxin transport. On the contrary, Rouphael et al. [38], found that SwE application does not significantly influence N concentration in tomato leaves. Thus, we may hypothesise that the plant N uptake and accumulation response to SwE supply, is significantly related to genotype and plant site. Results on minerals revealed that SwE treatment augmented P, K, Ca and Mg concentrations. These findings are partially coherent with those of Rouphael et al. [12], who found that *E. maxima* SwE application improve K and Mg concentrations in spinach plants. Furthermore, the outcomes concur with those of La Bella et al. [23], who evidenced that SwE treatment increase P, K and Mg concentration in spinach. As highlighted by Battacharyya et al. [29], this improved minerals uptake and build-up could be linked to a modification of the root architecture, resulting in an enhanced plant mineral uptake. Moreover, Soppelsa et al. [54] pointed out that commercial SwE includes a compound named kahydrin, which modifies plasma membrane proton pumps and elicit the H+ ions excretion into the apoplast determining rhizosphere acidification, resulting in a higher metal ions plant availability [52,53].

Among plant secondary metabolites, ascorbic acid and polyphenols provide benefits to human health and, furthermore, play a crucial role in numerous plant life aspects [55]. The current study showed that SwE application upgraded ascorbic acid and polyphenols concentrations in young shoots of bottle gourd. These results sustained the outcomes of Rouphael et al. [12] and La Bella et al. [23] on spinach and those of Abbas et al. [56] on onion. As reported by Ertani et al. [57] and Rouphael et al. [58], the enhancement of bioactive components, such as ascorbic acid and phenols, could be related to the chalcone isomerase activity, which is a key enzyme in phytochemical homeostasis [59].

5. Conclusions

In the current study, SwE application boosted plant growth traits, fruit and young shoot yield, NUE_{ys}, mineral profile and functional components of young shoots. Concurrently, Sicilian bottle gourd landraces showed a relevant range of genetic variability. Overall, the results suggested that combining G4 and G5 landraces with 3 mL L^{-1} SwE profoundly upgraded young shoot yield, plant height, number of leaves young shoot yield and NUE_{ys} index. Quality traits such as SSC, minerals concentration, ascorbic acid and polyphenols content were also improved.

Author Contributions: Conceptualisation, L.S. and G.I.; methodology, L.S., G.I., C.D.P. and S.L.B.; software, B.B.C., L.S., R.P.M. and C.N.; validation, L.S., G.I. and S.L.B.; formal analysis, B.B.C., L.S., R.P.M., C.N. and S.L.B.; investigation, B.B.C., L.S., C.D.P., G.I. and S.L.B.; resources, L.S., C.D.P. and S.L.B.; data curation, B.B.C., L.S., R.P.M. and C.N.; writing—original draft preparation, B.B.C., L.S. and G.I.; writing—review and editing, L.S., R.P.M., C.N., G.I. and S.L.B.; visualisation, L.S. and G.I.; supervision, L.S. and G.I.; project administration, L.S., G.I. and S.L.B.; funding acquisition, L.S., C.D.P. and S.L.B. All authors have read and agreed to the published version of the manuscript.

Funding: This research received no external funding.

Institutional Review Board Statement: Not applicable.

Informed Consent Statement: Not applicable.

Conflicts of Interest: The authors declare no conflict of interest.

References

1. Choudhary, B.L.; Katewa, S.S.; Galav, P.K. Plants in material culture of tribals and rural communities of Rajsamand district of Rajasthan. *Indian J. Trad. Knowl.* **2008**, *7*, 11–22.
2. Ghule, B.V.; Ghante, M.H.; Upaganlawar, A.B.; Yeole, P.G. Analgesic and anti-inflammatory activities of *Lagenaria siceraria* Stand. fruit juice extract in rats and mice. *Pharmacogn. Mag.* **2006**, *2*, 232–238.
3. Ghule, B.V.; Ghante, M.H.; Yeole, P.G.; Saoji, A.N. Diuretic activity of *Lagenaria siceraria* fruit extracts in rats. *Indian J. Pharm. Sci.* **2007**, *69*, 817–819. [CrossRef]
4. Ghule, B.V.; Ghante, M.H.; Saoji, A.N.; Yeole, P.G. Antihyperlipidemic effect of the methanolic extract from *Lagenaria siceraria* Stand. fruit in hyperlipidemic rats. *J. Ethnopharmacol.* **2009**, *124*, 333–337. [CrossRef] [PubMed]
5. Raimondo, F.M.; Bazan, G.; Troia, A. Taxa a rischio nella flora vascolare della Sicilia. *Biogeogr.-J. Integr. Biogeogr.* **2011**, *30*. [CrossRef]
6. D'Anna, F.; Sabatino, L. Morphological and agronomical haracterisation of eggplant genetic resources from the Sicily area. *J. Food Agric. Environ.* **2013**, *11*, 401–404.
7. Sabatino, L.; Palazzolo, E.; D'Anna, F. Grafting suitability of Sicilian eggplant ecotypes onto Solanum torvum: Fruit composition, production and phenology. *J. Food Agric. Environ.* **2013**, *11*, 1195–1200.
8. Sabatino, L.; Iapichino, G.; Vetrano, F.; D'Anna, F. Morphological and agronomical characterisation of Sicilian bottle gourd *Lagenaria siceraria* (Mol.) Standley. *J. Food Agric. Environ.* **2014**, *12*, 587–590.
9. Sabatino, L.; Iapichino, G.; Maggio, A.; D'anna, E.; Bruno, M.; D'Anna, F. Grafting affects yield and phenolic profile of Solanum melongena L. landraces. *J. Integr. Agric.* **2016**, *15*, 1017–1024. [CrossRef]
10. Virga, G.; Licata, M.; Consentino, B.B.; Tuttolomondo, T.; Sabatino, L.; Leto, C.; La Bella, S. Agro-morphological characterization of sicilian chili pepper accessions for ornamental purposes. *Plants* **2020**, *9*, 1400. [CrossRef]

11. Mauro, R.; Portis, E.; Acquadro, A.; Lombardo, S.; Mauromicale, G.; Lanteri, S. Genetic diversity of globe artichoke landraces from Sicilian small-holdings: Implications for evolution and domestication of the species. *Conserv. Genet.* **2009**, *10*, 431–440. [CrossRef]

12. Rouphael, Y.; Kyriacou, M.C.; Colla, G. Vegetable grafting: A toolbox for securing yield stability under multiple stress conditions. *Front. Plant Sci.* **2018**, *8*, 2255. [CrossRef] [PubMed]

13. Mauro, R.P.; Agnello, M.; Distefano, M.; Sabatino, L.; San Bautista Primo, A.; Leonardi, C.; Giuffrida, F. Chlorophyll fluorescence, photosynthesis and growth of tomato plants as affected by long-term oxygen root zone deprivation and grafting. *Agronomy* **2020**, *10*, 137. [CrossRef]

14. Expósito, A.; Pujolà, M.; Achaerandio, I.; Giné, A.; Escudero, N.; Fullana, A.M.; Conquero, M.; Loza-Alvarez, P.; Sorribas, F.J. Tomato and melon Meloidogyne resistant rootstocks improve crop yield but melon fruit quality is influenced by the cropping season. *Front. Plant Sci.* **2020**, *11*, 560024. [CrossRef]

15. Lee, J.M.; Oda, M. Grafting of herbaceous vegetable and ornamental crops. *Hortic. Rev.* **2003**, *28*, 61–124.

16. Karaca, F.; Yetişir, H.; Solmaz, İ.; Candir, E.; Kurt, Ş.; Sari, N.; Güler, Z. Rootstock potential of Turkish *Lagenaria siceraria* germplasm for watermelon: Plant growth, yield and quality. *Turk. J. Agric. For.* **2012**, *36*, 167–177.

17. Koike, S.T.; Gladders, P.; Paulus, A.O. *Vegetable Diseases: A Color Handbook*; Gulf Professional Publishing/Manson Publishing Ltd.: London, UK, 2007; pp. 220–228.

18. Colla, G.; Rouphael, Y. *Biostimulants in Horticulture*; Elsevier B.V.: Amsterdam, The Netherlands, 2015; pp. 1–134.

19. Du Jardin, P. Plant biostimulants: Definition, concept, main categories and regulation. *Sci. Hortic.* **2015**, *196*, 3–14. [CrossRef]

20. Consentino, B.B.; Virga, G.; La Placa, G.G.; Sabatino, L.; Rouphael, Y.; Ntatsi, G.; Iapichino, G.; La Bella, S.; Mauro, R.P.; D'Anna, F.; et al. Celery (*Apium graveolens* L.) Performances as Subjected to Different Sources of Protein Hydrolysates. *Plants* **2020**, *9*, 1633. [CrossRef]

21. Sabatino, L.; Iapichino, G.; Consentino, B.B.; D'Anna, F.; Rouphael, Y. Rootstock and Arbuscular Mycorrhiza Combinatorial Effects on Eggplant Crop Performance and Fruit Quality under Greenhouse Conditions. *Agronomy* **2020**, *10*, 693. [CrossRef]

22. Sabatino, L.; Consentino, B.B.; Rouphael, Y.; De Pasquale, C.; Iapichino, G.; D'Anna, F.; La Bella, S. Protein Hydrolysates and Mo-Biofortification Interactively Modulate Plant Performance and Quality of 'Canasta' Lettuce Grown in a Protected Environment. *Agronomy* **2021**, *11*, 1023. [CrossRef]

23. La Bella, S.; Consentino, B.B.; Rouphael, Y.; Ntatsi, G.; De Pasquale, C.; Iapichino, G.; Sabatino, L. Impact of Ecklonia maxima Seaweed Extract and Mo Foliar Treatments on Biofortification, Spinach Yield, Quality and NUE. *Plants* **2021**, *10*, 1139. [CrossRef] [PubMed]

24. Di Mola, I.; Conti, S.; Cozzolino, E.; Melchionna, G.; Ottaiano, L.; Testa, A.; Sabatino, L.; Rouphael, Y.; Mori, M. Plant-Based Protein Hydrolysate Improves Salinity Tolerance in Hemp: Agronomical and Physiological Aspects. *Agronomy* **2021**, *11*, 342. [CrossRef]

25. Khan, W.; Rayirath, U.P.; Subramanian, S.; Jithesh, M.N.; Rayorath, P.; Hodges, D.M.; Critchley, A.T.; Craigie, J.S.; Norrie, J.; Prithiviraj, B. Seaweed extracts as biostimulants of plant growth and development. *J. Plant Growth Regul.* **2009**, *28*, 386–399. [CrossRef]

26. Craigie, J.S. Seaweed extract stimuli in plant science and agriculture. *J. Appl. Phycol.* **2011**, *23*, 371–393. [CrossRef]

27. Colla, G.; Nardi, S.; Cardarelli, M.; Ertani, A.; Lucini, L.; Canaguier, R.; Rouphael, Y. Protein hydrolysates as biostimulants in horticulture. *Sci. Hortic.* **2015**, *196*, 28–38. [CrossRef]

28. Colla, G.; Hoagland, L.; Ruzzi, M.; Cardarelli, M.; Bonini, P.; Canaguier, R.; Rouphael, Y. Biostimulant action of protein hydrolysates: Unraveling their effects on plant physiology and microbiome. *Front. Plant Sci.* **2017**, *8*, 2202. [CrossRef]

29. Battacharyya, D.; Babgohari, M.Z.; Rathor, P.; Prithiviraj, B. Seaweed extracts as biostimulants in horticulture. *Sci. Hortic.* **2015**, *196*, 39–48. [CrossRef]

30. Branca, F.; La Malfa, G. Traditional vegetables of Sicily. *Chron. Horticult.* **2008**, *48*, 20–25.

31. Richards, L.A. Diagnosis and improvement of saline and alkali soils. In *Agricultural Hand Book*; U.S. Dept. of Agriculture: Washington, DC, USA, 1954; p. 160.

32. Baiamonte, G.; Crescimanno, G.; Parrino, F.; De Pasquale, C. Effect of biochar on the physical and structural properties of a sandy soil. *Catena* **2019**, *175*, 294–303. [CrossRef]

33. Walkley, A.J.; Black, I.A. Estimation of soil organic carbon by the chromic acid titration method. *Soil Sci.* **1934**, *37*, 29–38. [CrossRef]

34. Tesi, R. *Orticoltura Mediterranea Sostenibile*; Pàtron Editore: Bologna, Italy, 2010.

35. Meda, A.; Lamien, C.E.; Romito, M.; Millogo, J.; Nacoulma, O.G. Determination of the total phenolic, flavonoid and proline contents in Burkina Fasan honey, as well as their radical scavenging activity. *Food Chem.* **2005**, *91*, 571–577. [CrossRef]

36. Morand, P.; Gullo, J.L. Mineralisation des tissus vegetaux en vue du dosage de P, Ca, Mg, Na, K. *Ann. Agron.* **1970**, *21*, 229–236.

37. Fogg, D.N.; Wilkinson, A.N. The colorimetric determination of phosphorus. *Analyst* **1958**, *83*, 406–414. [CrossRef]

38. Rouphael, Y.; De Micco, V.; Arena, C.; Raimondi, G.; Colla, G.; De Pascale, S. Effect of *Ecklonia maxima* seaweed extract on yield, mineral composition, gas exchange, and leaf anatomy of zucchini squash grown under saline conditions. *J. Appl. Phycol.* **2017**, *29*, 459–470. [CrossRef]

39. Weiner, J.A.C.O.B. The influence of competition on plant reproduction. *Plant Reprod. Ecol. Patterns Strateg.* **1988**, *11*, 228–245.

40. Ali, N.; Farrell, A.; Ramsubhag, A.; Jayaraman, J. The effect of *Ascophyllum nodosum* extract on the growth, yield and fruit quality of tomato grown under tropical conditions. *J. Appl. Phycol.* **2016**, *28*, 1353–1362. [CrossRef]

41. Rolland, F.B.; Moore, B.; Sheen, J. Sugar sensing and signaling in plants. *Plant Cell* **2002**, *14* (Suppl. 1), S185–S205. [CrossRef] [PubMed]

42. Hussain, H.I.; Kasinadhuni, N.; Arioli, T. The effect of seaweed extract on tomato plant growth, productivity and soil. *J. Appl. Phycol.* **2021**, *33*, 1305–1314. [CrossRef]
43. Hassan, S.M.; Ashour, M.; Soliman, A.A.; Hassanien, H.A.; Alsanie, W.F.; Gaber, A.; Elshobary, M.E. The potential of a new commercial seaweed extract in stimulating morpho-agronomic and bioactive properties of *Eruca vesicaria* (L.) Cav. *Sustainability* **2021**, *13*, 4485. [CrossRef]
44. Stirk, W.A.; Novak, M.S.; Van Staden, J. Cytokinins in macroalgae. *Plant Growth Regul.* **2003**, *41*, 13–24. [CrossRef]
45. Wally, O.S.D.; Critchley, A.T.; Hiltz, D.; Craigie, J.S.; Han, X.; Zaharia, L.I.; Abrams, S.R.; Prithiviraj, B. Regulation of phytohormone biosynthesis and accumulation in Arabidopsis following treatment with commercial extract from the marine macroalga *Ascophyllum nodosum. J. Plant Growth Regul.* **2013**, *32*, 324–339. [CrossRef]
46. Di Mola, I.; Cozzolino, E.; Ottaiano, L.; Nocerino, S.; Rouphael, Y.; Colla, G.; El-Nakhel, C.; Mori, M. Nitrogen Use and Uptake Efficiency and Crop Performance of Baby Spinach (*Spinacia oleracea* L.) and Lamb's Lettuce (*Valerianella locusta* L.) Grown under Variable Sub-Optimal N Regimes Combined with Plant-Based Biostimulant Application. *Agronomy* **2020**, *10*, 278. [CrossRef]
47. Basile, B.; Brown, N.; Valdes, J.M.; Cardarelli, M.; Scognamiglio, P.; Mataffo, A.; Rouphael, Y.; Bonini, P.; Colla, G. Plant-Based Biostimulant as Sustainable Alternative to Synthetic Growth Regulators in Two Sweet Cherry Cultivars. *Plants* **2021**, *10*, 619. [CrossRef] [PubMed]
48. Cozzolino, E.; Di Mola, I.; Ottaiano, L.; El-Nakhel, C.; Rouphael, Y.; Mori, M. Foliar application of plant-based biostimulants improve yield and upgrade qualitative characteristics of processing tomato. *Ital. J. Agron.* **2021**, *16*, 1825. [CrossRef]
49. Hocking, B.; Tyerman, S.D.; Burton, R.A.; Gilliham, M. Fruit Calcium: Transport and Physiology. *Front. Plant Sci.* **2016**, *7*, 569. [CrossRef]
50. Rouphael, Y.; Schwarz, D.; Krumbein, A.; Colla, G. Impact of grafting on product quality of fruit vegetables. *Sci Hortic.* **2010**, *127*, 172–179. [CrossRef]
51. Rouphael, Y.; Cardarelli, M.; Bassal, A.; Leonardi, C.; Giuffrida, F.; Colla, G. Vegetable quality as affected by genetic, agronomic and environmental factors. *J. Food Agric. Environ.* **2012**, *10*, 680–688.
52. Krouk, G.; Lacombe, B.; Bielach, A.; Perrine-Walker, F.; Malinska, K.; Mounier, E.; Hoyerova, K.; Tillard, P.; Leon, S.; Ljung, K.; et al. Nitrate-regulated auxin transport by NRT1. 1 defines a mechanism for nutrient sensing in plants. *Dev. Cell* **2010**, *18*, 927–937. [CrossRef]
53. Castaings, L.; Marchive, C.; Meyer, C.; Krapp, A. Nitrogen signalling in Arabidopsis: How to obtain insights into a complex signalling network. *J. Exp. Bot.* **2011**, *62*, 1391–1397. [CrossRef] [PubMed]
54. Soppelsa, S.; Kelderer, M.; Casera, C.; Bassi, M.; Robatscher, P.; Matteazzi, A.; Andreotti, C. Foliar Applications of Biostimulants Promote Growth, Yield and Fruit Quality of Strawberry Plants Grown under Nutrient Limitation. *Agronomy* **2019**, *9*, 483. [CrossRef]
55. Di Matteo, A.; Sacco, A.; Anacleria, M.; Pezzotti, M.; Delledonne, M.; Ferrarini, A.; Frusciante, L.; Barone, A. The ascorbic acid content of tomato fruits is associated with the expression of genes involved in pectin degradation. *BMC Plant Biol.* **2010**, *10*, 163. [CrossRef]
56. Abbas, M.; Anwar, J.; Zafar-ul-Hye, M.; Iqbal Khan, R.; Saleem, M.; Rahi, A.A.; Danish, S.; Datta, R. Effect of Seaweed Extract on Productivity and Quality Attributes of Four Onion Cultivars. *Horticulturae* **2020**, *6*, 28. [CrossRef]
57. Ertani, A.; Pizzeghello, D.; Francioso, O.; Sambo, P.; Sanchez-Cortes, S.; Nardi, S. *Capsicum chinensis* L. growth and nutraceutical properties are enhanced by biostimulants in a long-term period: Chemical and metabolomic approaches. *Front. Plant Sci.* **2014**, *5*, 375. [CrossRef] [PubMed]
58. Rouphael, Y.; Kyriacou, M.C.; Petropoulos, S.A.; De Pascale, S.; Colla, G. Improving vegetable quality in controlled environments. *Sci. Hortic.* **2018**, *234*, 275–289. [CrossRef]
59. Kovalikova, Z.; Kubes, J.; Skalicky, M.; Kuchtickova, N.; Maskova, L.; Tuma, J.; Vachova, P.; Hejnak, V. Changes in Content of Polyphenols and Ascorbic Acid in Leaves of White Cabbage after Pest Infestation. *Molecules* **2019**, *24*, 2622. [CrossRef]

horticulturae

Article

Inferring the Potential Geographic Distribution and Reasons for the Endangered Status of the Tree Fern, *Sphaeropteris lepifera*, in Lingnan, China Using a Small Sample Size

Xueying Wei [1,2,†], AJ Harris [1,†], Yuwen Cui [1], Yangwu Dai [3], Hanjia Hu [4], Xiaoling Yu [1], Rihong Jiang [5] and Faguo Wang [1,6,*]

[1] Key Laboratory of Plant Resources Conservation and Sustainable Utilization, Guangdong Provincial Key Laboratory of Applied Botany, South China Botanical Garden, Chinese Academy of Sciences, Guangzhou 510650, China; hsdweixueying@163.com (X.W.); aj.harris@inbox.com (A.H.); capebulbs@gmail.com (Y.C.); yxl272614@163.com (X.Y.)
[2] College of Life Sciences, Anhui Normal University, Wuhu 241000, China
[3] Nan'ao County Lantian Gardens Virescence Co., Ltd., Shantou 515900, China; daiyangwu@126.com
[4] Houzhai Town People's Government of Nan'ao County, Shantou 515910, China; huhj2001@126.com
[5] Guangxi Forestry Research Institute, Nanning 530002, China; jiangrhg@163.com
[6] University of Chinese Academy of Sciences, Beijing 100049, China
* Correspondence: wangfg@scbg.ac.cn; Tel.: +86-2037093715
† Equal contribution.

Citation: Wei, X.; Harris, A.; Cui, Y.; Dai, Y.; Hu, H.; Yu, X.; Jiang, R.; Wang, F. Inferring the Potential Geographic Distribution and Reasons for the Endangered Status of the Tree Fern, *Sphaeropteris lepifera*, in Lingnan, China Using a Small Sample Size. *Horticulturae* **2021**, *7*, 496. https://doi.org/10.3390/horticulturae7110496

Academic Editors: Rosario Paolo Mauro, Carlo Nicoletto and Leo Sabatino

Received: 20 September 2021
Accepted: 10 November 2021
Published: 15 November 2021

Publisher's Note: MDPI stays neutral with regard to jurisdictional claims in published maps and institutional affiliations.

Abstract: In this study, we investigated suitable habitats for the endangered tree fern, *Sphaeropteris lepifera* (J. Sm. ex Hook.) R.M. Tryon, based on fieldwork, ecological niche modeling, and regression approaches. We combined these data with the characterization of spore germination and gametophytic development in the laboratory to assess the reasons why *S. lepifera* is endangered and to propose a conservation strategy that focuses on suitable sites for reintroduction and accounts for the ecology and biphasic life cycle of the species. Our methods represent an integration of process- and correlation-based approaches to understanding the distributional patterns of this species, and this combined approach, while uncommonly applied, is a more robust strategy than either approach used in isolation. Our ecological niche models indicated that cold temperature extremes, temperature stability over long- and short-terms, and the seasonality of precipitation were among the most important abiotic environmental factors affecting the distribution of *S. lepifera* among the variables that we measured. Moreover, distribution of this fern species is also strongly influenced by the timing of development of male and female gametes. Additionally, we observed that slope aspect, specifically south-facing slopes, facilitates more incoming sunlight for mature trees, and simultaneously, provides greater, much-needed shade for fiddleheads on account of the canopy being denser. We believe that our study can provide important guidance on the restoration of *S. lepifera* in the wild. Specifically, potential restoration areas can be screened for the specific environmental factors that we infer to have a critical impact on the survival of the species.

Keywords: MaxEnt; ecological niche modeling (ENM); endangered species; Cyatheaceae; environmental factors

1. Introduction

Since at least the beginning of the Anthropocene epoch (beginning between ca. 7000 BP to 1964) [1], human activities have profoundly threatened the survival of other, non-human species on Earth. Human activities include overuse of resources, which are not renewable quickly enough to meet demands, fragmentation of once-continuous habitats to make way for infrastructure, and intentionally or unintentionally transporting species around the globe into new environments where they may threaten the naturally occurring species [2]. These activities have had the outcome of fundamentally altering the local and global patterns of geographic distributions of many organisms [3].

Understanding the natural environmental tolerances and preferences of species is fundamental to developing in-situ and ex-situ conservation strategies that seek to increase the populations of species, especially those that have become rare due to anthropogenic impacts. Unfortunately, the breadth of environmental tolerances and preferences of species cannot always be inferred by examining populations in the field. This is because it is often difficult or impossible to examine all populations or measure all environmental dimensions of a habitat and because species do not always occupy the full extent of the habitat available to them due to various biological constraints and environmental barriers [4,5]. To overcome these challenges, several broad classes of approaches have been utilized to extrapolate from the existing, or realized, distributions, of species to infer their potential distributions [6]. Among these classes are process-based approaches and correlative approaches. Process-based approaches focus on aspects of species biology, such as dispersal capabilities and pollination syndrome (i.e., in the case of plants). In contrast, correlative approaches involve building models based on existing distributional sites and environmental factors at those sites to make inferences about the potential distribution. At present, one of the most common correlative approaches is ecological niche modeling (ENM), especially via maximum entropy in MaxEnt [7]. Process-based and correlative approaches have exceptional potential in combination [8] to inform conservation strategies because the former can predict the natural potential of a species, while the latter can elucidate in what environments and geographic locations human-mediated conservation efforts may protect and encourage that potential. However, unfortunately, process-based and correlative approaches are rarely used together.

In this study, we sought to combine process-based understanding with correlative modeling to infer the potential distribution of *Sphaeropteris lepifera* (J. Sm. ex Hook.) R.M. Tryon, a tree fern that mainly occurs in China [9]. *S. lepifera* is a large species, characterized by an erect trunk (often more than 6 m tall and up to 20 cm in diameter near base) bearing distinct leaf scars [10]. Its leaves are pale green to green, about 3 m long, with scales and spreading hairs on the abaxial sides. The sori are often on the back of the smallest leaf divisions, or pinnules, and lack indusia. *S. lepifera* differs from *S. guangxiensis* Y.F. Gu & Y.H. Yan by having sparse spiny warts on its laminae and pinnae rachides and small scales on the pinnae rachides and abaxially on pinnule rachides. *S. guangxiensis* lacks the scales and warts on these structures [11]. This tree fern is listed as endangered within China as a grade two protected plant [12]. *S. lepifera* is treated in the family Cyatheaceae and is valued medicinally [13], horticulturally [14], and as a scientific model for evolution and biogeography [15,16]. Despite its present conservation status, this fern represents an ancient lineage that originated ca. 200 million years ago, became relatively widespread, and, later, underwent extinctions leading to its present restricted distributional range [17]. Thus, it has survived and multiplied since the Mesozoic time period. Within China, *S. lepifera* is primarily distributed in Taiwan [10], but a few wild populations have been found in coastal areas of mainland China, such as in Guangxi and Yunnan, including on near-shore islands, like Hainan [18–20]. Moreover, the species has also been discovered in Japan (Ryukyu Islands) and the Philippines, where it is likely to be native [21–24]. In Guangdong Province of China, *S. lepifera* was first found on Nan'ao Island of Shantou City by the Second National Key Protected Wild Plant Resources Survey of Guangdong Province in 2018. At present, the Nan'ao Island population, while tiny, is the only population in Guangdong Province showing new plant recruitment. Thus, the population is presently protected in-situ [25].

The causes of a rarity of *S. lepifera* have been considered in several prior studies [26–28]. One set of possible causes are environmental constraints on reproduction; especially on the ability of flagellated, swimming sperm to reach the female reproductive apparatus, the archegonium. In *S. lepifera*, the sperm originates from antheridia, which separated in space on the gamete-producing body from archegonia, thus necessitating at least a thin film of water for fertilization to occur. Aside from water, other constraints on sperm reaching archegonia may especially arise due to the presence of heavy metal pollutants in the environment. For example, the rotation frequency and displacement speed of the flagella of sperm are negatively impacted by lead (Pb^{2+}), while cadmium (Cd^{2+}) may inhibit the ability of sperm to follow chemical signals that otherwise guide them to receptive archegonia [26,27]. However, heavy metals are likely not the main reason for limitations in reproduction asany species of Cyatheaceae have been shown to produce gametophytes but not sporophytes under experimental conditions where soil nutrients are free of heavy metals [28]. There are 19 species of Cyatheaceae in China with a protected conservation status [29], and only nine species, including *S. lepifera*, have been successfully bred under controlled conditions [28,30].

Protection of a species requires not only a detailed understanding of its unique biology but also accurate models of its environmental tolerances to support both meaningful in-situ and ex situ conservation actions. Recent developments in ENM have led to its application in diverse conservation issues, including prediction of suitable habitat and species ranges [31,32] and how human activities affect the distribution of species [33,34], but applications for the conservation of protected species are still somewhat limited, especially in combination with the biology of the species.

The main objective of our study of *S. lepifera* was to apply process-based understanding with correlative-based modeling to species distributional modeling to (1) determine the main environmental and biological factors affecting the wild distribution of the species, so as to better focus the scope of conservation and (2) assess suitable areas for ex situ conservation. We especially sought to find suitable areas in Guangdong, Hong Kong, and Guanxi Provinces for ex situ conservation because these places had or have wild populations of *S. lepifera*. To accomplish our study, we focused on the population on Nan'ao Island, which is the only one within the province that has been observed to undergo natural recruitment without human intervention. By comprehensively analyzing the environment of *S. lepifera* on Nan'ao Island and its environmental interactions, we expect to determine other similarly suitable areas for the species for reintroduction. We believe that our results provide a comprehensive framework for future endeavors at protection, reintroduction, and sustainable utilization of *S. lepifera*.

2. Study Area and Methods

2.1. Field Investigation

From 2018 to 2020, we carried out field surveys on Nan'ao Island of Guangdong Province. The island is located at 116°53′–117°19′ E and 23°11′–23°32′ N within the subtropical maritime climate zone, having four seasons, including mild winters and relatively cool summers. Within Nan'ao Island, *S. lepifera* occurs on the main island, which has an area of 111.44 km^2.

On Nan'ao Island, we investigated plant recruitment of populations of *S. lepifera* as well as diameter, height, crown breadth, and health status of adult individuals of the species. We regarded adults as those individuals with heights greater than 3 m.

To determine the vascular plant composition of the community where *S. lepifera* is located, we also performed a Drude scale analysis, which is a quantitative method of assessing taxonomic composition, abundance, and cover [35,36]. To accomplish this, we established 20 m × 20 m plots centered around adult individuals and surveyed all co-occurring vascular plants within four 10 × 10 m quadrats nested within each plot. Within each quadrat, we sought to assign each species to one of the seven levels of the Drude scale [37]: plants of high sociability (soc; *plantae sociales*), three levels of copious or numerous (cop1-3; *copiose intermixtae*), sparse–sporadic (sp; *sparsae, sporadice intermixtae*), solitary (sol; *plantae solitariae*), and low abundance and/or singular individuals (uni; *unicum*). Assignments to these groups were determined based on relative-cover (relative cover = species cover/quadrat area), and the specific classification was as follows: soc ($rc \geq 75\%$); cop3 ($75\% < rc \leq 50\%$); cop2 ($50\% < rc \leq 25\%$); cop1 ($15\% < rc \leq 25\%$); sp ($5\% < rc \leq 15\%$); sol ($5\% < rc \leq 3\%$); uni ($rc \leq 2\%$), where rc refers to relative cover. We performed the Drude analysis separately for the tree, shrub, and herb layers, which we delimited on the basis of height with shrubs being 3–6 m tall and trees and herbs being greater and lesser in height, respectively.

2.2. Reintroduction Experiments

From July to September 2020, we carried out reintroductions of *S. lepifera* to Nan'ao Island using plants initially grown in the greenhouse. We planted a total of 120 fiddleheads that had been growing in the greenhouse for about six months; 40 at the site of the existing population located in Nan'ao Island, and 80 others divided into two sites selected for translocation experiments. We also conducted preliminary translocation experiments at Maofeng Mountain in Guangzhou, Renhua County in Shaoguan, and Bajia Town in Yangchun, based on preliminary site assessments. At these sites, we planted 20 plants each, and at all sites, including Nan'ao Island, we planted the fiddleheads near a shady slope with a stream. We performed surveys at all sites of reintroduction and translocation in February 2021 to determine fiddlehead survival.

2.3. Gametophytic Development

In order to investigate possible biological reasons for the relative rarity of *S. lepifera*, we conducted a spore breeding experiment. Ferns have a relatively unique life cycle in which both gametophytes and sporophytes can live independently. However, compared with the sporophyte, the structure of the gametophyte of *S. lepifera* is small, fragile, and composed of only one cell layer, whereas the sporophyte is ultimately arborescent. Therefore, gametophytic development of *S. lepifera* might be involved in its rarity or its distributional patterns.

We studied gametophytic development using spores newly germinated under lab conditions. We obtained the spores in September 2018 at our field site on Nan'ao Island. Specifically, we collected 50 mature sporophylls from two individuals of *S. lepifera* and stored these in paper bags in a cool and ventilated place until the spores fell off naturally after about seven days. We separated the spores from the dry vegetative material and stored them at 4 °C before further processing.

For germination, we placed the spores on 1/2 concentration MS medium after disinfecting them by applying 4% NaClO solution for 5 min. We grew the spores in culture under 16 h/d of 5000 Lux at 25 °C following recommendations for this species in a prior study [38]. Beginning on the third day, we sampled spores for microscopic examination every other day of two weeks. We selected characteristic spores for making into temporary slides at different stages, which we photographed under an OlympusBX43 light microscope.

Following the observations of germination, we transplanted the four-week-old gametophytes to sterilized soil. We divided this cohort of plants into four treatment groups containing 90–100 gametophytes each to determine tolerances of the species to temperature and moisture conditions. For temperature, we grew one group at 18 °C and the other at 25 °C both with daily watering. For moisture, we watered one group daily and the other once per week, and grew these plants at 25 °C. We regarded 25 °C and daily watering as "normal" (or control) growth conditions based on field surveys and the available literature. In all other respects, factors between the groups, such as lighting, were maintained as constant. One month later, we counted the number of sporophytes as a measure of successful fertilization.

2.4. Ecological Niche Modeling

We generated ENMs in MaxEnt 3.4.1 to predict potentially suitable distributional areas for *S. lepifera*. For generating ENMs in MaxEnt, we obtained occurrence data from three sources: our field site where a natural population occurs, populations conserved ex situ in natural reserves, and georeferenced specimen records [9] from the Global Biodiversity Information Facility (GBIF) [9]. Using all obtained data records, we performed data thinning with resampling to only one occurrence per 1 km × 1 km area using the R package, dismo [39] in R 4.0.5 [40,41]. Following data thinning, 409 occurrences remained, including one point from our field site and five ex-situ conservation sites in Guangdong Province (Table 1, Table S1).

Table 1. Specimen data from our field site and ex situ reintroduction areas.

Source	Province	City	Longitude	Latitude
field site	Guangdong	Shantou	117.072	23.444
ex situ	Guangdong	Shaoguan	113.751	25.076
ex situ	Guangdong	Yangjiang	111.520	21.963
ex situ	Guangdong	Shantou	117.072	23.444
ex situ	Guangdong	Guangzhou	113.444	23.283
ex situ	Guangdong	Shantou	117.071	23.442

We constructed the ENMs using the 19 environmental BioClim variables from World-CLIM 2.0 (Bio1-Bio19) [42–44], as well as slope and aspect derived from the Digital Elevation Model (DEM) downloaded from the same source. We also obtained landcover data as a raster map from the literature [45], and human population data from the Center for International Earth Science Information Network (CIESIN) [46]. We did not directly include elevation, which is widely known to be tightly linked with temperature variables. We obtained or resampled all variables at 30 arc-second resolution and clipped them to the boundaries of the distributional map that we established using the Database of Global Administrative Areas (GADM) [45].

Within MaxEnt, we determined the contribution of each environmental variable to the model as a whole by performing a jackknife test as an extension of model building. At the same time, in order to test the robustness of the model, we repeated the modeling process 12 times through the crossvalidate method [47], and we used "random seed" to randomize each run. We applied the area under the curve (AUC) to determine the predictive ability of each model.

3. Results

3.1. Germination and Fertilization Observations and Experiments

S. lepifera exhibits a typical fern life cycle with generational alternation between the gametophyte and sporophyte stages. We observed that the sporangia of *S. lepifera* produce single-celled meiotic spores in April and October each year (Figure 1). The spores germinate in about 60–70 days via the vittaria-type germination pattern. The center of each spore (Figure 1A) contains oil droplets, and germination starts with the exertion of a rhizoid (Figure 1B–E). The apical chloroplast-bearing cells opposite the rhizoid development (Figure 1E) become enlarged, and the emerging prothallus (Figure 1F) develops several rhizoids on its ventral surface. The structure of the gametophyte is simple but with differentiation of the dorsal and ventral surfaces. In particular, when the gametophyte matures, there are many antheridia (Figure 1G) and archegonia (Figure 1H) on the ventral surface. Antheridia generate sperm that must swim in the water and sense the chemical attractants secreted by the archegonia to complete the fertilization.

Warm temperatures and high moisture appear essential for fertilization to occur in *S. lepifera*. Specifically, we found that fertilization (evidenced by production of a sporophyte) had visually-assessed higher success rate among plants grown at 25 °C than 18 °C. Moreover, we found that watering every day had an important impact on the formation of sporophytes, possibly related to maintaining high humidity (Figure 1I). In fact, we found that gametophytes that were watered only once weekly became malformed and yielded unproductive sporophytes.

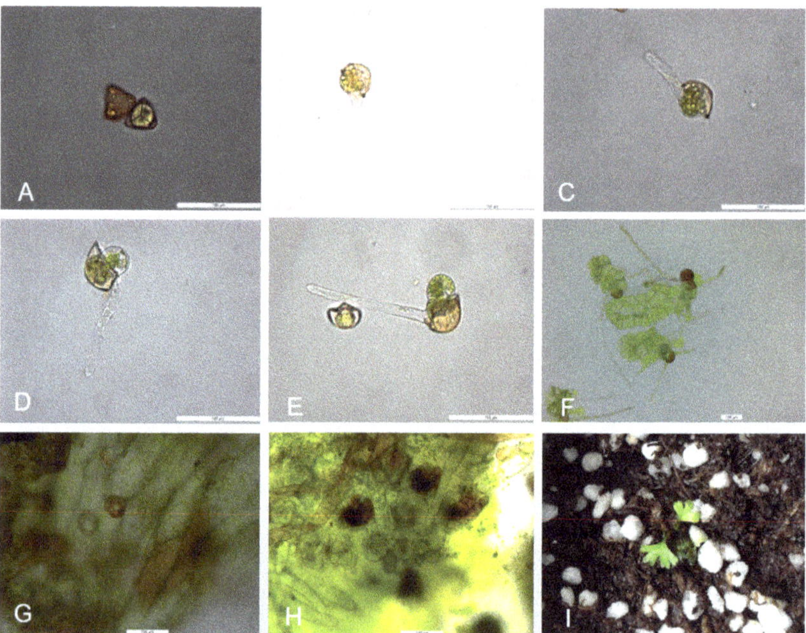

Figure 1. Spore germination process of *Sphaeropteris lepifera*. (**A**) Spore of *S. lepifera* (center of each spore contains oil droplets). Scale bar = 50 μm. (**B–E**) Spore germination. Scale bar = 100 μm. (**F**) Prothallus. Scale bar = 100 μm. (**G**) Antheridium. Scale bar = 100 μm. (**H**) Archegonium. Scale bar = 100 μm. (**I**) Young sporophyte. Scale bar = 10 cm.

3.2. Ecological Niche Modeling and Importance of Environmental Variables

Our ENM result for *S. lepifera* showed that the most suitable habitat was found on northern Taiwan, along the central and western coast of Guangdong, on coastal Guangxi and Yunnan, and on near shore islands south of these provinces, such as Nan'ao. Areas of moderate suitability included coastal mountains. The AUC was 0.983 (Table 2, Figure 2A) suggesting that our model was robust. Notably, among reintroduction sites, only Shaoguan was predicted as environmentally unsuitable.

Our estimates of the relative contributions of the sampled environmental variables to the MaxEnt model showed that mean diurnal range (Bio02) contributed the most to the model (29.3%) followed by temperature seasonality (Bio04, 24.7%), precipitation of the coldest quarter (Bio19, 13.5%), and mean annual precipitation (Bio12, 11.7%) (Table 2). Precipitation of the coldest quarter was also the most important variable based on permutation (66.6%).

The jackknife test of variable importance (Figure 2B–D) showed that, overall, models showed roughly equal gains when any one variable was excluded. However, the highest training and test gain (Figure 2B,C) for a variable used in isolation were for mean diurnal range (Bio02), followed by temperature seasonality (Bio04), mean temperature of the coldest quarter (Bio11), isothermality (Bio03) and precipitation of the warmest quarter (Bio18). The first two of these are in agreement with the measured percentage contributions, the mean temperature of the coldest quarter is consistent with the permutation result, and the others are broadly congruent with the inference that temperature and precipitation variability are critical for the species. The AUC calculations under the jackknife test (Figure 2D) showed a similar pattern as the training and test gain, except that the mean annual temperature (Bio01) also showed notable predictive capability in isolation.

Table 2. Contributions of the environmental variables to the MAXENT models using the 19 bioclimatic variables as well as slope, aspect, land cover, and population.

Code	Environmental Factor	Percent Contribution	Permutation Importance
Bio01	Annual Mean Temperature	0.4	0
Bio02	Mean Diurnal Range (Mean of monthly (max temp-min temp))	29.3	4.9
Bio03	Isothermality (Bio2/Bio7) ($\times$ 100)	3.7	7.4
Bio04	Temperature Seasonality (standard deviation $\times$ 100)	24.7	0.6
Bio05	Max Temperature of Warmest Month	1	1.2
Bio06	Min Temperature of Coldest Month	0.1	0.1
Bio07	Temperature Annual Range (Bio5-Bio6)	4.1	0.4
Bio08	Mean Temperature of Wettest Quarter	0.6	1.1
Bio09	Mean Temperature of Driest Quarter	0.1	0
Bio10	Mean Temperature of Warmest Quarter	0.5	0.2
Bio11	Mean Temperature of Coldest Quarter	2.1	5.7
Bio12	Annual Precipitation	11.7	3.7
Bio13	Precipitation of Wettest Month	0.1	0.3
Bio14	Precipitation of Driest Month	0.9	0.8
Bio15	Precipitation Seasonality (Coefficient of Variation)	0.2	0.5
Bio16	Precipitation of Wettest Quarter	0	0
Bio17	Precipitation of Driest Quarter	0.3	0.3
Bio18	Precipitation of Warmest Quarter	4.7	3.7
Bio19	Precipitation of Coldest Quarter	13.5	66.6
Slope	Slope	0.3	0.1
Aspect	Aspect	0.3	0.3
Landcover	Landcover	0.1	0.1
Human population	Human population	1.4	1.9

3.3. Community Composition

We observed that the forests of Nan'ao Island mainly comprised mixed stands of *Acacia confusa* Merr. and *Pinus massoniana* D. Don with wild shrubs from the families Rosaceae, Rutaceae, and Theaceae. According to the Drude abundance analysis of the community composition of *S. lepifera* (Table 3), the tree layer is mainly composed of *Machilus chinensis* (Benth.) Hemsl., *Schefflera heptaphylla* (L.) Frodin, *Cunninghamia lanceolata* (Lamb.) Hook. and *Acacia confusa* and is about 8–13 m in height. The shrub layer is dominated by *Psychotria rubra* (Lour.) Poir. and is about 1.2–1.5 m in height. The herb layer consists of *Eleutherococcus trifoliatus* (L.) S.Y. Hu, *Blechnum orientale* L., *Boehmeria nivea* (L.) Gaudich., *Cyclosorus parasiticus* (L.) Farw., *Deparia lancea* (Thunb.) Fraser-Jenk., *Dryopteris fuscipes* C. Chr., *Pteris fauriei* Hieron., *Alocasia odora* (Roxb.) K. Koch, *Adiantum flabellulatum* L., *Lygodium japonicum* (Thunb.) Sw., *Mussaenda pubescens* W.T. Aiton, and *Liriope spicata* (Thunb.) Lour.

3.4. Preliminary Reintroduction Experiment

The survey in February 2021 revealed that 14 of 40 introduced individuals of *S. lepifera* survived at the site of the original population on Nan'ao Island compared with 27 of 40 and three of 40 at the two nearby sites. In addition, Maofeng Mountain in Guangzhou, Renhua County in Shaoguan, and Bajia Town in Yangchun five, 10, and five survived, respectively, of the 20 planted. The environmental conditions of the site are shown in Table 4.

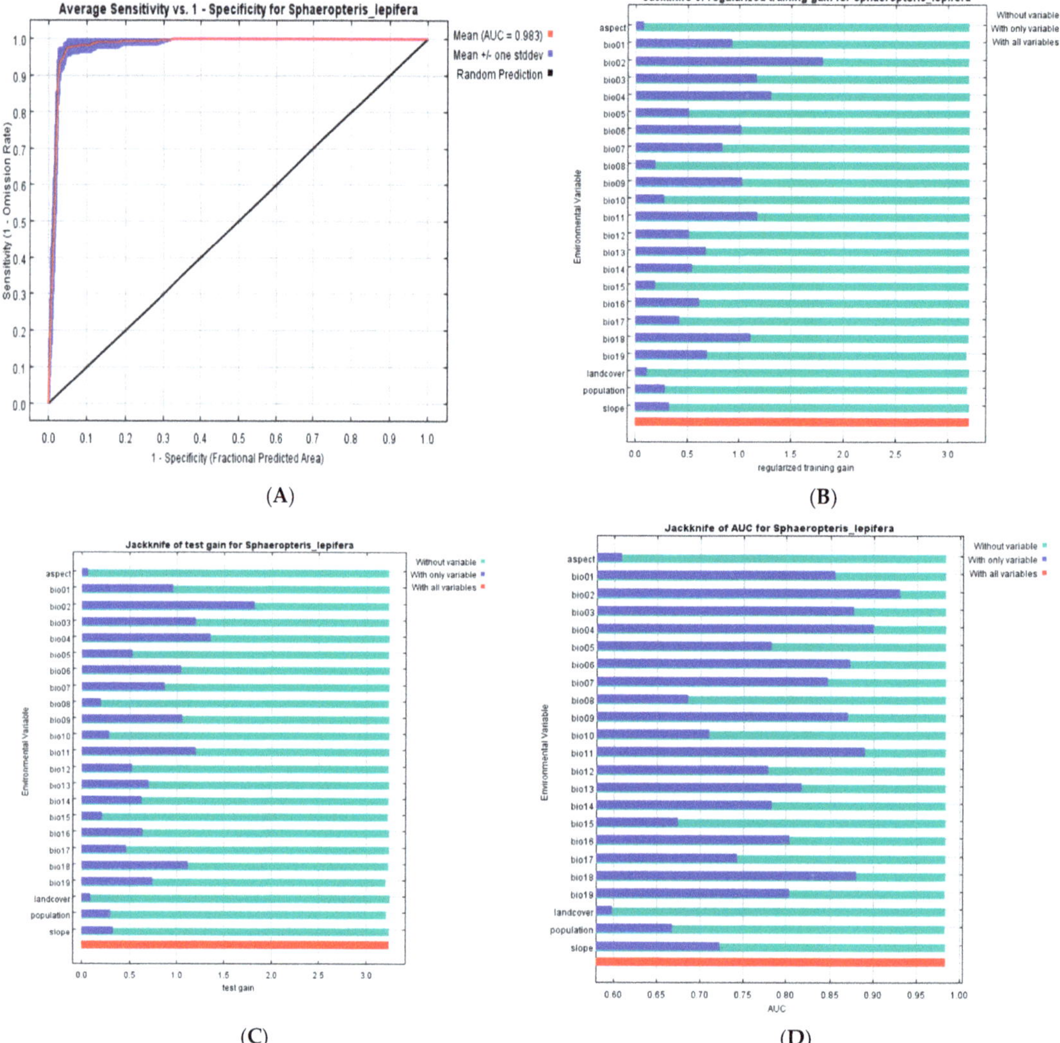

Figure 2. (**A**) ROC curve for MaxEnt prediction; (**B**) Result of the jackknife test of variable importance using training gain; (**C**) Result of the Jackknife test using test gain; (**D**) Result of the jackknife test using AUC on test data.

Although the sample size is small and the environmental variability limited due to geographic distance among reintroduction sites, it may be noteworthy that the site with the highest survival rate (after averaging among the three adjacent Shantao sites), Shaoguan (50%), represents the extremes of all values shown to be important based on the various modes of testing within the ENM framework (Figure 3). Moreover, in several cases, the site with the second-highest survival rate, Shantao (38% average rate), is usually closer to the same extremes as Shaoguan compared to the other key variables. This is the case with annual precipitation, isothermality, and precipitation of the warmest quarter. This may further highlight the importance of these specific four variables for the survival of reintroduced plants. However, in some cases Shaoguan and Shantao are at opposite extremes of the environmental variables surveyed. This includes for mean diurnal range, temperature seasonality, mean temperature of the coldest quarter, and precipitation of the coldest quarter. These four variables are all related to temperature. Notably, at Shantao, meticulous management practices included mitigating the effects of temperature, especially during winter, possibly suggesting the feasibility of successful reintroduction even under less than ideal temperature conditions.

Table 3. All species recorded in quadrats established around adult individuals of *Sphaeropteris lepifera* on Nan'ao Island.

Life Form	Frequency	Species	Drude Abundance	Relative Cover (%)
Tree	Dominant species	*Machilus chinensis* (Benth.) Hemsl.	Soc	211
		Schefflera heptaphylla (L.) Frodin	Soc	136.5
		Acacia confusa Merr.	Soc	90
		Cunninghamia lanceolata (Lamb.) Hook.	Soc	89.1
	Companion species	*Ilex viridis* Champ. ex Benth.	Cop3	52.8
		Litsea glutinosa (Lour.) C.B. Rob.	Cop2	36
		Aporosa dioica (Roxb.) Müll. Arg.	Sp	9
	Rare species	*Archidendron lucidum* (Benth.) I.C. Nielsen	Uni	1.28
Shrub	Dominant species	*Psychotria rubra* (Lour.) Poir.	Sp	9.6
	Companion species	*Camellia sinensis* (L.) Kuntze	Uni	2.92
		Diplospora dubia (Lindl.) Masam.	Uni	1.44
Herb	Dominant species	*Blechnum orientale* L.	Sol	4.5
		Boehmeria nivea (L.) Gaudich.	Sol	3.5
	Companion species	*Eleutherococcus trifoliatus* (L.) S.Y. Hu	Sol	3
		Cyclosorus parasiticus (L.) Farw.	Uni	2.2
		Alocasia odora (Roxb.) K. Koch	Uni	2
		Dryopteris fuscipes C. Chr.	Uni	1.2
		Pteris fauriei Hieron.	Uni	0.6
	Rare species	*Deparia lancea* (Thunb.) Fraser-Jenk.	Uni	0.5
		Adiantum flabellulatum L.	Uni	0.4
		Lygodium japonicum (Thunb.) Sw.	Uni	0.3
		Mussaenda pubescens W.T. Aiton	Uni	0.3
		Alpinia hainanensis K. Schum.	Uni	0.3
		Solanum Americanum Mill.	Uni	0.3
		Liriope spicata (Thunb.) Lour.	Uni	0.2

Note: Soc, society; Cop, copiosa; Sp, sparsal; Sol, solitariae; Uni, unicum.

According the MaxEnt results, we divided predictions for *S. lepifera* into five categories: currently present, high habitat suitability, moderate suitability, low suitability, and unsuitable (Figure 4). Much of the area throughout Guangdong Province is moderately to highly suitable for large-scale reintroduction. This is in contrast to Guangxi Province, which has a very similar flora to Guangdong, sharing ca. 92% of plant genera [48], and in both provinces, the flora has strongly tropical and subtropical characteristics. However, within Guangxi, mountainous terrain may affect critical environmental factors, such as temperature seasonality and precipitation of the coldest quarter, thus limiting the moderately suitable habitat for *S. lepifera* to areas along the border with Vietnam. Additionally, the MaxEnt model predicted that there are some places in Hong Kong that are highly suitable for introduction.

Table 4. Comparison of environmental conditions between the reintroduced area and Nanao Island.

	Nanao Island	Maofeng Mountain	Renhua County	Bajia Town
City	Shantou	Guangzhou	Shaoguan	Yangchun
Longitude	117.07190393	113.444498	113.751431	111.520311
Latitude	23.44393082	23.283165	25.07638	21.963179
Mean annual temperature	21.5 °C	21.9 °C	19.6 °C	22.3 °C
Average annual precipitation	1372.5 mm	1623.6 mm	1665 mm	2392.3 mm
Frost season	no	no	57 d	no
Annual minimum temperature	8 °C	3 °C	−3 °C	2 °C
Coldest month	January	January	January	January
Mean temperature of the coldest month	14 °C	15 °C	11 °C	16 °C

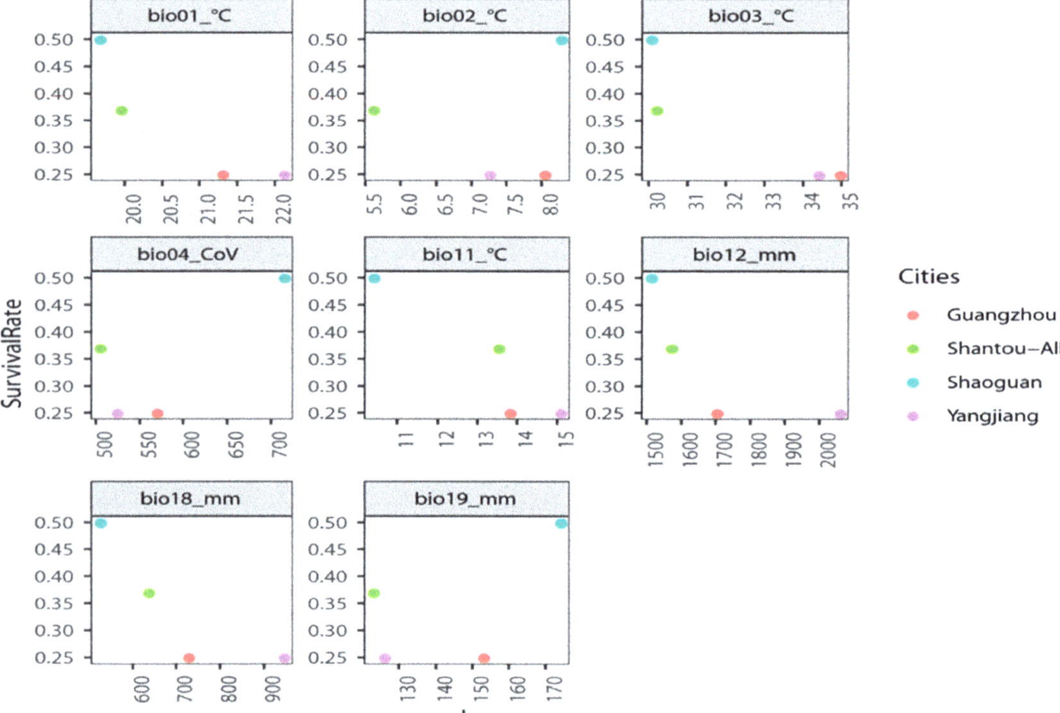

Figure 3. Survival rates of reintroduced seedlings at study sites based on the most important environmental variables according to assessments within the MaxEnt ENM framework. Survival rates were averaged across the three adjacent sites in Shantou, where the original population occurs.

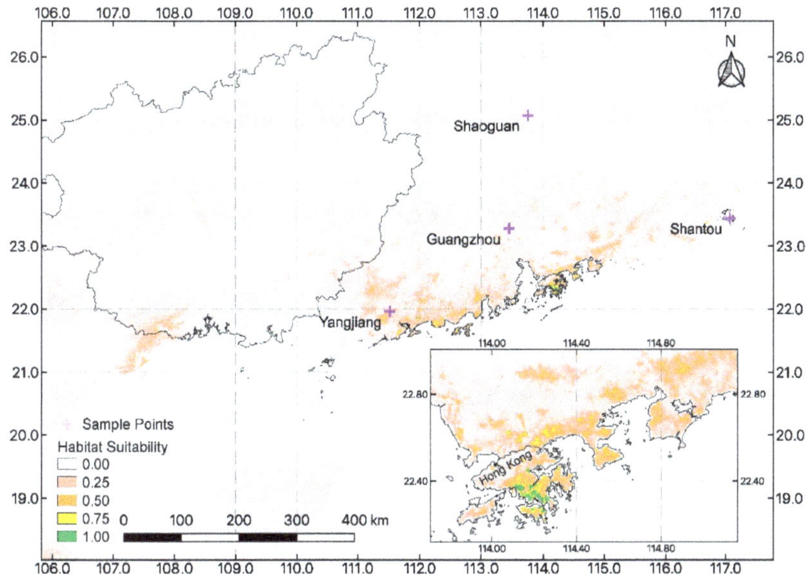

Figure 4. Prediction of habitat suitability for *Sphaeropteris lepifera* in eastern Asia.

4. Discussion

4.1. Factors and Reasons Affecting the Distribution of Sphaeropteris lepifera during Life History

We found that the rarity of *S. lepifera* may be due to the relatively large influence of environmental factors, especially temperature and available moisture, which appear critical to the transition between the gametophyte and sporophyte stages. *S. lepifera* generates spores twice yearly, at the end of April and the end of October. The process from spore dispersal to the formation of a sporophyte takes about three months so that the sporophytes should be formed in July-August and January-February. However, it happens that, within most of the native range of the species, January has the lowest temperature of the year (ca. about 14 °C) and represents the driest period. We observed in our germination experiments that lower temperatures and dry conditions led to lower rates of sporophytes production. Moisture is required for sperm to successfully fertilize archegonia, such that the moisture requirement is easily understood.

The role of temperature in the reproductive success of *S. lepifera* and its alteration of generations may be more complex. Although greenhouse-grown plants responded better to higher temperatures (25 °C) than lower ones (18 °C), we observed in the field that the Renhua area of Shaoguan, where the average monthly temperature is the lowest (Table 4), had more surviving plants than the other two places. This may be due to watering frequency by the management, as watering (and other maintenance) is well-known to improve plant growth of ferns introduced (or reintroduced) into the wild [49]. However, several species of the Cyatheaceae fern family, to which *S. lepifera* belongs, are known to be sensitive to phosphorus and have very specific requirements for this nutrient [50]. Phosphorus requires a higher energy investment to take up under higher temperatures [51], and while this may not be limiting for greenhouse-grown plants, it may have affected those reintroduced in wild areas. Thus, the interplay and balance between higher temperature preferences and investments in nutrient uptake at reintroduction should be examined in future studies.

4.2. Factors and Causes of Community Composition Affecting the Distribution of Sphaeropteris lepifera

Although we observed that *Acacia confusa* and *Pinus massoniana* were the dominant trees on Nan'ao Island, they were not the dominant species within the plot centered around mature individuals of *S. leipifera*. In fact, the 20 × 20 m plot contained no *Pinus massoniana* and only one *Acacia confusa*. Normally, *Acacia confusa* is 15 m in height while *Pinus massoniana* can reach 45 m. This is in contrast to *S. lepifera*, which reaches only ca. 8 m. The demand for light by mature *S. lepifera* would make it hard for these species to compete with tall trees. Overall, the average height of the community around the *S. lepifera* individuals was ca. 10 m, with the tallest species being *Machilus chinensis* and *Cunninghamia lanceolata*. These species are not normally so short in stature nor is the *Acacia confusa* also found in the plot. Their small stature may relate to the occurrence of the community on a south-facing slope. Notably, shrub species within the community, such as *Archidendron lucidum*, were taller than normal. This suggests that slope and aspect may also be critical factors affecting suitable habitat for *S. lepifera*, which may be better able to compete within environments that tend towards supporting lower canopy heights, either through recruitment of mainly short-stature trees or by limiting the growth of typically taller ones. This is also consistent with prior studies, which have shown that tree ferns commonly colonize light gaps within a canopy resulting from disturbances such as high wind, landslides, or other natural or anthropogenic causes [52,53].

4.3. Suitable Habitat for Reintroduction within the Chinese Mainland and Near-Shore Islands

In recent years, due to the rarity of wild *S. lepifera* on the China mainland, some plants were introduced at botanical gardens for protection. For example, South China Botanical Garden, Chinese Academy of Sciences and Xiamen Botanical Garden have successfully cultivated wild-sourced *S. lepifera* and possess thriving adults that are generating new fiddleheads and spores. Despite these successes with ex-situ conservation, in-situ conservation offers the advantages of facilitating long-term population health through natural selection and of supporting natural ecosystem services [54].

We base our recommendations for reintroduction of *S. lepifera* on our results and the prior work of Wei et al. [55] regarding the successful reintroduction of the endangered orchid, *Bletilla striata* (Thunb.) Rchb.f. This orchid has similar habitat requiremnets to *S. lepifera*, so we believe that protocols leading to its successful reintroduction may also be applicable to *S. lepifera*. Ideally, areas for reintroduction of *S. lepifera* in the wild should meet the following basic conditions: (1) be far away from villages and highways and relatively inaccessible to people to reduce anthropogenic disturbance; (2) have trees with average heights not be more than 10 m with small trees and shrubs as the dominant species and (3) be a moist environment, such as with a nearby water source, ideally at the foot of a shady slope. During reintroduction, we recommend the following steps to promote greater success: (1) plant in a scattered pattern and not too densely; (2) water frequently during dry periods from January to February or July to August or during drought, especially when drought occurs in the first six months after planting; and (3) perform preliminary in-situ experiments prior to large-scale transplanting, focusing on whether there are animal, disease, or other factors potentially affecting the survival of transplanted individuals in the local area.

Both ex-situ and in-situ conservation have advantages and disadvantages. For in-situ conservation, the disadvantages are that it requires a greater investment of time and money and, as evidenced from the recommendations above, the procedures are often cumbersome. However, given the aforementioned advantages, a model that utilizes both in situ and ex situ strategies for *S. lepifera* may be beneficial for the species. Considering the ornamental, economic, and medicinal value of *S. lepifera* [10,13–15], it could be widely promoted for cultivation in Guangdong Province, such as in parks or within urban landscaping, especially with orchids, for which its trunk is an important substrate for epiphytic growth.

The reason why many endangered plants cannot be cultivated at a large scales is that their unique, strict environmental requirements and the growing conditions lead to high maintenance costs. However, our work may not only provide a framework to support in-situ conservation of *S. lepifera*, but also finding suitable habitats for it and maintaining it within urban green spaces. Thus, our findings may help to reduce the costs of ex-situ conservation efforts.

Supplementary Materials: The following are available online at https://www.mdpi.com/article/10.3390/horticulturae7110496/s1, Table S1. Distribution information of *Sphaeropteris lepifera*.

Author Contributions: Conceptualization, X.W., A.H. and F.W.; methodology, Y.C. and A.H.; software, Y.C.; validation, A.H., F.W., Y.C. and X.W.; formal analysis, X.W. and A.H.; investigation, F.W., Y.D., H.H., X.Y. and R.J.; resources, F.W. and X.W.; data curation, X.W., A.H. and F.W.; writing—original draft preparation, X.W., A.H. and Y.C.; writing—review and editing, X.W., A.H. and F.W.; visualization, X.W.; supervision, F.W.; project administration, F.W.; funding acquisition, F.W. All authors have read and agreed to the published version of the manuscript.

Funding: This research was funded by grants from the Strategic Priority Research Program of the Chinese Academy of Sciences (grant number XDA13020601), Agricultural Biodiversity Investigation and Protection Technology System Construction Project (grant number 0835-210Z32104471), Guangdong Province Enterprise Science and Technology Special Commissioner Project (grant number GDKTP2020047500) and Guangdong Wildlife Protection and Management Project (grant number Y936061001).

Institutional Review Board Statement: Not applicable.

Informed Consent Statement: Not applicable.

Data Availability Statement: The data presented in this study are available on request from the corresponding author (F.-G.W.). It can be downloaded from the supporting files.

Acknowledgments: The authors would like to thank Jinfang Cai, Shaohui Wang and Jiyun Li of Shantou Nan'ao County Bureau of natural resources for their assistance with the field work, and Zhiqiang Song of South China Botanical Garden, Chinese Academy of Sciences for his assistance with the experiments observation.

Conflicts of Interest: The authors declare no conflict of interest.

References

1. Ehlers, E.; Krafft, T. The "Anthropocene". In *Earth System Science in the Anthropocene*; Springer: Berlin/Heidelberg, Germany, 2006; pp. 13–18.
2. Harris, A.; Stefanie, I.B.; Aarón, R. Long distance dispersal in the assembly of floras: A review of progress and prospects in north America. *J. Syst. Evol.* **2018**, *56*, 430–448. [CrossRef]
3. Xu, W.B.; Svenning, J.C.; Chen, G.K.; Zhang, M.G.; Huang, J.H.; Chen, B.; Ordonez, A.; Ma, K.P. Human Activities Have Opposing Effects on Distributions of Narrow—Ranged and Widespread Plant Species in China. *Proc. Natl. Acad. Sci. USA* **2019**, *116*, 266–274. [CrossRef]
4. Peterson, A.T. Uses and requirements of ecological niche models and related distributional models. *Biodivers. Inform.* **2006**, *3*, 59–72. [CrossRef]
5. Blonder, B. Hypervolume concepts in niche- and trait-based ecology. *Ecography* **2018**, *41*, 1441–1455. [CrossRef]
6. Peterson, A.T.; Pape, M.; Soberón, J. Mechanistic and correlative models of ecological niches. *Eur. J. Ecol.* **2015**, *1*, 28–38. [CrossRef]
7. Phillips, S.J.; Anderson, R.P.; Schapire, R.E. Maximum Entropy Modeling of Species Geographic Distributions. *Ecol. Model.* **2006**, *190*, 231–259. [CrossRef]
8. Carsten, F.D.; Stanislaus, J.S.; Juliano, C.; Isabelle, C.; Catherine, G.; Florian, H.; Michael, K.; Xavier, M.; Römermann, C.; Schröder, B.; et al. Correlation and process in species distribution models: Bridging a dichotomy. *J. Biogeogr.* **2012**, *39*, 2119–2131.
9. GBIF.org. GBIF Occurrence Download. Available online: https://doi.org/10.15468/dl.j8abq5 (accessed on 5 July 2021). [CrossRef]
10. Zhang, X.C.; Zhang, L.B. Cyatheaceae. In *Flora of China*; Wu, Z.Y., Raven, P.H., Hong, D.Y., Eds.; Science Press: Beijing, China, 2013; Volume 2, pp. 134–138.
11. Gu, Y.F.; Jiang, R.H.; Liu, B.D.; Yan, Y.H. Sphaeropteris guangxiensis Y. F. Gu & Y. H. Yan (Cyatheaceae), a new species of tree fern from Southern China. *Phytotaxa* **2021**, *518*, 69–74.
12. National Forestry and Grassland Administration, Ministry of Agriculture and Rural Affairs. List of National Key Protected Wild Plants in China. 2021-9-7. Available online: http://www.gov.cn/zhengce/zhengceku/2021-09/09/content_5636409.htm (accessed on 25 September 2021).
13. Ye, H.G. *Chinese Medicinal Plants*; Chemical Industry Press: Beijing, China, 2017; Volume 25, pp. 356–357.
14. Shi, L. *Ornamental Ferns of China*; China Forestry Publishing House: Beijing, China, 2002; p. 87.
15. Fujian Forestry Department. Sphaeropteris lepifera (J. Sm. ex Hook.). *For. China* **2003**, *11*, 29.

16. Morigengaowa, H.S.; Liu, B.; Kang, M.; Yan, Y. One or More Species? GBS Sequencing and Mmorphological Traits Evidence Reveal Species Diversification of Sphaeropteris brunoniana in China. *Biodivers. Sci.* **2019**, *27*, 36–44. [CrossRef]

17. Huang, C.H.; Qi, X.P.; Chen, D.Y.; Qi, J.; Ma, H. Recurrent genome duplication events likely contributed to both the ancient and recent rise of ferns. *J. Integr. Plant Biol.* **2020**, *62*, 433–455. [CrossRef]

18. Chen, X.X.; Pan, T.Z. The Report of a New Record Genera (*Sphaeropteris*) and a New Record Species (*Cyathea lepifera*) from Zhejiang, China. *J. Wenzhou Univ.* **2016**, *37*, 34–37.

19. Yin, H. *Rare and Endangered Plants in China*; China Forestry Publishing House: Beijing, China, 2013; pp. 13–26.

20. Zhao, X.Q. The Rare Wild Tree Fern Found in Xiamen: *Sphaeropteris lepifera* (J. Sm. ex Hook.) (Cycinaceae). *Wuyi Sci. J.* **1983**, *3*, 181.

21. Ebihara, A.; Yamaoka, A.; Mizukami, N.; Sakoda, A.; Nitta, J.H.; Imaichi, R. A survey of the fern gametophyte flora of japan: Frequent independent occurrences of noncordiform gametophytes. *Am. J. Bot.* **2013**, *100*, 735–743. [CrossRef] [PubMed]

22. Pelser, P.B.; Barcelona, J.F.; Nickrent, D.L. (Eds.) 2011 Onwards. Co's Digital Flora of the Philippines. Available online: www.philippineplants.org (accessed on 20 September 2021).

23. Barcelona, J.F. The Taxonomy and Ecology of the Pteridophytes of Mt. Iraya and Vicinity, Batan Island, Batanes Province, Northern Philippines. In *Pteridology in the New Millennium*; Springer: Dordrecht, The Netherlands, 2003. [CrossRef]

24. Holttum, R.E. Tree-Ferns of the Genus Cyathea Sm. in Asia (Excluding Malaysia). *Kew Bull.* **1965**, *19*, 463–487. [CrossRef]

25. Wang, B.S.; Yu, S.X.; Peng, S.L.; Li, M.G. *Handbook of Plant Community Experiments*; Guangdong Higher Education Press: Guangzhou, China, 1996.

26. Ji, S.B.; Xiang, J.Y.; Liu, B.D.; Ji, H.L. Behavior and Morphology of *Sphaeropteris lepifera* (J. Sm. ex Hook.) (Cyatheaceae) Swimming Sperm. *Acta Bot. Boreali-Occident. Sin.* **2018**, *38*, 92–97.

27. Zhang, X.; Xiang, J.J.; Xiao, X.; Ji, S.B.; Liu, B.D. Influence of Lead and Cadmium Stress on Sperm's Behavior and Morphology of *Sphaeropteris lepifera* (J. Sm. ex Hook.) (Cyatheaceae). *Acta Bot. Boreali-Occident. Sin.* **2019**, *5*, 776–783.

28. Cheng, M.C.; You, C.F. Sources of major ions and heavy metals in rainwater associated with typhoon events in southwestern taiwan. *J. Geochem. Explor.* **2010**, *105*, 106–116. [CrossRef]

29. Jiao, Y.; Wang, H.; Zhang, S.Z. *Fern Atlas*; China Forestry Publishing House: Beijing, China, 2014.

30. Wang, J.J.; Zhang, X.C.; Liu, B.D.; Cheng, X. Gametophyte Development of Three Species in Cyatheaceae. *J. Trop. Subtrop. Bot.* **2007**, *15*, 115–120.

31. Wani, I.A.; Verma, S.; Mustahq, S.; Alsahli, A.A.; Alyemeni, M.N.; Tariq, M.; Pant, S. Ecological analysis and environmental niche modelling of *D. hatagirea* (D. Don) Soo: A conservation approach for critically endangered medicinal orchid. *Saudi J. Biol. Sci.* **2021**, *28*, 2109–2122. [CrossRef]

32. Papeş, M.; Gaubert, P. Modelling ecological niches from low numbers of occurrences: Assessment of the conservation status of poorly known viverrids (Mammalia, Carnivora) across two continents. *Divers. Distrib.* **2010**, *13*, 890–902. [CrossRef]

33. Banks, S.C.; Finlayson, G.R.; Lawson, S.J.; Lindenmayer, D.B.; Paetkau, D.; Ward, S.J.; Taylor, A.C. The effects of habitat fragmentation due to forestry plantation establishment on the demography and genetic variation of a marsupial carnivore, *Antechinus agilis*. *Biol. Conserv.* **2005**, *122*, 581–597. [CrossRef]

34. Rhodes, J.R.; Wiegand, T.; McAlpine, C.A.; Callaghan, J.; Lunney, D.; Bowen, M.; Possingham, H.P. Modeling species' distributions to improve conservation in semiurban landscapes: Koala case study. *Conserv. Biol.* **2006**, *20*, 449–459. [CrossRef] [PubMed]

35. Li, X.; Fu, L.; Wang, F.G.; Xing, F.W. Michelia guang¬dongensis (Magnoliaceae), an Endangered Plant Species with Ex-tremely Small Population, should be Evaluated as CR C2a (i); D. *Biodivers. Sci.* **2017**, *25*, 91–93. [CrossRef]

36. Ren, Z.; Chen, L.; Peng, H. Orophea yunnanensis should be Listed as a Plant Species with Extremely Small Population. *Biodivers. Sci.* **2016**, *24*, 358–359. (In Chinese) [CrossRef]

37. Ptsch, R.; Jaková, A.; Chytr, M.; Kucherov, I.B.; Janssen, J. Making them visible and usable—vegetation lot observations from fennoscandia based on historical species quantity scales. *Appl. Veg. Sci.* **2019**, *22*, 465–473. [CrossRef]

38. Ma, H.N.; Li, Y.; Tan, L.Y.; Liu, B.D. Spore Propagation and Rejuvenation of *Sphaeropteris lepifera* (J. Sm. ex Hook.). *Acta Hortic. Sin.* **2010**, *37*, 1679–1684.

39. Robert, J.H.; Phillips, S.; Leathwick, J.; Elith, J. Dismo: Species Distribution Modeling. R Package Version 1.3-3. Available online: https://CRAN.R-project.org/package=dismo (accessed on 10 September 2020).

40. R Core Team. *R: A Language and Environment for Statistical Computing*; R Foundation for Statistical Computing: Vienna, Austria. Available online: https://www.R-project.org/ (accessed on 12 June 2021).

41. Tennekes, M. tmap: Thematic Maps in R. *J. Stat. Softw.* **2018**, *84*, 1–39. [CrossRef]

42. Fick, S.E.; Hijmans, R.J. Worldclim 2: New 1-km spatial resolution climate surfaces for global land areas. *Int. J. Climatol.* **2017**, *37*, 4302–4315. [CrossRef]

43. Booth, T.H.; Nix, H.A.; Busby, J.R.; Hutchinson, M.F. Bioclim: The first species distribution modelling package, its early applications and relevance to most current MaxEnt studies. *Divers. Distrib.* **2014**, *20*, 1–9.

44. Zuo, H.; Hutchinson, M.F.; Mcmahon, J.P.; Nix, H.A. Developing a mean monthly climatic database for China and Southeast Asia. In Proceedings of the International Workshop Held in Bangkok: Matching Trees and Sites, Bangkok, Thailand, 27–30 March 1995.

45. Buchhorn, M.; Smets, B.; Bertels, L.; Lesiv, M.; Tsendbazar, N.-E.; Masiliunas, D.; Linlin, L.; Herold, M.; Fritz, S. Copernicus Global Land Service: Land Cover 100 m: Collection 3: Epoch 2019: Globe (Version V3.0.1) [Data set]. Zenodo. 2020. Available online: https://land.copernicus.eu/global/products/lc (accessed on 10 July 2021).

46. WorldPop (www.worldpop.org-School of Geography and Environmental Science, University of Southampton; Department of Geography and Geosciences, University of Louisville; Departement de Geographie, Universite de Namur) and Center for International Earth Science Information Network (CIESIN), Columbia University. Global High Resolution Population Denominators Project-Funded by the Bill and Melinda Gates Foundation (OPP1134076). 2018. Available online: https://www.worldpop.org/geodata/summary?id=94 (accessed on 10 September 2020).

47. Chitta, R.D.; Jamir, N.S.; Zubenthung, P.K. Distribution Prediction Model of a Rare Orchid Species (Vanda bicolor Griff.) Using Small Sample Size. *Am. J. Plant Sci.* **2017**, *8*, 1388–1398.

48. Su, Z.R.; Zhang, H.D. The Relationship between the Flora of Guangxi and the Flora of Neighboring Areas. *J. South China Agric. Univ.* **1994**, *15*, 38–43.

49. Riano, K.; Briones, O. Sensitivity of three tree ferns during their first phase of life to the variation of solar radiation and water availability in a mexican cloud forest. *Am. J. Bot.* **2015**, *102*, 1472–1481. [CrossRef] [PubMed]

50. Brock, J.M.R.; Burns, B.R.; Perry, G.; Lee, W.G. Gametophyte niche differences among sympatric tree ferns. *Biol. Lett.* **2019**, *15*, 20180659. [CrossRef] [PubMed]

51. Fisher, J.B. Carbon cost of plant nitrogen acquisition: A mechanistic, globally applicable model of plant nitrogen uptake, retranslocation, and fixation. *Glob. Biogeochem. Cycles* **2010**, *24*. [CrossRef]

52. Brock, J.; Perry, G.; Lee, W.G.; Burns, B.R. Tree fern ecology in New Zealand: A model for southern temperate rainforests. *For. Ecol. Manag.* **2016**, *375*, 112–126. [CrossRef]

53. Bernabe, N.; Williams-Linera, G.; Palacios-Rios, M. Tree ferns in the interior and at the edge of a mexican cloud forest remnant: Spore germination and sporophyte survival and establishment1. *Biotropica* **1999**, *31*, 83–88.

54. Rotach, P. In situ conservation methods. In *Conservation and Management of Forest Genetic Resources in Europe*; Geburek, T., Turok, J., Eds.; Arbora Publishers: Zvolen, Slovakia, 2005; pp. 535–565.

55. Wei, K.H.; Liang, Y.; Yang, C.; Li, L.X.; Hu, Y.; Liao, J.H.; Gao, W.Y. Conservation Research of *Bletilla striata*. *Mod. Chin. Med.* **2019**, *21*, 689–693.

 *horticulturae*

Article

Fatty Acid Profile, Tocopherol Content of Seed Oil, and Nutritional Analysis of Seed Cake of Wood Apple (*Limonia acidissima* L.), an Underutilized Fruit-Yielding Tree Species

Shrinivas Lamani [1], Konerira Aiyappa Anu-Appaiah [2], Hosakatte Niranjana Murthy [1,*], Yaser Hassan Dewir [3,*] and Hail Z. Rihan [4,5]

1   Department of Botany, Karnatak University, Dharwad 580003, India; shrinivasgl92@gmail.com
2   Department of Microbiology and Fermentation Technology, CSIR-CFTRI, Mysuru 570020, India; anuappaiah@gmail.com
3   Plant Production Department, College of Food & Agriculture Sciences, King Saud University, Riyadh 11451, Saudi Arabia
4   School of Biological Sciences, Faculty of Science and Environment, University of Plymouth, Drake Circus, Plymouth PL4 8AA, UK; hail.rihan@plymouth.ac.uk
5   Phytome Life Sciences, Launceston PL15 7AB, UK
*   Correspondence: hnmurthy60@gmail.com (H.N.M.); ydewir@ksu.edu.sa (Y.H.D.)

Citation: Lamani, S.; Anu-Appaiah, K.A.; Murthy, H.N.; Dewir, Y.H.; Rihan, H.Z. Fatty Acid Profile, Tocopherol Content of Seed Oil, and Nutritional Analysis of Seed Cake of Wood Apple (*Limonia acidissima* L.), an Underutilized Fruit-Yielding Tree Species. *Horticulturae* 2021, 7, 275. https://doi.org/10.3390/horticulturae7090275

Academic Editors: Rosario Paolo Mauro, Carlo Nicoletto and Leo Sabatino

Received: 14 August 2021
Accepted: 27 August 2021
Published: 1 September 2021

Publisher's Note: MDPI stays neutral with regard to jurisdictional claims in published maps and institutional affiliations.

Abstract: The present study was aimed at analyzing the fatty acid composition, tocopherols, and physico-chemical characterization of wood apple (*Limonia acidissima* L.) seed oil and the nutritional profile of seed cake. The fatty acids in seed oil were analyzed by gas chromatography–mass spectrometry (GC-MS), and the total seed oil was 32.02 ± 0.08%, comprising oleic (21.56 ± 0.57%), alpha-linolenic (16.28 ± 0.29%), and linoleic acid (10.02 ± 0.43%), whereas saturated fatty acid content was 33.38 ± 0.60% including palmitic (17.68 ± 0.65%) and stearic acid (14.15 ± 0.27%). A greater amount of unsaturated fatty acids (52.37%) were noticed compared to saturated fatty acids (33.38%); hence the seed is highly suitable for nutritional and industrial applications. Gamma-tocopherol was present in a higher quantity (39.27 ± 0.07 mg/100 g) as compared to alpha (12.64 ± 0.01 mg/100 g) and delta (3.77 ± 0.00 mg/100 g) tocopherols, which are considered as natural antioxidants. The spectrophotometric technique was used for quantitative analysis of total phenolic content, and it revealed 135.42 ± 1.47 mg gallic acid equivalent /100 g DW in seed cake. All the results of the studied seed oil and cake showed a good source of natural functional ingredients for several health benefits.

Keywords: wood apple; fatty acid profile; tocopherol; nutritional; phenolics; GC-MS; HPLC

1. Introduction

Plant seeds are an important source of oils and fats to meet nutritional, industrial, and pharmaceutical needs [1]. Oils and fats are composed of neutral lipids, majorly triglycerides, which are the sources of nutraceutical compounds that are an essential part of the human diet and major constituents for the storage of energy, structural and functional composition of cells [2]. Seed oil is used for the preparation of soap and detergents, cosmetics, and also used as ingredients for paint and varnishes, lubricants, and organic pesticides [3]. Oils rich in polyunsaturated fats have been related to the prevention of coronary heart diseases, diabetes, cancer, and depression, whereas cholesterol and saturated fats cause chronic diseases [4]. Many seed oils are reported to contain tocopherols, and they are considered as effective fat-soluble antioxidants present in the oil, which helps to protect cell membranes, improvement in blood circulation, and treating various diseases [5]. Most seeds and vegetable oils, such as groundnut, sunflower, soybean, and peanut, are the most important oil sources for cooking, canning, and preparations of emulsions and margarine [4].

The increasing population of the world is creating a shortage of food sources which permits the interest toward underutilized fruits. Seed oil and cake is the major portion of the human diet due to abundant nutrients, protection from oxidative stress, and several diseases. Seeds have beengiven special attention throughout the world, particularly underutilized fruits. As a biodiversity country, India has been a habitat for thousands of wild underutilized fruit seeds, which could be exploited directly as foods or used to obtain valuable natural compounds and derivatives [6].

Wood apple (*Limonia acidissima* L.) is an underutilized fruit-yielding tree species native to India and Sri Lanka (Figure 1A). Wood apple (Figure 1B) is used by the tribal and rural population of the developing world, which contributes to the traditional health system [6]. Extracts of wood apple are used traditionally for curing various diseases, such as antimicrobial, antifungal, liver, and cardiac tonic, dysentery, hiccough, sore throat, and is a good antidote for snakebite and used as a face cream to remove small spots and lesions on the skin. The phytochemical analysis of *L. acidissima* plant parts showed the presence of alkaloids, flavonoids, phenols, terpenoids, tannins, fats, sterols, saponins, glycosides, gum, mucilage, and fixed oils [7,8]. The small, numerous, and white seeds scattered throughout fruits(Figure 1C) showed an abundance in protein, carbohydrate, amino acid content and high amount of iron (Fe), zinc (Zn), sodium (Na), potassium (K), phosphorus (P), copper (Cu), magnesium (Mg) and manganese (Mn) [9]. However, no detailed reports are available on wood apple seed oil (Figure 1D). Therefore, the present study was aimed atanalyzing the fatty acid composition by GC-MS, tocopherols by HPLC/FD, and physico-chemical properties of seed oil, and nutritional analysis of wood apple seed cake.

Figure 1. Wood apple (*Limonia acidissima* L.)-Tree (**A**), Fruit (**B**), Seeds (**C**), and Oil (**D**).

2. Materials and Methods

2.1. Samples

Ripened fruits of *Limonia acidissima* L. were collected from Itagi, Koppal District, Karnataka, India, during February 2019. The collected fruits were washed under running tap water to remove the pulp content. Seeds were dried in a hot air oven at $48 \pm 2\,°C$ to remove moisture and stored at $-20\,°C$ until future analysis.

2.2. Standards and Chemicals

n-Hexane, methanol, potassium hydroxide (KOH), hydrochloric acid, 14% methanolic boron trifluoride (w/v), helium, 5% ethanolic pyrogallol (w/v), diethyl ether, ethanol, milli-Q-water, acetone, phenolphthalein, chloroform, Hanus iodine solution, potassium iodide, sodium thiosulfate, sodium hydroxide, phosphate-buffered saline, bovine serum albumin, copper sulfate, Folin-Ciocalteu's reagent, anthrone reagent, and sodium carbonate were of analytical grade from Merck, Bangalore (India). Glucose, gallic acid, α, γ, and δ-tocopherols were purchased from Sigma–Aldrich, Bangalore, India.

2.3. Soxhlet Extraction of Wood Apple Seeds

Moisture-free wood apple seeds were pulverized into a fine powder in a mixer grinder and extracted with n-hexane using soxhlet apparatus for 8 h. The solvent was evaporated by using rotary evaporation, and the oil was collected in a glass vial. Concentrated oil was dried under a nitrogen air atmosphere. The extracted wood apple seed oil (WASO) was used for fatty acid composition, tocopherol, and physico-chemical characterizations, and defatted wood apple seed cake (DWASC) was used for nutritional analysis.

2.4. Investigation of Physico-Chemical Properties of WASO

Physico-chemical properties of WASO were achieved by different standard methods described by [10,11]. The viscosity of the oil was measured using a Brookfield viscometer (Model DV-III, Stoughton, MA, USA). Viscosity was reported in centipoise (cP) at a temperature of 28 ± 2 °C. The Refractive index was determined at 28 ± 2 °C using a hand refractometer (Atago Co. RX-500, Tokyo, Japan). The specific gravity was estimated using a 10 mL Pycnometer (Borosil, India) at 28 ± 2 °C temperature. The content of wax was determined gravimetrically using the standard method described [12]. Briefly, 25 mL of n-hexanewas added toseed oil (1 g) and vortexed. The acetone and n-hexane (3:1) were added to the mixture and refrigerated for 24 h at -20 °C. The liquid phase was separated and re-dissolved in n-hexane to removethe residue of the undissolved matter, and the wax content was expressed as mg/g oil.

The acid value (AV), iodine value (IV), free fatty acid value (FFA), saponification value (SV), and unsaponifiable matter (USM) of WASO were assayed by the method described by [11]. For AV determination, 0.1 g of oil was neutralized by ethyl alcohol. The mixture was incubated in a water bath until the oil dissolved completely. Using phenolphthalein, the mixture was titrated against 0.1 N potassium hydroxide. IV results were expressed as g I_2/100 g. Briefly, 0.1 g of oil was dissolved in 10 mL of chloroform, and 30 mL of Hanus iodine solution was added. Flasks were incubated in the dark for 30 min with occasional shaking. One milliliter of 15% potassium iodide was added to the incubated mixture and titrated against 0.1 N sodium thiosulfate. Once the yellow color existed, few drops of starch were added, and the titration was continued until the yellow color disappeared. SV was evaluated by adding 5% alcoholic potassium hydroxide to 0.1 g of oil and incubated in a water bath for 1 h to saponify. The solution was titrated against 0.1 N hydrochloric acid using a phenolphthalein indicator and expressed as mg KOH/g. The USM was assayed and expressed in percentage, where 15 mL of ethanol and 5 mL of 60% aqueous potassium hydroxide was added to 1 g of oil. The mixture was refluxed until the oil was saponified. The saponified matter was extracted with diethyl ether, andthe supernatant was collected. After evaporating to dryness, the residue was desiccated and weighed. The percentage of free fatty acid was measured by dissolving 1 g of oil in ethanol and diethyl ether (1:1). It was then mixed thoroughly and neutralized with 0.1 N potassium hydroxide, and the mixture was titrated against 0.1 N sodium hydroxide using a phenolphthalein indicator.

2.5. Fatty Acids Profiling

2.5.1. Preparation of Fatty Acid Methyl Esters (FAMEs)

FAME was prepared according to standard methods [13]. WASO (10 mg) wassaponified with 1 mL methanolic potassium hydroxide (0.5 M, w/v) at 60 °C for 1 h in a boiling water bath. The reaction was stopped by adding 1 mL of methanolic hydrochloric acid (0.5 N, v/v) and 2 mL of hexane. The hexane pool was collected (3 times) and evaporated to dryness under a nitrogen air atmosphere. Fatty acids were methylated with 0.7 mL of methanolic boron trifluoride (14%, w/v) and 0.3 mL of benzene. The extracted FAMEs were washed with hexane and water. The hexane pool was evaporated and re-dissolved in MS grade n-hexane.

2.5.2. GC-MS Characterization of Fatty Acids

Fatty acids analysis was achieved with Agilent 7890 B GC 5977 A MSD GC-MS (Agilent Technologies, Santa Clara, CA, USA). An Agilent DB-23 column (50%- Cyanopropyl-methylpolysiloxane; 60 m length, 0.25 μm, i.d. 250 μm) was used. Helium was used as a carrier gas with a flow rate of 1 mL/min. The injection volume was 1 μL, and the spilt ratio was 20:1. The GC oven program was initially set at 60 °C andheld for 1 min, and then increased at 20 °C/min to 130 °C, and 7.5 °C/min to 170 °C. Finally, the temperature reached 200 °C and was heldfor 3 min. MS data were collected in a 70 eV scanning electron ionization mode from m/z 40 to 500. Fatty acid methyl esters were identified with their mass spectrum data, and results were confirmed by a mass spectral library search (NIST 2.0 g).

2.6. Estimation of Tocopherols
2.6.1. Sample Preparation

The separation and quantification of tocopherols were carried out using High-performance liquid chromatography [14]. WASO (1 g) was saponified with 1 mL of potassium hydroxide (100%, w/v), 4 mL of ethanolic Pyrogallol (5%, w/v), and incubated in a water bath for 3 min at 80 °C. The reaction was neutralized by sudden cooling, and 30 mL of distilled water was added. Diethyl ether fractions were separated and collected in another tube and repeated the same thrice. The pooled extract was washed and evaporated to dryness under a vacuum at 40 °C. The residue was dissolved in 1 mL of ethanol and 4 mL of benzene and evaporated to dryness under nitrogen air. The dried extract was dissolved with 1 mL of n-hexane and used for the characterization of tocopherols.

2.6.2. Tocopherol Estimation

The quantification of tocopherols was carried out by using the Shimadzu LC 8A-HPLC system, SCL-10Avp system controller coupled with a RF-20A fluorescence detector. Separation was achieved with a reverse-phase Kinetex C18 column (250 × 4.6 mm, 5 μm), and the injection volume was 5 μL. The excitation wavelength used was 290 nm, whereas the emission wavelength was 330 nm. The mobile phase consisted of methanol and water (95:5, v/v) at a flow rate of 1mL per minute in an isocratic mode of elution [15]. Tocopherols were identified and quantified on retention time bases of known standards.

2.7. Nutritional Characterization of Defatted Seed Cake
2.7.1. Nutritional Analysis

The proximate analysis, including total soluble solids, titratable acidity, and pH was analyzed using the [16] methods. Briefly, the Defatted seed cake was homogenized with 10 mL of distilled water at room temperature, and the mixture was used to quantify. Titratable acidity was expressed as a percentage of citric acid. The pH was measured by directly immersing the electrode into the homogenized cake using a pH meter (CD Instrumental Pvt. Ltd., Bangalore, India). Total soluble solids were measured directly using a refractometer (Erma handheld refractometer, Japan) and expressed in °Brix.

Total moisture and ash content were assayed according to the methods described in the [17]. The moisture content was evaluated by drying 100 g of whole seeds at 102 °C in an air circulating oven (Serwell Instrument Incorporation, Bangalore, India) until it reached constant weight. The defatted seed cake (1 g) was taken into a preheated and weighed crucible, and it was kept in a muffle furnace at 555 °C for 6 h to calculate the total ash content. The crucible was cooled in a desiccator, and the final weight was noted.

Crude fat, protein, and carbohydrate contents were quantified bythe method reported by [18]. Defatted seed cake (1 g) was ground with 10 mL of phosphate-buffered saline and used for protein estimation. Aliquots of each sample were placedinto test tubes, and 4.5 mL of alkaline copper sulfate reagent was added. Tubes were mixed well and incubated at room temperature for 10 min. Then 0.5 mL of 2N Folin-Ciocalteau's reagent was added and incubated again for 30 min, and the optical density was measured at 660 nm. Bovine serum albumin was used as a standard, and the amount of the total protein content was calculated by regression analysis. Total carbohydrate content was achieved using the anthrone reagent method. Defatted seed cake (1 g) was hydrolyzed in a boiling water bath for 3 h with 5 mL of 2.5 N HCl and cooled at room temperature. It was then neutralized with sodium carbonate until effervesces ceased. The supernatant was centrifuged and collected for analysis. Aliquots of each were taken, and 4 mL of Anthrone reagent was added to each tube. Tubes were mixed well and incubated in aboiling water bath for 10 min, cooled rapidly, and the optical density was read at 630 nm. Glucose was used as a standard, and the amount of total carbohydrate content was calculated by regression analysis.

2.7.2. Extraction and Quantification of Total Phenolic Content

Total phenolic content was carried out as per apreviously described method [19]. Defatted WASC powder (2 g) was extracted with 20 mL of aqueous methanol (80%, v/v) and filtered the extract. Three milliliters of milli-Q-water and 0.5 mL of 2N Folin-Ciocalteu reagent were added toan aliquot of filtrate (0.5 mL) and mixed thoroughly. Then, 2mL of 20% (w/v) sodium carbonate was added, and the tubes were incubated for 30 min in the dark at room temperature. Absorbance was recorded at 760 nm against the blank using a UV-VIS Spectrophotometer. Results were expressed as mg gallic acid equivalent/100 g dry weight.

2.8. Statistical Analysis

All the determined values were estimated in triplicate using three lots; each contained 100 g of seeds (n = 3). Results are presented as mean $\pm$ standard deviation (SD). Statistical analysis was carried out by using SPSS software.

3. Results and Discussion

3.1. Soxhlet Extraction of Wood Apple Seed Oil

Oilseed crops that yield more than 15% (*w/w*) on a dry weight basis will be considered as a predominant source for edible oil production and/or industrial applications [20]. The total content of oil in wood apple seeds was 32.02 $\pm$ 0.08 g/100 g dry weight (Table 1). The oil content was relatively equal to commercially available oil sources, for example, palm (*Elaeis guineensis*) 37.19%, sunflower (*Helianthus annuus*) 32%, and kokum (*Garcinia indica*) 16.80% [12,21].

Table 1. Physico-chemical properties of wood apple (*Limonia acidissima* L.) seed oil *.

Sl. No.	Parameters (at 30 °C)	Composition
1	Color	Golden yellow
2	Physical state at 4 °C	Liquid
3	Viscosity (cP)	42.33 $\pm$ 0.57
4	Wax content (mg/g)	0.13 $\pm$ 0.00
5	Refractive index	1.44 $\pm$ 0.01
6	Specific gravity	0.92 $\pm$ 0.00
7	Free fatty acid (%)	1.38 $\pm$ 0.02
8	Acid value (mg KOH/g)	2.12 $\pm$ 0.01
9	Iodine value (g I_2/100 g)	116.16 $\pm$ 0.28
10	Saponification value (mg KOH/g)	186.50 $\pm$ 0.86
12	Unsaponifiable matter (%)	1.15 $\pm$ 0.01

* Mean $\pm$ standard deviation of triplicate determinations.

3.2. Investigation of Physico-Chemical Properties of WASO

Physico-chemical characterization of WASO is depicted in Table 1. Physical and chemical properties of oil indicated the overall acceptance with respect to oil purity, quality, stability, flavor, degree of unsaturation, and natural antioxidants. The quality of the oil was decided based on the physico-chemical properties, including viscosity, iodine value, saponification value, specific gravity, and peroxide value of edible oil [22]. WASO showed interesting results as compared to the other accepted edible oils on the market. The color of the oil was clear golden yellow. This indicated that oil didnot consist of chlorophyll and carotenoid pigments, and the physical state at 4 °C was semi-liquid. Viscosity was 42.33 ± 0.57 cP, which is in line with sunflower (48.20 cP), soybean (48.70 cP), and cotton (53.50 cP). This indicated that the degree of saturated fatty acids was less compared to unsaturated fatty acids, and the quality of the oil is suitable for cooking and industrial purpose [23]. WASO showed a lower content of wax (0.13 ± 0.00%) which agreed well with reported sunflower oil (0.12%). A higher percentage of waxes in oil tended to crystallize, and the oil becameturbid when cooled at room temperature [24]. WASO exhibited a refractive index of 1.44 ± 0.01, and values were comparably similar to soybean and sunflower, which are1.46 and 1.46,respectively [25] and more than palm oil, 1.40 [26]. The specific gravity was 0.92 ± 0.00, comparatively similar to palm (0.99) and soybean (0.95) [25,26]. Free fatty acid content was 1.38 ± 0.02% as oleic acid, whereas the other edible oils contain 0.81% in soybean, 1.36% in peanut [27]. Any oil consisting of less than 5% of free fatty acid content will be considered as edible oil, whereas high FFA content in oils, directly permitting oxidation, will become rancid in a short period of time [28]. The acid value was 2.12 ± 0.01 mg KOH/g, and the result obtained was less than the maximum acceptance level (4 mg KOH/g) [29]. Low AV represents the stability of oil at room temperature, whereas high acid value in oil leads to unpleasant flavor and odor generation. IV represents the total unsaturated fatty acids. The presence of double/triple bonds in fatty acids was the cause for high IV in the oil samples. Seed oil tested for IV showed 116.16 ± 0.28 g I_2/100 g, whereas palm oil has low IV (50.0 g I_2/100 g) and soybean have ahigh IV (123.42 g I_2/100 g) [26]. This represents that the percentage of unsaturated fatty acid content was more in WASO. SV in WASO showed 186.50 ± 0.86 mg KOH/g, which is approximately similar to soybean, groundnut, and palm oil, which are 188, 198, and 214.17 mg KOH/g, respectively [25,26]. More SV is useful in industrial applications, such as the preparation of soap, shampoo, and paints [26]. When the SV of oil is less than 190 mg KOH/g, it is an indication of the presence of high molecular weight of triglycerides, such as linoleic and linolenic acids [30]. The unsaponifiable matter was 1.15 ± 0.01%, and the other reported oils showed 1.23% in olive oil and 0.81% in sunflower [27]. The presence of lignans, crude fiber, proteins, and mineral elements are causes for the high unsaponifiable matter [31].

3.3. Fatty Acid Profiling

The fatty acid profile of WASO ispresented in Table 2 andshown in Figure 2. This showed the five major fatty acids, namely palmitic acid, stearic acid, oleic acid, linoleic, and linolenic acids. The monounsaturated (MUFA) and polyunsaturated fatty acid (PUFA) content was almost equal, 25.51 ± 0.67% and 26.86 ± 0.13%, respectively. At the same time, saturated fatty acid (SFA) wasrelatively higher (33.38 ± 0.60%) among the total fatty acid content. The saturated fatty acid content was approximately equal to the oil-yielding tree species, for example, African mangosteen (*Garcinia livingstonei*) 38.23%, palm oil 49.9%, and less than kokum 61.20% [12,32,33]. Palm oil shares 38% of marketing in India because it constitutes a higherpercentage of saturated fatty acids (50.70%) [34]. Palmitic acid (17.68 ± 0.65%) was the highest among the saturated fatty acid content; subsequently, stearic acid (14.15 ± 0.27%), behenic acid (1.24 ± 0.10%), and margaric acid (0.30 ± 0.04%). Oils rich in saturated fatty acids had high oxidative stability and are suitable for frying and baking [35]. These fatty acids areconsidered a major component of cell membranes, secretary and transport lipids, with crucial roles in protein palmitoylation and palmitoylated signal molecules [36]. Hence, WASO can be used as a substitute for palm oil because it has comparably more or less saturated fatty acids.

Table 2. Fatty acid profiling of wood apple (*Limonia acidissima* L.) seed oil (Mean ± standard deviation, *n* = 3).

Peak	t_R (min)	Identified Compounds		Common Name	Rel. Percentage (%)
1	14.66	Hexadecanoic acid, methyl ester	C16:0	Palmitic acid	17.68 ± 0.65
2	15.61	9-Hexadecenoic acid, methyl ester, (Z)-	C16:1	Palmitoleic acid	0.38 ± 0.14

Monounsaturated fatty acids, especially oleic acid, is well known for human health benefits and also important for lowering the effectiveness of low-density lipoprotein (LDL) cholesterol level which reduces the risk for coronary heart diseases. Further, it is documented as an insulin resistance contrary promoter to the polyunsaturated fatty acids with the protection against insulin resistance [37]. The oleic acid (21.56 ± 0.57%) is the dominant fatty acid, and this could be comparably more or less than in sunflower, soybean, and orange at 28.0%, 21.3%, and 26.1%, respectively [38,39]. Next to this, vaccenic acid (1.70 ± 0.04%), elaidic acid (1.36 ± 0.4%), palmitoleic, and paullinic acids were present in traces. Oil-rich in monounsaturated fatty acids makes the oil advisable in terms of nutrition and provides enough stability to be used forbaking purposes [40]. Hence, the higher oleic acid content in oil, which has high oxidative stability, can be used as a healthy alternative to partially hydrogenated vegetable oils.

Polyunsaturated fatty acid compositions of the obtained WASO werecomparatively equal to saturated and monounsaturated fatty acids. Linoleic and alpha-linolenic acids are long-chain polyunsaturated fatty acids, which are considered to be essential fatty acids because they cannot be synthesized endogenously [41]. α-Linolenic acid (16.28 ± 0.29%) is a major omega-3 fatty acid, whereas soybean (9.4%), rapeseed (8.6%), and flaxseed (52.46%) are close to the total content of alpha-linolenic acid [38,42]. Next to this, a loweramount of gamma-linolenic acid was detected. These fatty acids are well known for preventing heart and blood vessels related disease treatments and for selective antitumor properties with negligible systemic toxicity [43]. Linoleic acid (10.02 ± 0.43%) was the next highest among the total polyunsaturated fatty acids, whereas flaxseed, olive, and palm oil are 16.16%, 8.50%, and 9.3%, respectively [27,43]. WASO wasa good source for essential fatty acids, whereas high content of omega-3 and omega-6 fatty acids in the diet increases HDL-cholesterol and decreases LDL-cholesterol. In contrast, a higher rate of oleic acid decreases LDL-cholesterol in the diet and does not affect HDL-cholesterol [44].

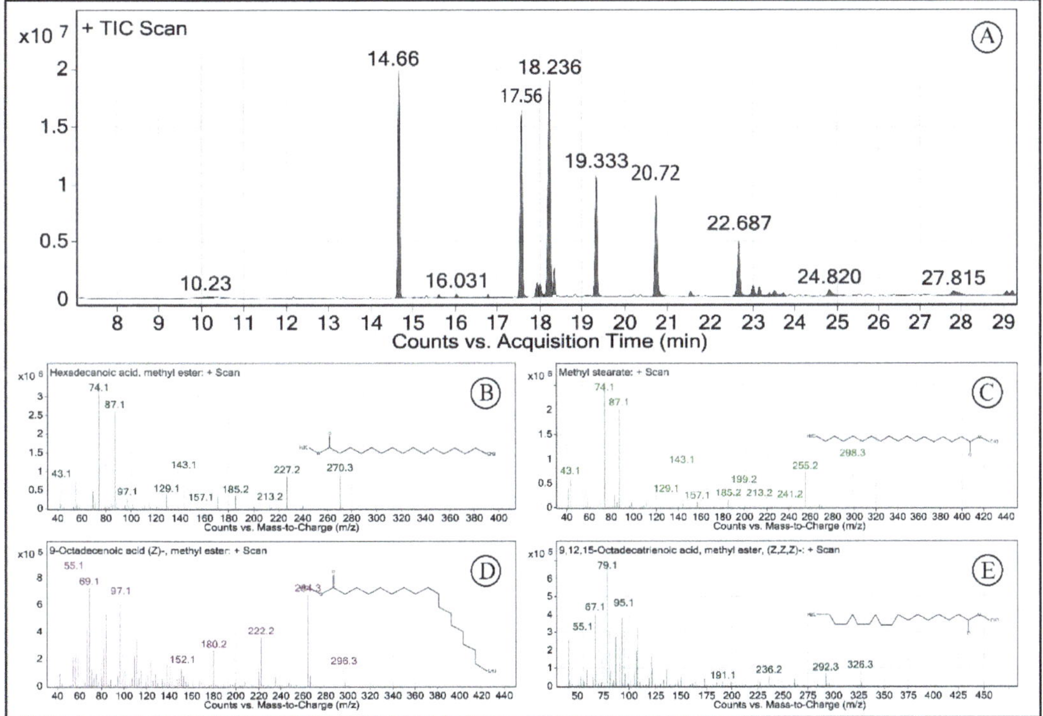

Figure 2. GC-MS chromatograms of wood apple (*Limonia acidissima* L.) seed oil fatty acid methyl esters (FAMEs) (**A**) and MS spectra of major FAMEs of analyzed fatty acids. Methyl palmitate (**B**), methyl stearate (**C**), methyl oleate (**D**), methyl linolenate (**E**).

3.4. Estimation of Tocopherols

Tocopherols are nothing but vitamin E, a fat-soluble vitamin and essential for human health. Tocopherols have beengiven much interest because of their overall health impact and arrest of oxidation at cellular levels [45]. Tocopherols are important natural antioxidants [46]. The total tocopherol content in WASO was (55.68 ± 0.08 mg/100 g) which showed as a comparatively good source to obtain vitamin E. Gamma-tocopherol content was present in greaterquantity, whereas the content of alpha-tocopherol and delta-tocopherol was 12.64 ± 0.01 mg/100 g and 3.77 ± 0.00 mg/100 g, respectively (Table 3, Figure 3). High gamma-tocopherol content in the oil indicates the abundance of α-linolenic acid [47]. The result obtained in WASO was approximately more or less in other plant sources, such assunflower and flaxseed oil, which are 57.97, 42.24 mg/100 g, respectively [48].

Table 3. Tocopherol composition (mg/100 g) in wood apple seed oil *.

Sl. No.	t_R	Tocopherols	Composition
1	9.54	δ-Tocopherol	12.64 ± 0.01
2	10.92	γ-Tocopherol	39.27 ± 0.07
3	12.25	α-Tocopherol	3.77 ± 0.00
	Total tocopherol content		55.68 ± 0.08

* Mean ± standard deviation of triplicates.

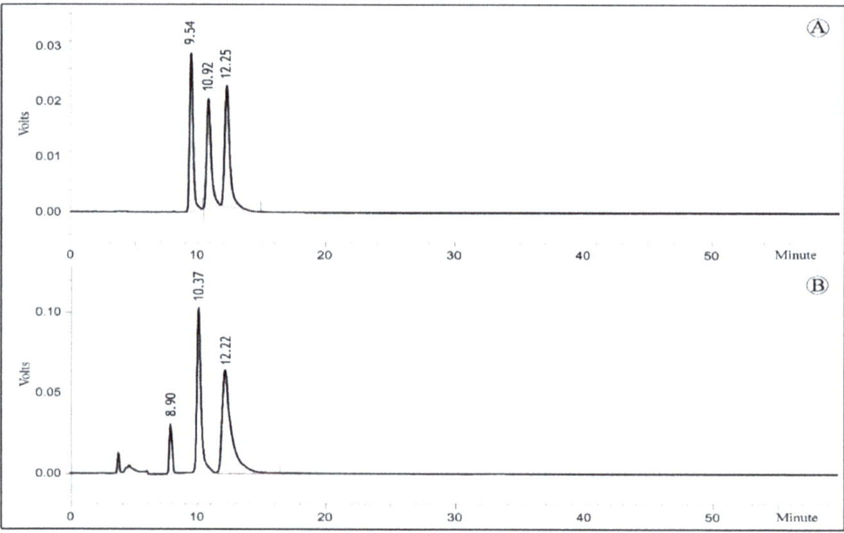

Figure 3. HPLC/FD chromatograms of tocopherol standards (**A**) and wood apple (*Limonia acidissima* L.) seed oil (**B**). Standard peak at RT-9.54 is δ-Tocopherol, peak RT-10.92 is γ-Tocopherol, and peak RT-12.25 is α-Tocopherol.

3.5. Nutritional Characterization of Defatted Seed Cake

The nutritional composition of defatted seed cake is presented in Table 4. Total moisture, fat, ash, protein, and carbohydrate contents agreed with the earlier report [49], whereas the total soluble solids, pH, acidity, and total carbohydrate content were reported for the first time in this present study. Total soluble solids, pH, and acidity reported in WASC were 1.52 ± 0.29°Brix, 1.76 ± 0.19, and 1.83 ± 0.02%, respectively. The composition of total carbohydrate content was (31.66 ± 0.04%), quite high when compared to groundnut (25.41%) and palm (23.1%) seeds [50]. Moisture content in seeds was slightly lower (4.83%) as compared to the other edible oils, such as sunflower (5.50%), groundnut (4.45%), and palm kernel (14.26%) [50,51]. Low moisture content suggests the ability to store for a longer period of time, whereas high moisture value leads to spoilage [52]. Ash content was a major indication of the presence of mineral elements. The ash content of seed cake (3.76 ± 0.18%) was comparable to reported values of groundnut (2.77%), palm (1.50%), and kokum (2.62%) [12,50]. The total 25.24 ± 0.07% crude protein present in seed cake was comparable to groundnut and African mangosteen, which are26.5% and 31.76%, respectively [32,50]. The proximate results of defatted cake showed an excellent source for nutritional properties. The high content of crude fat, protein, and carbohydrate in defatted WASC is an excellent source and ingredient for food products to meet the requirements of a growing population. The °Brix content of WASC was very low compared to other seed cakes. However, the seed cakes with low °Brix values could be used for biogas production [53]. Phenolic acids are the major bioactive constituents that are available from different plant sources. WASC showed 135.42 mg gallic acid equivalent/100 g of total phenolic content, and the results of African mangosteen, linseed, and rapeseed were 206.39, 102.0, and 830 mg gallic acid equivalent/100 g DW, respectively [32,54]. Food that consists of phenolic acids, which are principal constituents to protect our body from free radicals and chronic diseases [55], are beneficial to human health.

Table 4. Nutritional and nutraceutical composition of wood apple (*Limonia acidissima* L.) seeds *.

Sl. No.	Parameters	Composition
1	Moisture (%)	4.83 ± 0.01
2	Oil (%)	32.02 ± 0.08
3	TSS (°Brix)	1.52 ± 0.29
4	pH	1.76 ± 0.19
5	Acidity (%)	1.83 ± 0.02
6	Ash (%)	3.76 ± 0.18
7	Total Protein (%)	25.24 ± 0.07
8	Total Carbohydrate (%)	31.66 ± 0.04
9	Total phenolics (mg GAE/100 g)	135.42 ± 1.47

* Mean ± standard deviation of triplicate determinations.

4. Conclusions

The quality of edible oil was decided based on physico-chemical properties. Wood apple seed oil values were represented an overall acceptance with respect to oil purity, stability, flavor, degree of unsaturation, and natural antioxidants. The viscosity of oil indicated that the degree of saturated fatty acids was approximately equal to the degree of unsaturated fatty acids and is assessed as of suitable quality for cooking and industrial purposes. GCMS characterization resulted in a greater amount of unsaturated fatty acids compared to saturated fatty acids. Hence, the seed is a remarkably good source to obtain essential fats and for cooking and frying. The presence of tocopherol in the seed oil indicated its potential as a good source for scavenging the free radicals. Nutritional and nutraceutical analysis of seed cake revealed a rich content of protein, carbohydrate, and total phenols, which are helpful in maintaining good health. The utilization of such underutilized fruit seeds facilitates an extra economic benefit to the local people.

Author Contributions: Conceptualization, H.N.M. and K.A.A.-A.; methodology, H.N.M., K.A.A.-A. and Y.H.D.; Formal analysis, H.N.M., K.A.A.-A., Y.H.D. and H.Z.R.; investigation and data curation, S.L. and H.N.M.; validation, S.L., H.N.M., K.A.A.-A., Y.H.D. and H.Z.R.; writing—original draft preparation, S.L., H.N.M., K.A.A.-A. and Y.H.D.; writing—review and editing, S.L., H.N.M., K.A.A.-A., Y.H.D. and H.Z.R. All authors have read and agreed to the published version of the manuscript.

Funding: Researchers Supporting Project number (RSP-2021/375), King Saud University, Riyadh, Saudi Arabia.

Institutional Review Board Statement: Not applicable.

Informed Consent Statement: Not applicable.

Data Availability Statement: Not applicable.

Acknowledgments: The authors acknowledge Researchers Supporting Project number (RSP-2021/375), King Saud University, Riyadh, Saudi Arabia. The authors wish to thank CSIR-CFTRI, Mysuru, for providing laboratory facilities with constant support. The first author also thanks DST-KSTePS-Ph.D. fellowship, Karnataka, India for financial support.

Conflicts of Interest: The authors declare no conflict of interest.

References

1. Saidani, S.; Dhifi, W.; Marzouk, B. Lipid evaluation of some Tunisian citrus seeds. *J. Food Lipids* **2004**, *1*, 242–250. [CrossRef]
2. Yoshida, H.; Hirakawa, Y.; Abe, S. Roasting influences on molecular species of triacylglycerols in sunflower seeds (*Helianthus annuus* L.). *Food Res. Int.* **2001**, *34*, 613–619. [CrossRef]
3. Malacrida, C.R.; Kimura, M.; Jorge, N. Phytochemicals and antioxidant activity of citrus seed oils. *Food Sci. Technol. Res.* **2012**, *18*, 399–404. [CrossRef]
4. Timilsena, Y.P.; Vongsvivut, J.; Adhikari, R.; Adhikari, B. Physicochemical and thermal characteristics of Australian chia seed oil. *Food Chem.* **2017**, *228*, 394–402. [CrossRef]
5. Theriault, A.; Chao, J.; Wang, Q.; Gapor, A.; Adeli, K. Tocotrienol: A review of its therapeutic potential. *Clin. Biochem.* **1999**, *32*, 309–319. [CrossRef]
6. Murthy, H.N.; Dalawai, D. Bioactive compounds of wood apple (*Limonia acidissima* L.). In *Bioactive Compounds in Underutilized Fruits and Nuts*; Murthy, H.N., Bapat, V.A., Eds.; Springer Nature: Cham, Switzerland, 2020; pp. 544–569.
7. Nithya, N.; Saraswathi, U. In vitro antioxidant and antibacterial efficacy of *Feronia elephantum* Correa fruit. *Indian J. Nat. Prod. Resour.* **2010**, *1*, 301–305.
8. Vijayvargia, P.; Choudhary, S.; Vijayvergia, R. Preliminary phytochemical screening of *Limonia acidissima* Linn. *Int. J. Pharm. Pharm. Sci.* **2014**, *6*, 134–136.
9. Sonawane, S.K.; Bagul, M.B.; LeBlanc, J.G.; Arya, S.S. Nutritional, functional, thermal and structural characteristics of *Citrullus lanatus* and *Limonia acidissima* seed flours. *J. Food Meas. Charact.* **2016**, *10*, 72–79. [CrossRef]
10. AOCS. *Official Methods and Recommended Practices of the American Oil Chemists' Society*, 5th ed.; AOCS Press: Champaign, IL, USA, 1998.
11. AOCS. *Official Methods and Recommended Practices of the American Oil Chemists' Society*, 6th ed.; AOCS Press: Urbana, IL, USA, 2009.
12. Patil, M.M.; Muhammed, A.M.; Anu-Appaiah, K.A. Lipids and fatty acid profiling of major indian garcinia fruit: A comparative study and its nutritional impact. *J. Am. Oil Chem. Soc.* **2016**, *93*, 823–836. [CrossRef]
13. Morison, W.R.; Smith, L.M. Preparation of fatty acid methyl esters and dimethylacetals from lipids with boron fluoride-methanol. *J. Lipid Res.* **1964**, *5*, 600–608. [CrossRef]
14. Prasad, P.; Savyasachi, S.; Reddy, L.P.A.; Sreedhar, R.V. Physico-chemical characterization, profiling of total lipids and triacylglycerol molecular species of Omega-3 fatty acid rich *B. arvensis* seed oil from India. *J. Oleo Sci.* **2019**, *68*, 209–223. [CrossRef] [PubMed]
15. Manasa, V.; Chaudhari, S.R.; Tumaney, A.W. Spice fixed oils as a new source of γ-oryzanol: Nutraceutical characterization of fixed oils from selected spices. *RSC Adv.* **2020**, *10*, 43975–43984. [CrossRef]
16. AOAC. *Official Methods of Analysis of Association of Analytical Chemists*, 15th ed.; AOAC: Washington, DC, USA, 1998.
17. AOAC. *Official Methods of Analysis of Association of Analytical Chemists*, 17th ed.; AOAC: Arlington, VA, USA, 2000.
18. Sadasivam, S.; Manickam, A. *Biochemical Methods*, 3rd ed.; New Age International (P) Limited, Publishers: New Delhi, India, 2008.
19. Singleton, V.L.; Orthofer, R.; Lamuela-Raventos, R.S. Analysis of total phenols and other oxidation substrates and antioxidants by means of Folin Ciocalteau Reagent. *Methods Enzymol.* **1999**, *299*, 152–178.
20. Paul, A.A.; Southgate, D.A.T. *McCance and Widdowson's the Composition of Foods*, 4th ed.; Her Majesty's Stationery Office: London, UK, 1980.
21. Tan, C.H.; Ghazali, H.M.; Kuntom, A.; Tan, C.P.; Ariffin, A.A. Extraction and physicochemical properties of low free fatty acid crude palm oil. *Food Chem.* **2009**, *113*, 645–650. [CrossRef]
22. Khaneghah, M.; Shoeibi, S.; Ameri, M. Effect of storage conditions and PET packing on quality of edible oils in Iran. *Adv. Environ. Biol.* **2012**, *6*, 694–701.
23. Yalcin, H.; Toker, O.S.; Dogan, M. Effect of oil type and fatty acid composition on dynamic and steady shear rheology of vegetable oils. *J. Oleo Sci.* **2012**, *61*, 181–187. [CrossRef] [PubMed]
24. Carelli, A.A.; Frizzera, L.M.; Forbito, P.R.; Crapiste, G.H. Wax Composition of sunflower seed oils. *J. Am. Oil Chem. Soc.* **2002**, *79*, 763–768. [CrossRef]
25. Rossell, J.B. Vegetable oil and fats. In *Analysis of Oil Seeds, Fats and Fatty Foods*; Rossell, J.B., Pritchard, J.L.R., Eds.; Elsevier Applied Sciences: New York, NY, USA, 1991; pp. 261–319.
26. Eze, S.O.O. Physico-chemical properties of oil from some selected underutilized oil seeds available for biodiesel preparation. *Afr. J. Biotechnol.* **2012**, *11*, 10003–10007.
27. Konuskan, D.B.; Arslan, M.; Oksuz, A. Physicochemical properties of cold pressed sunflower, peanut, rapeseed, mustard and olive oils grown in the Eastern Mediterranean region. *Saudi J. Biol. Sci.* **2019**, *26*, 340–344. [CrossRef]

28. Cornelius, J.A. Some technical aspects influencing the quality of palm kernels. *J. Sci. Food Agric.* **1966**, *17*, 57–61. [CrossRef]
29. Codex Alimentarius. Standard for Named Vegetable Oils (CODEX STAN 210). 1999. Available online: http://www.fao.org/input/download/standards/336/CXS210e2015.pdf (accessed on 25 July 2021).
30. Aremu, M.O.; Adebisi, O. Compositional studies and physicochemical characteristics of cashew nut (*Anarcadium occidentale*) flour. *Pak. J. Nutr.* **2006**, *5*, 328–333.
31. Mathews, S.; Singhai, R.S.; Kulkarni, P.R. Some physico-chemical characteristics of *Lepidium sativum* (haliv) seeds. *Nahrung* **1993**, *37*, 69–71. [CrossRef]
32. Joseph, K.S.; Bolla, S.; Joshi, K.; Bhat, M.; Naik, K.; Patil, S.; Bendre, S.; Brunda, G.; Haibatti, V.; Payamalle, S.; et al. Determination of chemical composition and nutritive value with fatty acid compositions of African Mangosteen (*Garcinia livingstonei*). *Erwerbs-Obstbau* **2016**, *59*, 195–202. [CrossRef]
33. Edem, D.O. Palm oil: Biochemical, physiological, nutritional, hematological and toxicological aspects: A review. *Plant. Foods Hum. Nutr.* **2002**, *57*, 319–341. [CrossRef] [PubMed]
34. Johnson, S.; Saikia, N. *Fatty Acids Profile Of Edible Oils and Fats in India*; Centre for Science and Environment: New Delhi, India, 2009.
35. Matthaus, B. Use of palm oil for frying in comparison with other high-stability oils. *Eur. J. Lipid Sci. Technol.* **2007**, *109*, 400–409. [CrossRef]
36. German, J.B. Dietary lipids from an evolutionary perspective: Sources, structures and functions. *Matern. Child. Nutr.* **2011**, *7*, 2–16. [CrossRef]
37. FAO/WHO. *Fats and Fatty Acids in Human Nutrition*; Report of an Expert Consultation; FAO/WHO: Geneva, Switzerland, 2010.
38. Gordon, M.H.; Miller, L.A.D. Development of steryl ester analysis for the detection of admixtures of vegetable oils. *J. Am. Oil Chem. Soc.* **1997**, *74*, 505–510. [CrossRef]
39. Ajewole, K.; Adeyeye, A. Characterisation of Nigerian citrus seed oils. *Food Chem.* **1993**, *47*, 77–78. [CrossRef]
40. Petukhov, I.; Malcolmson, L.J.; Przybylski, R.; Armstrong, L. Frying performance of genetically modified canola oils. *J. Am. Oil Chem. Soc.* **1999**, *76*, 627–632. [CrossRef]
41. Simopoulos, A.P. Omega-3 fatty acids in health and disease and in growth and development. *Am. J. Clin. Nutr.* **1991**, *54*, 438–463. [CrossRef]
42. Tan, Z.; Yang, Z.; Yi, Z.; Wang, H.; Zhou, W.; Li, F.; Wang, C. Extraction of oil from flaxseed (*Linum usitatissimum* L.) using enzyme-assisted three-phase partitioning. *Appl. Biochem. Biotechnol.* **2016**, *179*, 1325–1335. [CrossRef]
43. Kenny, F.S.; Pinder, S.E.; Ellis, I.O.; Gee, J.M.; Nicholson, R.I.; Bryce, R.P.; Robertson, J.F. Gamma linolenic acid with tamoxifen as primary therapy in breast cancer. *Int. J. Cancer* **2000**, *85*, 643–648. [CrossRef]
44. Przybylski, R.; Mcdonald, B.E. *Development and Processing of Vegetable Oils for Human Nutrition*; AOCS Press: Champaign, IL, USA, 1995.
45. Malekbala, M.R.; Soltani, S.M.; Hosseini, S.; Babadi, F.E.; Malekbala, R. Current technologies in the extraction, enrichment and analytical detection of tocopherols and tocotrienols: A review. *Crit. Rev. Food Sci. Nutr.* **2017**, *57*, 2935–2942. [CrossRef] [PubMed]
46. Paz, S.M.; Aguilar, F.M.; Garcia-Gimenez, M.D.; Fernandez-Arche, M.A. Hemp (*Cannabis sativa* L.) seed oil: Analytical and phytochemical characterization of the unsaponifiable fraction. *J. Agric. Food Chem.* **2014**, *62*, 1105–1110.
47. Eldin, A.K.; Andersson, R. A multivariate study of the correlation between tocopherol content and fatty acid composition in vegetable oils. *J. Am. Oil Chem. Soc.* **1997**, *74*, 375–380. [CrossRef]
48. Gornas, P.; Siger, A.; Junhevica, K.; Lacis, G.; Sne, E.; Seglina, D. Cold-pressed Japanese quince (*Chaenomeles japonica* (Thunb.) Lindl. ex Spach) seed oil as a rich source of α-tocopherol, carotenoids and phenolics: A comparison of the composition and antioxidant activity with nine other plant oils. *Eur. J. Lipid Sci. Technol.* **2014**, *116*, 563–570. [CrossRef]
49. Ramakrishna, G.; Azeemoddin, G.; Ramayya, D.A.; Rao, S.D.T. Characteristics and composition of Indian wood apple seed and oil. *J. Am. Oil Chem. Soc.* **1979**, *56*, 870–871. [CrossRef]
50. Onyeike, E.N.; Acheru, G.N. Chemical composition of selected Nigerian oil seeds and physiochemical properties of oil extracts. *Food Chem.* **2002**, *77*, 431–437. [CrossRef]
51. Srilatha, K.; Krishnakumari, K. Proximate composition and protein quality evaluation of recipes containing sunflower cake. *Plant. Foods Hum. Nutr.* **2003**, *58*, 1–11. [CrossRef]
52. Onyeike, E.N.; Olungwe, T.; Uwakwe, A.A. Effect of heat-treatment and defatting on the proximate composition of some Nigerian local soup thickeners. *Food Chem.* **1995**, *53*, 173–175. [CrossRef]
53. Ancuta, P.; Sonia, A. Oil press-cakes and meals valorization through circular economy approaches: A review. *Appl. Sci.* **2020**, *10*, 7432. [CrossRef]
54. Terpinc, P.; Ceh, B.; Ulrih, N.P.; Abramovic, H. Studies of the correlation between antioxidant properties and the total phenolic content of different oil cake extracts. *Ind. Crop. Prod.* **2012**, *39*, 210–217. [CrossRef]
55. Pulido, R.; Bravo, L.; Calixto, F.S. Antioxidant activity of dietary polyphenols as determined by a modified ferric reducing/antioxidant power assay. *J. Agric. Food Chem.* **2000**, *48*, 3396–3402. [CrossRef] [PubMed]

Article

Ginkgo biloba Seeds—An Environmental Pollutant or a Functional Food

Teodora Tomova [1,*], Iva Slavova [1], Desislav Tomov [2], Gergana Kirova [1] and Mariana D. Argirova [1]

1   Department of Chemical Sciences, Faculty of Pharmacy, Medical University of Plovdiv,
    4000 Plovdiv, Bulgaria; iva.slavova@mu-plovdiv.bg (I.S.); gergana.kirova@mu-plovdiv.bg (G.K.);
    mariyana.argirova@mu-plovdiv.bg (M.D.A.)
2   Department of Bioorganic Chemistry, Faculty of Pharmacy, Medical University of Plovdiv,
    4000 Plovdiv, Bulgaria; desislav.tomov@mu-plovdiv.bg
*   Correspondence: teodora.tomova@mu-plovdiv.bg

Abstract: *Ginkgo biloba* has been cultivated in Bulgaria since the end of the 19th century. Ividual specimens can be seen in almost every park. Females of the tree are considered contaminants of the landscape because their ripe seeds have a strong odor and are not utilized. We undertook this study to clarify whether ginkgo seeds of local origin can be converted from an unwanted and unused environmental pollutant into a source of beneficial compounds. Various analytical and chromatographic methods were used to quantify the major constituents and ten biologically active compounds in methanol seed extract. The results showed that the seeds are low in proteins (5%) and fats (1%); the seeds were also rich in unsaturated fatty acids and tocopherols. About 44% of nut starch was resistant to in vitro enzymatic hydrolysis. The amount of terpene trilactones in an aqueous-methanol seed extract was significantly higher than the number of flavonoids. Ginkgotoxin and ginkgolic acid were also found. The extract demonstrated weak antimicrobial activity against thirteen microorganisms. This study revealed that seeds of locally grown Ginkgo trees can be used as a source of biologically active substances. The chemical composition show similarity to those of seeds from other geographical areas.

Keywords: *Ginkgo biloba*; trace elements; starch; flavonoids; terpene trilactones; ginkgotoxin; ginkgo-lik acid; antimicrobial

Citation: Tomova, T.; Slavova, I.; Tomov, D.; Kirova, G.; Argirova, M.D. *Ginkgo biloba* Seeds—An Environmental Pollutant or a Functional Food. *Horticulturae* **2021**, 7, 218. https://doi.org/10.3390/horticulturae7080218

Academic Editors: Rosario Paolo Mauro, Carlo Nicoletto and Leo Sabatino

Received: 20 May 2021
Accepted: 29 July 2021
Published: 2 August 2021

Publisher's Note: MDPI stays neutral with regard to jurisdictional claims in published maps and institutional affiliations.

1. Introduction

Ginkgo biloba seeds do not have the popularity of the leaves and leaf extract; the latter is used as multi-target phytochemical drug to improve memory, blood circulation, and to reduce symptoms of psychiatric disorders [1–3]. Ginkgo nuts have been used as a delicious food after cooking or fermentation in several Asian countries for millennia. Although Ginkgo nuts have a longer history than leaves of uses in traditional medicine, including treatment of frequent urination, enuresis, asthmatic response, and lung tuberculosis the chemical composition and pharmacological potential of this natural product have been the subject of more systematic studies in only the last two decades.

While the *Ginkgo biloba* tree is native to China, and its kernels are used in national cuisine, in Bulgaria, the seeds have not been used as food. The tree has been cultivated since the end of the 19th century and several varieties are available for landscaping. During the period of 2011–2014, a partial inventory of the available Bulgarian genetic resources of *Ginkgo biloba* was made in the areas where it is distributed (southern Bulgaria and the Black Sea coast). Data from this study have shown that Plovdiv was the city with the most specimens (182) of this tree [4]. Most of the cultivars were *Ginkgo biloba* L. and only two specimens—*Ginkgo biloba* var. *laciniata*. Only 7.2% of these plants were female. Due to the unpleasant smell of ripened fleshy seed coats, sometimes female trees have been intentionally uprooted after planting. Most of the female plants investigated in the abovementioned study were 40–50 years old; perhaps in the past, no attempts were made to avoid the female plants that create problems in landscape maintenance.

We undertook this study to clarify whether ginkgo seeds of local origin can be converted from an unwanted and unused environmental pollutant into a source of beneficial compounds. Thus, the first objective of the study was to quantify the substances in the seeds to determine their nutritional value and the second, to quantify some secondary metabolites in the seeds that are thought to be responsible for most of the pharmacological properties of ginkgo leaf extracts. These studies may reveal an opportunity to increase the utilization of Ginkgo seeds in Bulgaria, to reduce environmental pollution, and to open new perspectives for local cultivation of this species.

2. Materials and Methods

2.1. Plant Material

Most of the seeds were collected from a tree located in the yard of the Medical College of the Medical University of Plovdiv (N 42.120501, E 24.751327) in the first half of November 2017–2020. The soft outer skin was removed, and the seeds were washed thoroughly with tap water and air-dried. Then, they were shelled, and the endosperm was homogenized in a high-speed tissue homogenizer and stored in plastic packages at $-20\ ^\circ$C until use. For comparative studies, seeds from the North (N 42.157510, E 24.732101) and Central (N 42.134982, E 24.733517) regions of Plovdiv were used. In this text, the terms seed, nut, and kernel are used interchangeably.

Dry matter of the seeds was determined gravimetrically, according to European Pharmacopeia 9.6, by heating at 105 $^\circ$C to constant weight. The average value of two parallel samples was used when needed.

2.2. Reagents and Instruments

All reagents and chromatographic reference compounds used in these experiments were obtained from Merck-Sigma-Aldrich (Germany) and were of the highest purity available. Solvents used for chromatographic analyses were of LC-MS grade. Deionized water (18.2 mΩ/cm^2) was used thoroughly. Spectronic Camspec M550 (Spectronic Camspec Ltd., Garforth, Leeds, UK) spectrophotometer was used for absorbance readings.

2.3. Micro Elemental Analysis

Before analysis, plant material was dried at 105 $^\circ$C to constant weight. After grinding in a glass mortar and sieving through a polyamide sieve, the powder obtained was subjected to microwave digestion (Multiwave Go, Anton-Paar, Ashland, VA 23005, USA). One hundred mg of the sample was mineralized with 2 mL of conc. HNO_3 and 1 mL deionized water at 190 $^\circ$C. The following oven program was used: ramp for 20 min, hold for 20 min, and cool for 10 min. Thereafter, the samples were brought up to 30 mL with water and analyzed by inductively coupled plasma mass spectrometry (ICP-MS) (Thermo Fisher Scientific, Waltham, MA 02451, USA, iCAPQ) in kinetic energy discrimination (KED) mode using 103Rh as an internal standard. Reference material NCS DC 73348 was analyzed in parallel for evaluating the accuracy of the digestion process. Standard multi-element acid solutions were used for calibration and accurate quantitative analysis.

2.4. Fat and Protein Analyses

The method approved by the American Oil Chemists Society [5] was used for protein quantification. Fat content was determined according to ISO 659, 2009 [6]. Fatty acids profile was analyzed according to ISO 5509, 2000 and ISO 5508, 2004 [7,8] using HP 5890 gas chromatography system with flame ionization detector, capillary column Supelco SPTM 2390 (30 m $\times$ 0.25 mm), and carrier gas H_2 (20 mL/min). Injector and detector temperature was set at 250 $^\circ$C; the column temperature program was as follows: 70 $^\circ$C (1 min), increase to 190 $^\circ$C with a rate 6 $^\circ$C/min, then increase to 250 $^\circ$C with a rate

10 °C/min. Individual fatty acids were identified and quantified based on the retention times of reference compounds and their peak areas.

Determination of tocopherols in kernel oil was performed by high-performance liquid chromatography (Merck–Hitachi, Darmstadt, Germany) and detected by emission at 330 nm and excitation at 290 nm with a fluorescent detector according to ISO 9936, 2006 [9]. Separation was carried out on Nucleosyl Si 50-5 column (25 × 0.4 cm) and mobile phase hexane: dioxane (96:4, v/v) delivered with rate 1.5 mL/min and pressure 50 bar. The volume of the standard solution injected (mixture of pure α-, β-, γ- and δ-tocopherols with exact concentrations) was 20 µL. Individual tocopherols in Ginkgo seed oil were identified and quantified based on the retention times of standards and peak area, respectively.

2.5. Analyses of Ginkgo Starch

The residue (10 g) obtained after triple extraction of seeds with 70% methanol was homogenized in deionized water (100 mL) using a laboratory blender at full speed for 4 min. The slurry was filtered successively through a sieve (0.25 cm) to remove the fiber fraction and a nylon screen (150 µm) to obtain homogeneous starch particle size. After triple washing with water followed by centrifugation at 6000× g for 20 min the pellet was collected and dried at 40 °C for 12 h. The dried starch was ground, passed through a nylon screen, and extracted in Soxhlet apparatus with 75% 1-propanol for 4 h. The cellulose capsule was then air-dried for 12 h; defatted starch was transferred to a Petri dish and dried for an additional 24 h at 30 °C. For comparative studies, commercial wheat and maize starch samples obtained from local vendors were sieved through the same size nylon screen and used further without any treatment.

Nicolet iS10 instrument fitted with Smart iTR™ Attenuated Total Reflectance (ATR) Sampling Accessory with a diamond crystal (Thermo Fisher Scientific, Waltham, MA 02451, USA) was used to collect infrared spectra of the starch samples; 128 scans were applied at resolution 4 cm^{-1}.

Amylose content of the samples was measured according to the iodine binding method [10]. Amylose from potato was used for the calibration plot.

Differential scanning calorimeter B DSC Q200 (TA Instruments, New Castle, DE 19720, USA) was used to study phase transformations of Ginkgo starch; calibration of the instrument was done using indium as a standard and the results were processed with specialized software Universal V4.1D, TA Instruments. Samples (5–10 mg) were moistened with water in an aluminum dish keeping a ratio of 30% starch (dry matter) and 70% water. The dish was sealed and left at room temperature for at least 30 min before analysis. The heating was performed at a linear rate of 10 °C/min. During the analysis, the weight of the analyzed sample was continuously compared to the weight of an empty aluminum dish.

The rate of in vitro starch hydrolysis was analyzed using K-DSTRS kit (Megazyme, Ireland) according to the manufacturer's assay procedure, and calculations were done using Mega-Calc™ program. Briefly, starch samples were enzymatically hydrolyzed with pancreatic α-amylase and amyloglucosidase at pH 6.0 and 37 °C with continuous stirring. Aliquots were taken at 20 min to measure rapidly digestible starch (RDS), at 120 min to measure slow digestible starch (SDS), and at 240 min to measure total digestible starch (TDS). Another aliquot was taken after 240-min incubation and subjected to enzymatic hydrolysis with amyloglucosidase to measure resistant starch (RS).

2.6. Analyses of Secondary Metabolites in Ginkgo Seeds

Low molecular mass secondary metabolites were extracted from Ginkgo kernels by stirring plant material with 70% methanol (1:10 w/v) for 24 h in dark. After centrifugation and filtration, the insoluble residue was processed in the same way two more times. Dry matter of the combined aqueous-methanolic extracts was determined gravimetrically prior to further analyses.

Total phenol content and flavonoid content of the extracts obtained were determined as described elsewhere [11]. Tannins were quantified according to the protein precipitation method [12]. Total carbohydrate content was determined using the anthrone/sulfuric acid method [13]. Qualitative and quantitative determination of phenolic acids was performed using HPLC chromatography with UV detection, as previously reported [14].

Analyses of flavonoids (free aglycones), terpenes, and ginkgotoxin were carried out by LC-MS as previously described [15]. Quantification of ginkolic acid was performed on LC-MS (Thermo Dionex Ultimate 3000 chromatographic system and a Thermo TSQ Quantum Access MAX triple quadrupole mass detector with a Heated Electrospray Ionization ionizer) using Syncronis C18 column (150 × 4.6, 5 µm) in gradient mode with mobile phases A: 0.1% formic acid in acetonitrile and B: 0.1% formic acid in acetonitrile-water (10:90, *v/v*) at a flow rate 1 mL/min. Transition 319.2 → 275.3 was observed at collision energy 23 V. Ginkgolic acid C13:0 was used as a standard.

2.7. Antimicrobial Analysis

Twenty microorganisms including seven Gram-positive bacteria, six Gram-negative bacteria, two yeasts, and five fungi strains from the collection of the Department of Microbiology at the University of Food Technologies, Plovdiv, Bulgaria, were selected for antimicrobial tests. Antimicrobial activity of methanol extracts obtained from *Ginkgo biloba* seeds was determined using the standard agar well diffusion method in Luria–Bertani glucose agar medium (LBG agar medium) as previously described [16]. Methanol was used as a control.

The test bacteria (except for *Bacillus subtilis, B. cereus, B. amyloliquefaciens, Micrococcus luteus* and yeast *Saccharomyces cerevisiae*) were cultured on LBG agar medium at 30 °C for 24 h. The other pathogenic bacteria and yeast *Candida albicans* were cultured on LBG agar medium at 37 °C for 24 h. The test fungi were grown on malt extract agar (MEA) at 28 °C for 7 days or until sporulation.

Ginkgo biloba extract and control were pipetted in a quantity of 60 µL into the agar wells (d = 6 mm) in triplicates. The inoculated Petri dishes were incubated as shown above. Antimicrobial activity was determined by recording the diameter of the inhibition zones around the wells on the 24th and 48th h of incubation.

Minimum inhibitory concentration (MIC) was determined applying serial dilution technique—a series of two-fold dilutions of the extracts, ranging from 3.5 mg/mL to 0.014 mg/mL. The samples were pipetted in a quantity of 60 µL per well in a preliminarily inoculated with the test microorganisms LBG agar media. The Petri dishes were incubated at the conditions described above. The MIC values were defined as the lowest concentration of the extracts inhibiting completely the growth of each test microorganism around the agar well.

3. Results

3.1. Microelements in Ginkgo biloba Seeds

Different harvests of *Ginkgo biloba* kernels were analyzed for the presence of important trace elements—iron, copper, zinc, and selenium. The highest variation was found for iron: between 18 and 64 ppm, copper was in the range 6.5–11.4 ppm, and zinc was 8.8–24 ppm; the amount of selenium in all analyzed samples was below 30 ppb (Table 1).

Table 1. Microelement content of different *Ginkgo biloba* seed samples. The results are mean of four scans ± standard deviation (SD).

Sample	Year of Collection	Fe, ppm	Cu, ppm	Zn, ppm
North region	2013	29.5 ± 0.2	7.5 ± 0.1	8.8 ± 0.2
	2014	23.8 ± 0.2	9.3 ± 0.4	23.7 ± 0.5
	2015	31.4 ± 1.2	9.0 ± 0.6	16.7 ± 1.1
	2016	64.4 ± 2.2	11.4 ± 0.3	20.4 ± 0.8
Central region	2017	41.8 ± 0.8	9.67 ± 0.2	24.07 ± 0.9
Medical college (South region)	2017	18.3 ± 0.2	6.5 ± 0.2	13.5 ± 0.05
	2018	26.0 ± 1.4	6.8 ± 0.4	17.2 ± 0.8

3.2. Proteins and Lipids in Ginkgo biloba Seeds

The protein content determined by the Kjeldahl method was 5% of the raw seed weight or about 11% of the dry matter and comparable to the data summarized in [17].

The lipid content of raw *Ginkgo biloba* nuts was 1% (2.04% dry weight basis). Individual fatty acids quantified as methyl esters by gas-chromatography are shown in Table 2.

Table 2. Fatty acid composition of triacylglycerides obtained from *Ginkgo biloba* nuts (crop 2019).

Fatty Acids	%
C10:0	0.1
C12:0	0.1
C14:0	0.6
C14:1	0.5
C15:0	0.4
C16:0	14.9
C16:1	3.5
C17:0	0.2
C17:1	0.3
C18:0	2.0
C18:1	24.5
C18:1 trans	11.6
C18:2 (ω-6)	32.2
C18:2 trans	1.8
C18:3 (ω-3)	0.9
C20:0	0.5
C20:1	0.3
C20:2 (ω-6)	0.3
C20:3 (ω-3)	3.4
C20:4 (ω-6)	1.3
C22:0	0.1
C22:1	0.1
C22:2 (ω-6)	0.2
C20:5 (ω-3)	0.2
C20:1	0.3
C20:2 (ω-6)	0.3
C20:3 (ω-3)	3.4
C20:4 (ω-6)	1.3
C22:0	0.1
C22:1	0.1
C22:2 (ω-6)	0.2
C20:5 (ω-3)	0.2
C20:3 (ω-3)	3.4

Figure 1a shows the ratio between saturated, monounsaturated, and polyunsaturated fatty acids with different nutritional values.

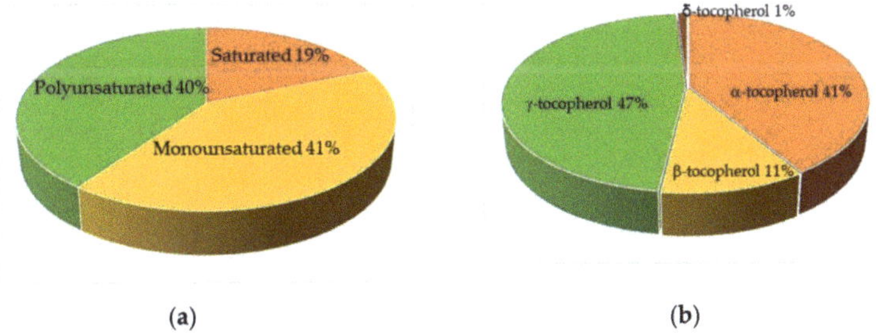

Figure 1. Lipid constituents of *Ginkgo biloba* nuts (crop 2019): (**a**) Relative part of saturated, monounsaturated, and polyunsaturated fatty acids in oil; (**b**) Individual tocopherol composition.

The total amount of tocopherols was 192.6 mg/100 g oil. Chromatographic quantification of the individual tocopherols showed that, as with most nuts, γ- and α-tocopherol dominated and were approximately equally represented (Figure 1b).

3.3. Properties of Ginkgo Starch

The starch was isolated as water-insoluble residue of seeds after extraction with organic solvents and water. The average yield of product after extraction with 70% methanol and 70% propanol was about 70% of the Ginkgo seeds dry matter.

The product obtained was identified as starch by recording its infrared spectrum with attenuated total reflection (FTIR-ATR) (Figure 2). The characteristic bands for O–H and C–H oscillations can be seen at 3340 and 2930 cm^{-1}, respectively; the triad between 1000 and 1150 cm^{-1} is characteristic for C–O, C–C stretching, and C–O–H bending. Spectra of maize and wheat starch were identical; differences that were more significant were observed in the 1550–1750 cm^{-1} region, which are probably due to the presence of minor components bound to starch (proteins, phosphate esters, etc.).

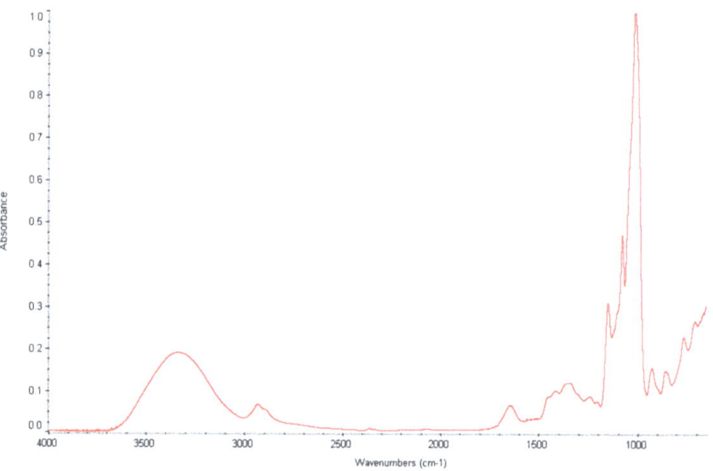

Figure 2. FTIR-ATR spectrum of *Ginkgo biloba* starch.

The percentage of amylose fraction was highest in *Ginkgo biloba* starch: 36.9 ± 5.0 compared to maize: 23.6 ± 4.3 and wheat starch: 22.2 ± 4.8 (data are means $\pm$ standard deviation of three parallel samples). The phase transitions in *Ginkgo biloba* starch begun at about 62 °C (T_0), reached their maximum at 79.5 °C (T_p), which is assumed gelling temperature, and the final transition temperature was 90 °C (T_c). The change in enthalpy determined by instrumental software was 833.1 J/g. For comparison, the parameters of wheat and maize starches (T_0, T_p and T_c) transitions were very close to those of *Ginkgo biloba* starch but the changes in enthalpy were 20–23% lower.

In vitro kinetic study on Ginkgo starch digestion was based on the presumption that the average time of residence of food in the human small intestine is 4 ± 1 h. Ginkgo starch demonstrated a lower digestion rate and a higher percentage of resistible starch compared to the types found in Bulgaria wheat and maize starch (Table 3).

Table 3. Rate of in vitro enzymatic hydrolysis of four types of starch. RDS—rapid digestible starch; SDS—slow digestible starch, TDS—total digestible starch, RS—resistible starch.

Starch	RDS	SDS	TDS	RS	Total Starch
Ginkgo biloba (crop 2019)	7.3	26.3	42.1	33.4	75.5
Wheat	26.3	43.0	67.1	1.7	68.8
Maize	12.17	55.1	71.9	0.8	72.7
Standard (manufacturer's data)	6.9	16.2	35.9	47.4	83.3
Standard (values found)	6.0	16.8	32.8	48.9	81.7

3.4. Secondary Metabolites in Ginkgo Kernels

Organic extracts containing low molecular mass compounds from Ginkgo seeds were obtained from four crops: 2017–2020. Typically, the yield of dry matter in the methanol extract obtained by cold maceration was 5–8% of seeds dry matter. Although the extraction procedure was standardized, spectrophotometric and chromatographic methods showed some variation in the content of the major classes of secondary metabolites with, in some cases, concentrations below the limit of detection. Preliminary phytochemical screening of the extracts showed the presence of phenolic compounds (26.4–118 gallic acid equivalents/g dry matter) including flavonoids (from not detectable to 20.0 quercetin equivalents/g) and tannins (1.9–9.4 tannic acid equivalents/g). The extracts also contained carbohydrates (426–716 glucose equivalents/g).

The extracts were further analyzed for some individual bioactive compounds. Among the 10 typically found in plants—hydroxybenzoic and hydroxycinnamic acids (gallic acid, 3,4-dihydroxy benzoic acid, chlorogenic acid, caffeic acid, ferulic acid, p-coumaric acid, sinapic acid, rosmarinic acid, chicoric acid, and cinnamic acid)—only p-coumaric acid was detected at level 0.478 mg/g dry matter.

Ten other phytochemicals typical for the *Ginkgo biloba* species were quantified. Variations in the content of flavonoids (quantified as aglycones), terpene trilactones, ginkgotoxin, and ginkgolic acid are shown in Table 4. The selection of reference compounds was based on the mass-spectral studies of Zhou et al. [18], results of which are also included for comparison in Table 4. Recently, data about chemical constituents in plant material (leaves, seed coats, and embryoids) collected from Ginkgo trees aged between 25 and 2000 years have been reported [19], also shown in Table 4.

Table 4. Amounts of bioactive compounds in *Ginkgo biloba* seed extracts. The results are presented in µg/g of dry extract.

Compound	Own Results	[18]	[19]
Quercetin	1.4–14.4	na	nd
Kaempferol	nd–20.0	na	nd
Isorhamnetin	4.5–30.0	0.45	nd–3.5
Rutin	2–20	na	nd–1.9
Ginkgolide A	242	165.17	0.92–3.47
Ginkgolide B	388 *	199.67	575–1988
Ginkgolide J	–	na	15–47
Ginkgolide C	143	136.3	152–600
Bilobalide	122	72.52	22–46
Ginkgotoxin	335	na	na
Ginkgolic acid	18	51.7	56.5–172.2

* Ginkgolide B and ginkgolide J are isomers and could not be separated upon the chromatographic conditions used; na—no data available; nd—not detected.

3.5. Antimicrobial Studies

Initially, the antimicrobial activity of Ginkgo seed extract was evaluated by the agar well diffusion method, which measures the diameter of the area around which the test agent has diffused, and inhibited germination and growth of the microbial strain tested. The highest applied concentration—3.5 mg/mL for the time and conditions of incubation resulted in zones of inhibition between 0 and 13 mm. No inhibition was caused by the control. Minimum inhibitory concentrations (MIC) of Ginkgo seed extract with respect to the microorganisms tested are shown in Table 5.

Table 5. Minimum inhibitory concentrations (MIC, mg/mL) of 70% methanol extract from *Ginkgo biloba* kernels.

Strain	MIC
Gram (+) bacteria	
Bacillus subtilis ATCC 6633	0.875
Bacillus cereus	0.875
Bacillus amyloliquefaciens	1.75
Staphylococcus aureus ATCC 25923	0.875
Listeria monocytogenes ATCC 8632	–
Enterococcus faecalis	3.5
Micrococcus luteus	3.5
Gram (-) bacteria	
Salmonella enteritidis	1.75
Salmonella abony	–
Klebsiella sp.	–
Escherichia coli ATCC 8739	1.75
Proteus vulgaris ATCC 6380	–
Pseudomonas aeruginosa ATCC 9027	1.75
Yeasts	
Candida albicans NBIMCC 74	–
Saccharomyces cerevisiae	–
Fungi	
Aspergillus niger ATCC 1015	0.875
Aspergillus flavus	–
Penicillium sp.	1.75
Rhizopus sp.	0.109
Fusarium moniliforme ATCC 38932	3.5

4. Discussion

Ginkgo biloba seeds have an advantage over the leaves in that they can not only be a source of phytochemicals, but also a source of macro-(proteins, fats, and carbohydrates) and micronutrients (trace elements, vitamins, amino acids, etc.). In this study, we quantified some constituents related to the nutritional value of Ginkgo kernels of Bulgarian origin. As the biosynthesis of plant primary and secondary metabolites shows significant variability due to the influence of population (location), season, climatic and agronomic factors, post-collection treatment, and other factors, we compared our findings with available data for other populations as well and with similar foods used in Bulgaria. Furthermore, we analyzed 70% methanol seed extract, which contains a significant proportion of polar and moderately polar secondary metabolites that are considered as principal carriers of the biological activity of plants.

Microelements are necessary for plant growth and development but also essential for humans. Most of them are part of enzymes; participate in antioxidant protection, metabolism of hormones, proteins, immune processes, etc. Plant food is a major supplier of these microelements. In the earth's crust, they are most often found as insoluble minerals; plants through their root system release organic compounds capable of chelating metal ions, bringing them into a soluble form that helps their movement between plant cells. In such soluble form, they are also delivered to humans through food intake.

The only available data for comparison regarding microelement content of seeds were those reported by Goh and Barlow [20]. They found 1.2 mg iron per 100 g dry weight in nuts purchased from China, and those purchased from the US contained 1.4 mg/100 g (12–14 ppm), which is lower than the values we found.

The amount of protein in Ginkgo nuts was relatively low, given the significantly higher protein content of other nuts popular in Bulgaria: walnuts (26.1%), peanuts (25.8%), and almonds (21.9%) [21]. Despite the low protein content, the proteins in *Ginkgo biloba* seeds seem to have other beneficial properties [17].

The fat content of seeds was much lower not only compared to other types of nuts (typically above 50% raw weight basis) but also compared to Ginkgo nuts originating in Korea: 2.2% on raw weight basis [22] or Massachusetts—USA: 3.75% of dried nuts [23]. However, seed fat contains fatty acids, which are beneficial for humans, albeit in small amounts, and provide a good proportion of omega-3 and omega-6 fatty acids and tocopherols. This may explain the observation that the lipid fraction of Ginkgo seeds applied to an animal model was capable of modulating liver and serum apolipoproteins and cholesterol [23].

Starch, the reserve food for the embryo, was the constituent present in the largest amount in the seeds—about 70% of the dry matter. Due to the specificity of the enzymatic reaction, which always hydrolyzes the glycosidic bonds starting from the non-reducing end of the polysaccharide, the presence of significant amounts of amylose in the starch composition reduces the rate of glucose release. Thus, amylose content is an important but not the only factor influencing the rate of starch hydrolysis. The ratio of crystalline/amorphous parts of the polysaccharide, the degree of swelling, the length, folding, and number of branches of amylopectin also have an impact. Some of these factors also affect the enthalpy of gelling recorded by DSC. The amylose content of ginkgo seeds usually varies between 28.5% [24,25] and 33.2% [26]. The transition temperatures (T_0, T_p, and T_c) of starches obtained from three Chinese Ginkgo cultivations [24] were close to those found for Ginkgo starch of Bulgarian origin. The high amylose content and high enthalpy of gelling of *Ginkgo biloba* starch were preliminary indicators that it may be more resistant to enzymatic hydrolysis than the other two starches studied. This assumption was confirmed by an in vitro study of its digestibility. The data obtained showed that *Ginkgo biloba* starch contains a smaller amount of digestible starch and a higher proportion of resistant starch compared to wheat and maize starch. A good negative correlation was found between the content of amylose and total digestible starch (r = −0.875) and a positive correlation between amylose and resistant starch (r = 0.808), which confirms that the content of starch linear fraction is a significant factor in the degree of enzymatic hydrolysis of starch to glucose.

The flavonoids quercetin, kaempferol, and isorhamnetin and their glycosides represent about a quarter of the composition of the standardized *Ginkgo biloba* leaf extract EGb 761®. In the extract of Ginkgo kernels, they are in significantly lower amounts. Although both aglycones and their glycosides have chromophores, we could not detect them by chromatographic separation with a UV detector. The quantification of the above-mentioned flavanols as free aglycones using LC-MS revealed that their content in the seed extract was indeed very low.

The unique constituents of *Ginkgo biloba* are the terpene trilactones, ginkgolides, and bilobalide. Specific beneficial effects of standardized leaf extract for patients with ischemic cardiovascular and cerebrovascular diseases are most often attributed to this class of bioactive constituents [27]. Our data from chromatographic analysis, as well as data from other authors, confirm that the terpene trilactones in seed extract are in greater quantities than flavonoids, and it could have healing properties similar to leaf extract. In a recent study, we confirmed that this extract enhances learning and memory functions in rats and has an effect comparable to that of the nootropic medication Piracetam [28].

Among the unfavorable components of the seed extract are ginkgotoxin (antivitamin B_6) and ginkgolic acid. While the presence of ginkgotoxin in leaves is still uncertain, and perhaps is the reason why the European Pharmacopoeia does not set a limit for its content in the standardized leaf extract, the ginkgolic acid level is restricted in EGb 761 to 5 ppm because of allergenic and genotoxic effects. Some approaches to seed detoxification have recently been summarized [17]; for example, pre-cooking the seeds significantly reduces the amount of ginkgotoxin due to its good water solubility. In addition, our experiment showed that after further treatment of the aqueous methanolic extract with hexane, the level of ginkgolic acid was reduced from 19 ppm to 4 ppm with minor changes in the concentrations of other bioactive constituents. However, the pharmacological properties of ginkgolic acids family are not limited to those mentioned above [29] and undoubtedly deserve further research.

Studies on the antimicrobial activity of seeds and their extracts are sporadic. Park and Cho have studied the antimicrobial properties of *Ginkgo biloba* leaf, seed, and seed coat extracts [30]. While leaf and seed coat extracts have shown very good antimicrobial activity, aqueous, methanol and ethanol seed extracts have proved to be inactive against the microorganisms tested. Based on an ancient Chinese recipe, Chassagne and colleagues have found that an 80% ethanol seed extract has been effective against pathogens involved in skin and soft tissue infections [31]. The authors attribute this antimicrobial activity to the presence of ginkgolic acid.

The extract obtained by cold maceration of Ginkgo seeds demonstrated weak antimicrobial activity against 13 of 20 tested microorganisms with MIC ranging from 0.109 to 1.75 mg/mL. Gram-negative bacteria were less sensitive toward the constituents of extract, which is usually attributable to the permeability barrier posed by their outer membrane, which makes them more resistible to antimicrobial agents. The two yeasts tested were resistant to this extract. These antimicrobial properties might be more pronounced after removal of some constituents considered as inert, for example, carbohydrates, which constitute a significant part of the extract dry matter.

This study demonstrates that seeds of locally grown Ginkgo trees, that up to now have been considered an environmental pollutant, can be used as a source of biologically active substances. The chemical composition shows similarity to those of seeds from other geographical areas.

5. Further Perspectives

While the *Ginkgo biloba* tree is native to China and its nuts are used in Far Eastern national cuisines, in Bulgaria, Ginkgo seed is still unknown as food. The pursuit of healthy and organic eating has led to the entry of exotic—unknown until 10–20 years ago—foods into the country. Maybe it is time to promote Ginkgo seeds as well because they fully meet the definition of a functional food product, which in addition to providing nutrients and energy, favorable modulates one or more targeted functions in the body by enhancing a particular physiological response and/or reduce the risk of disease [32]. They can provide essential microelements, tocopherols, omega-3 and omega-6 fatty acids albeit in low concentrations, carbohydrates with low digestibility, high content of terpenoids (ginkgolides and bilobalide) with proven health benefits. These are all arguments in favor of the prospect of promoting Ginkgo seeds as healthy functional food. The limitation to this perspective is the content of ginkgotoxin and ginkgolic acid in nuts, so consumption of more than 10–20 nuts a day might pose health risks.

The other perspective is the separate use of detoxified organic seed extract, its standardization and use of its pharmacological potential as a nootropic agent, like the leaf extract, as well as purification of residual starch and its use in food and pharmaceutical industry. Both approaches could lead to a reduction of environmental pollution, better maintenance of park areas, utilization of this undoubtedly useful natural product, and potentially its future cultivation in Bulgaria.

Author Contributions: Conceptualization, T.T. and M.D.A.; formal analysis, D.T. and G.K.; data curation, I.S.; writing—original draft preparation, T.T. and I.S.; writing—review and editing, M.D.A.; project administration, M.D.A. All authors have read and agreed to the published version of the manuscript.

Funding: This work was supported by the Bulgarian Ministry of Education and Science (Grant D01-217/30.11.2018) under the National Research Programme "Innovative Low-Toxic Bioactive Systems for Precision Medicine (BioActiveMed)" approved by DCM # 658/14.09.2018.

Institutional Review Board Statement: Not applicable.

Informed Consent Statement: Not applicable.

Conflicts of Interest: The authors declare no conflict of interest. The funder had no role in the design of the study; in the collection, analyses, or interpretation of data; in the writing of the manuscript, or in the decision to publish the results.

References

1. DeFeudis, F.V.; Drieu, K. *Ginkgo biloba* extract (EGb 761) and CNS functions: Basic studies and clinical applications. *Curr. Drug Targets* **2000**, *1*, 25–58. [CrossRef]
2. Wieland, L.S.; Feinberg, T.M.; Ludeman, E.; Prasad, N.K.; Amri, H. *Ginkgo biloba* for cognitive impairment and dementia (Protocol). *Cochrane Database Syst. Rev.* **2002**, *4*, CD003120. [CrossRef]
3. Belwal, T.; Giri, L.; Bahukhandi, A.; Tariq, M.; Kewlani, P.; Bhatt, I.D.; Rawal, R.S. Chapter 3.19. *Ginkgo biloba*. In *Nonvitamin and Nonmineral Nutritional Supplements*; Nabavi, S.M., Silva, A.S., Eds.; Elsevier Inc.: Amsterdam, The Netherlands, 2019; pp. 241–250. [CrossRef]
4. Ivanova, V.; Natcheva, L.; Gercheva, P. Use of *Ginko biloba*, L. in green system of the town of Plovdiv. *Sci. Works USB Plovdiv. Ser. C Tech. Technol.* **2015**, *13*, 241–244.
5. American Oil Chemists Society (AOAC). Kjeldahl's method for protein determination in cereals and feed. In *Official Methods of Analysis^{SM}*, 16th ed.; Method 945; AOAC International: Rockville, MD, USA, 1996; p. 18-B.
6. International Organization for Standardization (ISO Standard No. 659:2009). *Oilseeds—Determination of Oil Content (Reference Method)*; International Organization for Standardization: Geneva, Switzerland, 2009.
7. International Organization for Standardization (ISO Standard No. 5509:2000). *Animal and Vegetable Fats and Oils. Preparation of Methyl Esters of Fatty Acids*; International Organization for Standardization: Geneva, Switzerland, 2000.
8. International Organization for Standardization (ISO Standard No. 5508:2004). *Animal and Vegetable Fats and Oils. Analysis by Gas Chromatography of Methyl Esters of Fatty Acids*; International Organization for Standardization: Geneva, Switzerland, 2004.
9. International Organization for Standardization (ISO Standard No. ISO 9936). *Animal and Vegetable Fats and Oils. Determination of Tocopherols and Tocotrienols Contents. Method Using High-Performance Liquid Chromatography*; International Organization for Standardization: Geneva, Switzerland, 2006.
10. Knutson, C.A.; Grove, M.J. Rapid Method for Estimation of Amylose in Maize Starches. *Cereal Chem.* **1994**, *71*, 469–471.
11. Morado-Castillo, R.; Quintanilla-Licea, R.; Gomez-Flores, R.; Blaschek, W. Total Phenolic and Flavonoid Contents and Flavonoid Composition of Flowers and Leaves from the Mexican Medicinal Plant Gymnosperma glutinosum (Spreng.) Less. *Eur. J. Med. Plants* **2016**, *15*, 1–8. [CrossRef]
12. Hagerman, A.E.; Butler, L.G. Protein precipitation method for the quantitative determination of tannins. *J. Agric. Food Chem.* **1978**, *26*, 809–812. [CrossRef]
13. Haldar, D.; Sen, D.; Gayen, K. Development of Spectrophotometric Method for the Analysis of Multi-component Carbohydrate Mixture of Different Moieties. *Appl. Biochem. Biotechnol.* **2017**, *181*, 1416–1434. [CrossRef] [PubMed]
14. Ivanov, I.G.; Vrancheva, R.Z.; Marchev, A.S.; Petkova, N.T.; Aneva, I.Y.; Denev, P.P.; Georgiev, V.G.; Pavlov, A.I. Antioxidant activities and phenolic compounds in Bulgarian Fumaria species. *Int. J. Curr. Microbiol. App. Sci.* **2014**, *3*, 296–306.
15. Feodorova, Y.; Tomova, T.; Minchev, D.; Turiyski, V.; Draganov, M.; Argirova, M. Cytotoxic effect of *Ginkgo biloba* kernel extract on HCT116 and A2058 cancer cell lines. *Heliyon* **2020**, *6*, e04941. [CrossRef]
16. Tumbarski, Y.; Lincheva, V.; Petkova, N.; Nikolova, R.; Vrancheva, R.; Ivanov, I. Antimicrobial activity of extract from aerial parts of potentilla (*Potentilla reptans* L.). *Ind. Technol.* **2017**, *4*, 37–43.
17. Wang, H.-Y.; Zhang, Y.-Q. The main active constituents and detoxification process of *Ginkgo biloba* seeds and their potential use in functional health foods. *J. Food Comp. Anal.* **2019**, *83*, 103247. [CrossRef]
18. Zhou, G.; Yao, X.; Tang, Y.; Qian, D.; Su, S.; Zhang, L.; Jin, C.; Qin, Y.; Duan, J. An optimized ultrasound-assisted extraction and simultaneous quantification of 26 characteristic components with four structure types in functional foods from ginkgo seeds. *Food Chem.* **2014**, *158*, 177–185. [CrossRef]
19. Chen, X.; Zhong, W.; Shu, C.; Yang, H.; Li, E. Comparative analysis of chemical constituents and bioactivities of the extracts from leaves, seed coats and embryoids of *Ginkgo biloba* L. *Nat. Prod. Res.* **2020**, *35*, 1–4. [CrossRef]
20. Goh, L.M.; Barlow, P.J. Antioxidant capacity in *Ginkgo biloba*. *Food Res. Int.* **2002**, *35*, 815–820. [CrossRef]
21. Brufau, G.; Boatella, J.; Rafecas, M. Nuts: Source of energy and macronutrients. *Br. J. Nutr.* **2006**, *96*, 24–28. [CrossRef] [PubMed]
22. Hierro, M.T.G.; Robertson, G.; Christie, W.W.; Joh, Y.-G. The fatty acid composition of the seeds of *Ginkgo biloba*. *J. Am. Oil Chem. Soc.* **1996**, *73*, 575–579. [CrossRef]
23. Mahadevan, S.; Park, Y.; Park, Y. Modulation of cholesterol metabolism by *Ginkgo biloba* L. nuts and their extract. *Food Res. Int.* **2008**, *41*, 89–95. [CrossRef]
24. Miao, M.; Jiang, H.; Jiang, B.; Cui, S.W.; Jin, Z.; Zhang, T. Structure and functional properties of starches from Chinese ginkgo (*Ginkgo biloba* L.) nuts. *Food Res. Int.* **2012**, *49*, 303–310. [CrossRef]
25. Cai, J.; Cai, C.; Man, J.; Xu, B.; Wei, C. Physicochemical Properties of Ginkgo Kernel Starch. *Int. J. Food Prop.* **2015**, *18*, 380–391. [CrossRef]
26. Zheng, Y.; Zhan, H.; Yao, C.; Hu, L.; Peng, Y.; Shen, J. Study on physicochemical and in-vitro enzymatic hydrolysis properties of ginkgo (*Ginkgo biloba*) starch. *Food Hydrocoll.* **2015**, *48*, 312–319. [CrossRef]

27. Liu, X.-W.; Yang, J.-L.; Niu, W.; Jia, W.-W.; Olaleye, O.E.; Wen, Q.; Duan, X.-N.; Huang, Y.-H.; Wang, F.-Q.; Du, F.-F.; et al. Human pharmacokinetics of ginkgo terpene lactones and impact of carboxylation in blood on their platelet-activating factor antagonistic activity. *Acta Pharmacol. Sin.* **2018**, *39*, 1935–1946. [CrossRef] [PubMed]
28. Tomova, T.; Doncheva, N.; Mihaylova, A.; Kostadinov, I.; Peychev, L.; Argirova, M. An Experimental Study on Phytochemical Composition and Memory Enhancing Effect of *Ginkgo biloba* Seed Extract. *Folia Med.* **2021**, *63*, 203–212. [CrossRef] [PubMed]
29. Gerstmeier, J.; Seegers, J.; Witt, F.; Waltenberger, B.; Temml, V.; Rollinger, J.M.; Stuppner, H.; Koeberle, A.; Schuster, D.; Werz, O. Ginkgolic Acid is a Multi-Target Inhibitor of Key Enzymes in Pro-Inflammatory Lipid Mediator Biosynthesis. *Front Pharmacol.* **2019**, *10*, 797. [CrossRef] [PubMed]
30. Park, S.-B.; Cho, G.-S. Antimicrobial Activity of Extracts and Fractions of *Ginkgo biloba* Leaves, Seed and Outer Seedcoat. *J. Korean Soc. Food Sci. Nutr.* **2011**, *40*, 7–13. [CrossRef]
31. Chassagne, F.; Huang, X.; Lyles, J.T.; Quave, C.L. Validation of a 16th Century Traditional Chinese Medicine Use of *Ginkgo biloba* as a Topical Antimicrobial. *Front. Microbiol.* **2019**, *10*, 775. [CrossRef]
32. Bigliardi, B.; Galati, F. Innovation trends in the food industry: The case of functional foods. *Trends Food Sci. Technol.* **2013**, *31*, 118–129. [CrossRef]

 *horticulturae*

Article

Productive and Morphometric Traits, Mineral Composition and Secondary Metabolome Components of Borage and Purslane as Underutilized Species for Microgreens Production

Giandomenico Corrado [1,†], Christophe El-Nakhel [1,†], Giulia Graziani [2], Antonio Pannico [1], Armando Zarrelli [3], Paola Giannini [2], Alberto Ritieni [2], Stefania De Pascale [1], Marios C. Kyriacou [4] and Youssef Rouphael [1,*]

[1]   Department of Agricultural Sciences, University of Naples Federico II, 80055 Portici, Italy; giandomenico.corrado@unina.it (G.C.); christophe.elnakhel@unina.it (C.E.-N.); antonio.pannico@unina.it (A.P.); depascal@unina.it (S.D.P.)
[2]   Department of Pharmacy, University of Naples Federico II, 80131 Naples, Italy; giulia.graziani@unina.it (G.G.); paola.gian704@gmail.com (P.G.); alberto.ritieni@unina.it (A.R.)
[3]   Department of Chemical Sciences, University of Naples Federico II, 80126 Naples, Italy; zarrelli@unina.it
[4]   Department of Vegetable Crops, Agricultural Research Institute, 1516 Nicosia, Cyprus; m.kyriacou@ari.gov.cy
*   Correspondence: youssef.rouphael@unina.it
†   These authors equally contributed to the work.

Citation: Corrado, G.; El-Nakhel, C.; Graziani, G.; Pannico, A.; Zarrelli, A.; Giannini, P.; Ritieni, A.; De Pascale, S.; Kyriacou, M.C.; Rouphael, Y. Productive and Morphometric Traits, Mineral Composition and Secondary Metabolome Components of Borage and Purslane as Underutilized Species for Microgreens Production. *Horticulturae* **2021**, *7*, 211. https://doi.org/10.3390/horticulturae7080211

Academic Editor: Juan A. Fernández

Received: 4 June 2021
Accepted: 21 July 2021
Published: 23 July 2021

Abstract: Neglected and underutilized species (NUS) offer largely unexplored opportunities for providing nutritious plant food, while making agro-ecosystems more diverse and resilient to climate change. The aim of this work was to explore the potential of two typical Mediterranean underutilized species, purslane and borage, as novel vegetable product (microgreens). Micro-scale production of edible plants is spreading due to the simplicity of their management, rapid cycle, harvest index, and phytochemical value of the edible product. Microgreens, therefore, represent an opportunity to link NUS, nutrition, and agricultural and dietary diversification. By analyzing yield, antioxidants activities, mineral composition, and main phenolic acids and flavonoids, our work indicated that the two species provide interesting results when compared with those reported for crops and horticultural species. Specifically, purslane should be considered highly nutritional due to the amount of phenolic compounds and ascorbic acid, and to potential good β-carotene bioavailability. Borage microgreens have a very high fresh yield and a more composite and balanced phenolic profile. In conclusion, our work provided evidence for implementing new ways to expand the NUS market-chains and for developing added-value food products.

Keywords: *Portulaca olearacea*; *Borago officinalis*; yield; antioxidants; phenolics; flavonoids

1. Introduction

Neglected and underutilized endogenous plants are an appealing option to increase food variety [1]. The Mediterranean basin has a plenty of wild and semi-domesticated plants that have been exploited by rural communities for centuries and even during food crises, thus earning the name of "famine food" [2]. A declining consumption of non-cultivated edible plants is occurring due to social factors and the diffusion of the so-called "Western-style diet" [3,4]. Nonetheless, NUS are receiving increased scientific attention due to nutritional benefits, richness of bioactive components, and suitable micronutrient contents [3,5]. Numerous non-cultivated species are considered weeds in intensive crop systems and therefore, are largely ignored by researchers, farmers, and consumers [2]. These non-commodity plants offer the opportunity to enrich urban-style diets at an affordable cost [2,5]. For instance, several wild and semi-domesticated plants are consumed raw in salads, such as purslane and borage [6]. These halophytes are commonly richer in bioactive compounds than typical salad crops and can provide nutritious food while ensuring a more diversified food basket and a sustainable diet [2].

Purslane (*Portulaca oleracea* L.) is generally seen as a common species among summer crops [7,8] but it was already considered a medicinal and food source by the Ancient Egyptian civilization [9]. In Southern Italy, this herbaceous plant is typically harvested in the wild, although some family farms leave an area for this species to grow, with the harvest typically sold in small markets serving a specific community or area. Purslane is potentially suitable for hydroponics due to the ease of harvest (i.e., stems and leaves are edible), mechanical properties (i.e., the succulent leaves and stems are suitable for motorized harvest), and amenability to the cut-and-come again strategy (i.e., ability to regrow and to produce roots). In floating systems, the cultivation cycle can last from a minimum of 13 days in a summer cycle [10] to approximately 3 months from sowing to the final harvest [7,8]. Purslane is consumed in the Mediterranean basin fresh, cooked or as a dried vegetable [8]. In Southern Italy, it is employed mainly as fresh salad, often with wild rocket. It is rich in molecules with antioxidant potentials such as alfa-tocopherol, beta-carotene, and ascorbic acid [7–9,11]. This species has a high content of proteins, carbohydrates, and minerals (iron, phosphorus, magnesium, calcium, and potassium) [1,7]. On the other hand, mature leaves of purslane frequently contain a high amount of oxalates (between spinach and tea) [12] and nitrates (similar to spinach and celery) [13], which make this species more suitable for occasional consumption. Nonetheless, it has been pointed out that these traits have sufficient intraspecific variability for the selection of improved lines for ready-to-eat products [14].

Borage (*Borrago officinalis* L.) is an annual herb [15–17], probably native to Syria [15], naturalized in the Mediterranean basin and common in Asia Minor, Europe, North Africa, and America [15,18]. In several countries, this herb is cultivated in open-field conditions mainly to extract oil from seeds. Especially in Europe, *B. officinalis* is also grown for culinary and medicinal uses and often harvested in the wild [17,18]. Nonetheless, a recent study indicated the suitability of borage to produce ready-to-eat fresh cut leaves [19]. To our knowledge, information on the cultivation of borage in soilless systems is limited to the study of the effect of salinity on yield and seed characteristics [20]. Borage smells as cucumber, while leaves, stems, and sometimes flowers, are eaten (cooked or raw) in soups and salads, as well as in vegetable and meat dishes [17]. The same authors mentioned that in Northern Spain, borage leaves, petioles, and stems are eaten fresh or moderately fried in salads, whereas in Italy, borage flowers and leaves are eaten in omelets, stews, soups, condiments or pickled and in oil. Borage is rich in fatty acids and is consumed under the belief that it is a treatment for various diseases such as diabetes, arthritis, multiple sclerosis, and eczema [16]. In addition, borage is characterized by tannins, saponins, flavonoids (kaempferol, quercetin, and isorhamnetin), and phenolic acids (p-coumaric, vanilic, chlorogenic, rosmarinic, and caffeic) [17].

In recent years, the production of microgreens to complement that of mature plants has become a trendy market opportunity for novel foods, due to their rich phytochemical content and sensory value [21]. These young leafy greens enhance the human diet by representing not only a different source of bioactive compounds but also by combining vivid colors and tastes. Currently, several horticultural species have been evaluated and exploited to produce microgreens [22].

There is a wide consensus that we need to reverse the abandonment of NUS by changing their reputation (e.g., old-fashioned and associated with the rural lifestyle) and microgreens offer interesting nutritional and social benefits. Producing and selling NUS microgreens to populations who already are accustomed to specific plant species is a practical way to actively encourage people to improve the nutritional value of their diet by harnessing traditional biodiversity. Moreover, microgreens also represent a way to promote NUS in an urban cultural context. The exploitation of the NUS qualities as microgreens is strongly limited by insufficient awareness of their nutritional value. The current study aimed to characterize two underutilized species, purslane and borage, grown as microgreens, with the goal of promoting their value as a novel and sustainable food complement. Specifically, we assessed yield, macro- and micronutrients using an inductively coupled plasma mass spectrometer (ICP-OES), carotenoids by a high-performance liquid chromatographic method with diode-array detection (HPLC-DAD), and polyphenols by ultra-high-performance liquid chromatography coupled to quadrupole orbitrap high-resolution mass spectrometry (UHPLC-Q-Orbitrap HRMS).

2. Materials and Methods

2.1. Plant Material and Growth Condition

Seeds of borage (*Borrago officinalis* L.), also known as starflower, and of purslane (*Portulaca oleracea* L.), also known as duckweed, were obtained from "Pagano Costantino & F.lli S.R.L" (Scafati, Italy) and "Nehme Establishment for Trade & Agriculture" (Batroun, Lebanon), respectively. Sowing density was 40,000 seeds m^{-2} for borage and 80,000 seeds m^{-2} for purslane. Plants were sown and grew in a climatic chamber (KBP-6395F, Termaks, Bergen, Norway) in 204 cm^2 plastic trays filled with 650 cm^3 of peat-based substrate (Special Mixture, Floragard, Vertriebs-GmbH, Oldenburg, Germany). The macronutrient supply of the substrate (electric conductivity (EC): 282 μS cm^{-1}; pH: 5.48) is described elsewhere [23]. Fertigation was applied daily using a modified (quarter-strength) Hoagland's nutrient solution (NS) prepared with osmotic water (EC: 100 ± 25 μS cm^{-1}). The NS had an EC of 500 ± 50 μS cm^{-1}, a pH of 6 ± 0.2, and the following mineral composition: 2.0 mM NO_3^--N, 0.25 mM S, 0.20 mM P, 0.62 mM K, 0.75 mM Ca, 0.17 mM Mg, 0.25 mM NH_4^+-N, 20 μM Fe, 9 μM Mg, 0.3 μM Cu, 1.6 μM Zn, 20 μM B, and 0.3 μM Mo. Light was provided by light-emitting diode (LED) panels (K5 XL 750, Kind LED Grow Light, Santa Rosa CA, USA) with a 12 h photoperiod. The photosynthetic photon flux density was 300 ± 15 μmol m^{-2} s^{-1} measured at the tray level. The temperature and relative humidity were set at 24 °C day and 18 °C night ($\pm$ 2 °C) and 65% day and 75% night ($\pm$5%), respectively. Each species was replicated three times and randomly placed on the shelf of the climate chamber. Trays were relocated daily across the shelf to avoid time-invariant position effects among experimental units.

2.2. Sampling and Morphometric Measurements

At the emergence of the first true leaf, 23 days after sowing (DAS), borage and purslane microgreens were harvested with scissors by cutting at the substrate level (Supplementary Figure S1). Fresh weight (fw) was measured, and yield expressed in kg fw m^{-2}. Each replicate was divided into homogeneous sub-samples for the destructive analyses. A pool was immediately frozen and stored at -80 °C for the determination of total ascorbic acid and chlorophylls. Another pool was weighed, placed in a forced-air oven (65 °C) until constant weight, for dry weight (dw) assessment and the subsequent analysis of the mineral composition. An additional pool was first snap-frozen in liquid nitrogen and then cold-lyophilized (Christ, Alpha 1–4, Osterode, Germany) for the quantification of phenolic compounds, carotenoids (lutein and β-carotene), and antioxidant activities.

2.3. Antioxidant Activity Measurements

The free radical scavenging activity was quantified with a 2,2-diphenyl-1-picryl-hydrazyl (DPPH)-based method using a previously described procedure with few modifications [24]. The DPPH radicals were formed by dissolving 4 mg in 10 mL of methanol. Samples were diluted with the same solvent to obtain a DPPH radical working solution (DRWS) with an absorbance of 0.90 ($\pm$0.02) at 517 nm. A mixture of 1 mL of DRWS and 200 μL of sample was incubated for 10 min at room temperature, and absorbance was spectrophotometrically read at 517 nm. The activity was expressed as TEAC (mmol Trolox equivalents kg^{-1} dw of sample).

The ferric reducing antioxidant activity was measured using a FRAP assay [25] with few modifications. Briefly, the FRAP working solution was prepared by mixing 1.25 mL of 10 mM 2,4,6- tripyridyl-striazine (TPTZ) in HCl (40 mM), 1.25 mL of $FeCl_3$ (20 mmol) in water, and 12.5 mL of 0.3 M sodium acetate buffer (pH 3.6). The reaction of the FRAP solution (2.850 mL) and samples (150 μL) was incubated at room temperature for 4 min and then, absorbance was read at 593 nm. The results were expressed as TEAC (mmol Trolox equivalents kg^{-1} dw of sample).

The ABTS-scavenging activity was evaluated according to the previously published procedures with minor modifications [26]. The 2,2-azinobis-(3-ethylbenzothiazoline-6-sulfonate) (ABTS$^{\bullet+}$) radicals were generated by mixing 5 mL of 7 mM ABTS aqueous stock solution with 88 μL of 2.45 mM aqueous potassium persulfate, diluted with ethanol to a working solution with an absorbance of 0.700 ± 0.002 at 734 nm. Subsequently, 100 μL of sample and 1 mL of the above resulting solution were mixed and incubated for 3 min at room temperature. Absorbance was then read at 734 nm. The results were expressed as TEAC (mmol Trolox equivalents kg^{-1} dw of sample).

2.4. Quantification of Chlorophylls, Catotenoids, and Total Ascorbic Acid

Chlorophylls were spectrophotometrically quantified according to a previously published procedure [27]. Briefly, samples were weighed, and pigments extracted in 90% acetone. Aliquots were read at 662 and 645 nm using a Hach DR 4000 spectrophotometer (Hach Co., Loveland, CO, USA). Total chlorophyll was estimated as the sum of chlorophyll a and b and expressed in mg kg^{-1} fw. Total ascorbic acid was determined as previously described [28] and expressed as mg ascorbate equivalents 100 g^{-1} fw.

Carotenoids (β-carotene and lutein) were quantified by HPLC-DAD essentially as reported [29]. The apparatus comprised a 1200 Series quaternary pump and a 1260 Diode Array Detector Separation (Agilent Technologies, Santa Clara, CA, USA), equipped with Gemini C18 (Phenomenex, Torrance, CA, USA) reverse phase columns (250 × 4.6 mm, 5 μm). Calibration curves were built with using β-carotene and lutein commercial standards (Sigma-Aldrich, Milan, Italy) in the 5 to 100 μg mL^{-1} range. Results were expressed in μg g^{-1} dw.

2.5. Analysis of Macro- and Micro-Minerals by ICP-OES

Minerals (P, K, Ca, Mg, Na, Mn, Fe, Zn, Cu, Se, B, Cr, Mo, Ni, Al, Ba, Cd, and Pb) were quantified by inductively coupled plasma-optical emission spectrometry (Spectroblue, Spectro Ametek, Berwyn, PA, USA) [30]. Briefly, 1 g of dried plant tissue was processed by microwave-assisted digestion (MLS-1200, Microwave Laboratory Systems, Milestone, Shelton, CT, USA) in 10 mL of a 3:1 (*v/v*) solution of nitric acid and fuming hydrochloric acid. The slurry was brought to a final volume of 50 mL with ultra-pure water (Merck Millipore, Darmstadt, Germany). For non-alkaline elements (Fe, Mn, Mo, Se, and Zn), the calibration curve was built in the 1.0 to 100 μg L^{-1} interval and the quantity of the minerals expressed in μg g^{-1} dw. For alkaline elements (P, K, Ca, Mg, and Na), the calibration curve was built in the 100 μg L^{-1} to 10 mg L^{-1} range and the quantity of the minerals expressed in mg g^{-1} dw. For the determination of the accuracy, we used standard reference material (BCR CRM 142R-Commission of the European Communities, 1994). The recovery range was in the 86% to 98% interval.

2.6. Analysis of Polyphenols by UHPLC-Q-Orbitrap HRMS

Lyophilized plant tissue (100 mg) was extracted using 5 mL of a methanol/water (60:40, *v/v*) solution by sonication for half an hour. The mixture was centrifuged (4000 rpm, 15 min), the supernatant filtered through Whatman paper, and then aliquots (10 μL) were used for anthocyanins and polyphenols quantification with a UHPLC system (Dionex UltiMate 3000, Thermo Fisher Scientific, Waltham, MA, USA) coupled to a Q-Exactive Orbitrap mass spectrometer (UHPLC, Thermo Fisher Scientific, Waltham, MA, USA) essentially as described [29,31]. Chromatographic separation was carried out in a Luna Omega PS 1.6 μm column (50 × 2.1 mm, Phenomenex, Torrance, CA, USA) and identification was performed with a Q-Exactive Quadrupole-Orbitrap mass spectrometer (Thermo Fisher Scientific, Waltham, MA, USA) operating in fast negative/positive ion switching mode. Two scan events (full scan MS and all ion fragmentation, AIF) were set for the compounds. Data processing was carried out with the Xcalibur software 3 (Xcalibur, Thermo Fisher Scientific, Waltham, MA, USA). Polyphenols were expressed in $\mu g\ g^{-1}$ dw. The individual phenolic compounds were identified and quantified by comparison with the available standards as described [29,31].

2.7. Statistical Analysis

Morphometric measurements were independently carried out on the three replications. All instrumental determinations for each of the three biological replications were performed in two technical replicates. For all variables, the equality of the means between the two species was evaluated with an unpaired two-tailed Student's *t*-test.

3. Results

3.1. Morphometric and Yield Characteristics

Microgreens were harvested 23 days after sowing. Borage provided the highest amount of fresh product per cultivated surface and a largely higher dry biomass (Table 1). In our controlled conditions, yield was highly correlated with the harvested dry mass ($r = 0.99$; $p < 0.001$; Pearson two-tails). On the other hand, the percentage of dry matter was higher for purslane.

Table 1. Fresh yield, dry weight, dry matter, and hypocotyl length of the two microgreens growing in controlled conditions. Means were statistically separated using a two-tailed Student's *t*-test. **: $p < 0.01$; ***: $p < 0.001$.

Species	Yield ($kg\ fw\ m^{-2}$)	Dry Weight ($g\ m^{-2}$)	Dry Matter (%)	Hypocotyl Length (cm)
Borage	6.44 ± 0.09	307.4 ± 5.93	4.77 ± 0.05	5.32 ± 0.11
Purslane	1.19 ± 0.01	74.26 ± 0.68	6.23 ± 0.04	4.35 ± 0.07
t-test	***	***	***	**

3.2. Mineral Content

Regardless of the higher percentage of dry matter, purslane had a higher amount of 13 mineral elements out of the 18 analyzed, with only Se and Cu (not detected in purslane) present in higher percentage in borage (Table 2). In relative terms, the largest difference was observed for Mo (+461%), followed by Zn (+297%) for the microelements, and for K (+127%) and Mg (+98%) among the macro-elements. B, Al, and Pb did not display a significant variation between the species. As expected, K was the most abundant mineral element for both species. In addition, the Na/K ratio was not significantly different between purslane and borage (not shown), being K approximately $10 \times$ higher than Na.

Table 2. Mineral composition of the microgreens growing in controlled conditions. Means were statistically separated using a two-tailed Student's *t*-test. ns: Not significant; *: $p < 0.05$; **: $p < 0.01$; ***: $p < 0.001$; nd: Not detected.

Species	P ($mg\ g^{-1}$ dw)	K ($mg\ g^{-1}$ dw)	Ca ($mg\ g^{-1}$ dw)	Mg ($mg\ g^{-1}$ dw)	Na ($mg\ g^{-1}$ dw)	Mn ($\mu g\ g^{-1}$ dw)	Fe ($\mu g\ g^{-1}$ dw)	Zn ($\mu g\ g^{-1}$ dw)	Cu ($\mu g\ g^{-1}$ dw)
Borage	11.95 ± 0.26	16.03 ± 0.15	11.51 ± 0.26	2.70 ± 0.07	1.72 ± 0.11	32.57 ± 0.06	27.42 ± 0.44	20.62 ± 0.49	3.95 ± 0.52
Purslane	18.93 ± 0.60	36.43 ± 1.12	17.44 ± 0.49	5.34 ± 0.08	3.63 ± 0.11	76.03 ± 1.76	32.07 ± 0.88	81.86 ± 2.32	nd
t-test	***	***	***	***	***	***	**	***	-

Species	Se ($\mu g\ g^{-1}$ dw)	B ($\mu g\ g^{-1}$ dw)	Cr ($\mu g\ g^{-1}$ dw)	Mo ($\mu g\ g^{-1}$ dw)	Ni ($\mu g\ g^{-1}$ dw)	Al ($\mu g\ g^{-1}$ dw)	Ba ($\mu g\ g^{-1}$ dw)	Cd ($\mu g\ g^{-1}$ dw)	Pb ($\mu g\ g^{-1}$ dw)
Borage	0.89 ± 0.05	35.06 ± 0.31	0.08 ± 0.01	0.13 ± 0.03	0.48 ± 0.04	2.52 ± 0.08	23.23 ± 1.10	0.07 ± 0.002	0.10 ± 0.02
Purslane	0.68 ± 0.04	32.16 ± 1.32	0.15 ± 0.02	0.73 ± 0.03	0.72 ± 0.03	2.78 ± 0.17	67.28 ± 0.64	0.15 ± 0.001	0.10 ± 0.03
t-test	*	ns	*	***	**	ns	***	***	ns

3.3. Radical Scavenging Activity

Purslane microgreens had significantly higher antioxidant properties as indicated by the activities measured in the plant extracts with the three different assays (Table 3).

Specifically, the different methodologies gave a comparable ranking of antioxidant activities, consistently higher (around 40%) in purslane.

Table 3. Antioxidant activities in microgreen extracts. Means were statistically separated using a two-tailed Student's *t*-test. **: $p < 0.01$; ***: $p < 0.001$.

Species	DPPH (mmol Trolox eq. kg^{-1} dw)	ABTS (mmol Trolox eq. kg^{-1} dw)	FRAP (mmol Trolox eq. kg^{-1} dw)
Borage	31.09 ± 1.59	62.60 ± 1.17	31.09 ± 1.59
Purslane	45.40 ± 1.70	82.83 ± 0.82	45.40 ± 1.70
t-test	**	***	**

3.4. Ascorbic Acid, Chlorophyll, Lutein, and β-Carotene

The total ascorbic acid differed significantly between the two microgreen genotypes studied (Table 4). Interestingly, purslane had significantly more total ascorbic acid (more than 3 times higher than that of borage), while determinants of the leaf color (i.e., chlorophylls and carotenoids) were not different between the two species (Table 4). The data implied that the different colors of the microgreens (Supplementary Figure S1) are mainly due to other components, such as chromogenic phenolic compounds (see below). In absolute terms, β-carotene was more abundant than lutein for both species (Table 4). Nonetheless, the β-carotene/lutein ratio was significantly higher for purslane ($p < 0.01$; *t*-test).

Table 4. Total ascorbic acid, total chlorophylls, and carotenoids (lutein and β-carotene) in microgreen extracts. Means were statistically separated using a two-tailed Student's *t*-test. ns: Not significant; ***: $p < 0.001$.

Species	Total Ascorbic Acid (mg Ascorbate 100 g^{-1} fw)	Total Chlorophyll (mg kg^{-1} fw)	Lutein (µg g^{-1} dw)	β-Carotene (µg g^{-1} dw)
Borage	85.27 ± 3.97	612.0 ± 7.11	23.67 ± 1.22	132.0 ± 15.6
Purslane	276.9 ± 19.6	635.9 ± 17.5	25.99 ± 2.83	197.4 ± 26.4
t-test	***	ns	ns	ns

3.5. Polyphenols

The phenolic profiling of the microgreens indicated the presence of qualitative and quantitative differences (Table 5). Purslane had more than double the amount of total phenols per dry matter compared to borage, with a larger difference present for the flavonoid class (25.6 × higher) compared to phenolic acids (2.2 × higher). These strong dissimilarities are likely to reflect the species-specific biochemical profiles of the microgreens, considering the short cycle and the controlled environmental conditions. Specifically, purslane presented one largely predominant flavonoid (rutin), while other members of this chemical class were essentially present in a much lower amount. On the other hand, borage had a more balanced flavonoid profile, with kaempferols and luteolins being the two equally represented major flavonoid forms in microgreens. Species-specific differences were more complex for phenolic acids.

Seven of the 11 of the detected compounds were present in only one species. Moreover, of the four shared compounds, two (ferulic and caffeic acids) were present in a higher quantity in borage and two (Caffeoyl quinic acid and sinapinic acid hexose) in purslane.

Table 5. Phenolic profiles and phenolic acids and flavonoid composition of the microgreen species grown in controlled conditions. Means were statistically separated using a two-tailed Student's *t*-test. ***: $p < 0.001$; nd: Not detected.

	Phenolic Acids										
Species	Caffeic Acid (μg g^{-1} dw)	Caffeoyl Feruloyl Tartaric Acid (μg g^{-1} dw)	Caffeoyl Quinic Acid (μg g^{-1} dw)	Dihydroferulic Acid (μg g^{-1} dw)	Ferulic Acid (μg g^{-1} dw)	Feruloyl Hexoside (μg g^{-1} dw)	Feruloyl Quinic Acid (μg g^{-1} dw)	Rosmarinic Acid (μg g^{-1} dw)	Salicylic Acid Glucoside (μg g^{-1} dw)	Sinapinic Acid Hexose (μg g^{-1} dw)	Vanillic Acid (μg g^{-1} dw)
Borage	313.1 ± 14.6	nd	4.43 ± 0.42	5.27 ± 0.81	1124 ± 95.9	nd	nd	257.3 ± 12.2	1218 ± 73.4	1.06 ± 0.06	102.4 ± 11.3
Purslane	2.26 ± 0.17	701.7 ± 26.5	4926 ± 43.6	nd	284.4 ± 8.69	862.0 ± 23.4	17.98 ± 0.85	nd	nd	16.85 ± 0.21	nd
t-test	***	-	***	-	***	-	-	-	-	***	-

	Flavonoids							SUMS		
Species	Catechin-Glucoside (μg g^{-1} dw)	Kaempferol Trimethyl Ether (μg g^{-1} dw)	Kaempferol-3-glucoside (μg g^{-1} dw)	Luteolin trimethyl eher (μg g^{-1} dw)	Luteolin-7-*O*-glucoside (μg g^{-1} dw)	Quercetin rhamnoside (μg g^{-1} dw)	Rutin (μg g^{-1} dw)	Total Phenolic Acids (μg g^{-1} dw)	Total Flavonoids (μg g^{-1} dw)	Total Phenols (μg g^{-1} dw)
Borage	2.47 ± 0.11	7.66 ± 0.16	3.27 ± 0.17	7.50 ± 0.16	3.67 ± 0.16	3.72 ± 0.17	nd	3026 ± 183	28.28 ± 0.39	3054 ± 183
Purslane	nd	nd	0.25 ± 0.02	nd	0.28 ± 0.02	0.31 ± 0.03	721.2 ± 29.9	6812 ± 49.6	722.1 ± 29.9	7534 ± 36.6
t-test	-	-	***	-	***	***	-	***	***	***

4. Discussion

The two species provided very different yields that, according to the literature, place borage among the most productive species for microgreens (grown in a peat-based substrate) and purslane among the less productive [32,33]. Specifically, the fresh yield of borage was higher compared to plant species that were sown at a similar density (e.g., purple and green basil, tatsoi, mibuna) or that were harvested after a similar time (e.g., jute, kohlrabi, basil) [32,34]. If fresh yield is standardized considering the plant density (the number of seeds per unit of surface), borage still ranked among the most productive species, following radish and Swiss chard [32,34]. The fresh weight and hypocotyl length of borage was also higher than in another study on the same species, most likely due to an almost double growing cycle (23 vs. 12 days) [35]. On the other hand, purslane's yield was low, only higher than chicory [33], slightly below jute, basil, and rocket [32,34,36,37]. NUS are often disregarded due to their modest yield, but this is also related to the cultivation in marginal and poor soils, and to low or absent agricultural input [38]. The good performance of purslane, which deserves further consideration, is likely to be related to the succulent nature of the plant, as also implied by the high percentage of dry matter of the edible product [31,39].

Our evaluation also explored the mineral composition of the microgreen since nutritional parameters are considered one of the most interesting assets of this class of horticultural products, thus essential for promoting their commercial value [40]. There were significant differences between the two species, with the low-yield purslane providing the highest accumulation for the mineral elements of major dietary importance, such as K, Ca, Mg, Mn, Fe, and Zn. In absolute terms, both species presented a predominant presence of K and Ca, as also previously seen in purslane [41] and other microgreens [39]. It is significant that P was present in amounts similar to K, to an extent higher than previously observed in other species [32,39]. In addition, both species presented a low Na/K ratio, an index of dietary importance [42]. The observed ratio is close to the lowest level recorded in microgreens [32,39]. Purslane demonstrated a higher accumulation efficiency (on dry matter basis) also of the non-essential plant elements. This was shown for some (i.e., Ba and Cd) of the non-essential elements for humans (i.e., Al, Ba, Cd, and Pb). Consistent with the literature on other species, Mn, Zn, and Fe were among the most abundant micro-elements [39,43].

The high antioxidant activity is thought to be one the potential dietary advantages provided by microgreens [44]. Purslane resulted as the most valuable species also having a higher ascorbate content. Ascorbates are usually the most abundant water-soluble antioxidant molecules in plant cells [45] and it has been previously noted that faster growing microgreens have a higher quantity of ascorbic acid than sprouts [32]. The total ascorbate concentration of borage and purslane was within the range of a study on 12 species of microgreens (between 25.1 and 147.0 mg 100 g^{-1} fw) [46]. Differences between borage and purslane were not present in the main pigments of the plants, but purslane had a higher β-carotene/lutein ratio. This represents a useful feature since it is important that vitamins are present in plants in a form that can be easily absorbed. Specifically, it has been reported that lutein has an inhibitory effect on β-carotene absorption [47].

The antioxidant activity measured with the three methods consistently indicated the better performance of purslane compared to borage, coherent with the higher content of total polyphenols. The analysis of the phenolic compounds revealed the large diversity that exists between the two analyzed species. Regarding the absolute content, while the amount of total phenols in borage was comparable to that of several microgreens, purslane can be ranked among the top-producing microgreen species of phenolic compounds [32,39]. The polyphenols in borage and purslane were higher than in rocket, lettuce, mustard, and tatsoi [37,43], and lower than in basil and coriander [43]. Purslane extracts were characterized by a predominant amount of rutin [48]. This flavonol glucoside is common in the plant kingdom, especially in the Polygonaceae and Fabaceae families [49]. Rutin was not detected in several Brassica species, while it was highly concentrated in coriander microgreens [32]. A major source of rutin in the Mediterranean diet is represented by capers, olives, and asparagus. The amount detected in purslane microgreens is higher than in most vegetables and fruits, except for some berries, such as red raspberry and black currant [50,51]. Rutin has several beneficial properties, and is used in pharmacology for its vasoprotective and capillary stabilizing activities [52]. On the other hand, borage had a more composite phenolics' profile evident for both the phenolic acids and flavonoids. Among phenolic acids, ferulic acid for borage and caffeoyl quinic acid for purslane were confirmed as being major compounds in these species also at the microgreen stage [53]. Considering the literature relative to the adult plants, borage had a significant amount of salicylic acid glucoside. Salicilates (e.g., salts or esters of salicylic acids) are known to be involved in the plant response to biotic stress. Specifically, inactive forms of the plant hormone salicylic acid are made through conjugation with glucose (or other small organic molecules), to be stored in plant vacuoles [54]. Moreover, the borage flavonoid profile was characterized by a more balanced amount of kaempferols and luteolins, among the most widely distributed flavones in the plant kingdom [55].

5. Conclusions

Our work characterized main features and demonstrated the potential of two underutilized species for microgreens production. These species are expected to provide yield and phytochemical compositions comparable or higher to that of microgreens from largely cultivated and genetically improved horticultural varieties [29,32,39]. Compared with other microgreens, purslane should be considered highly nutritional due to the amount phenolic compounds with known beneficial impacts on human health, the high amount of ascorbic acid, and possibly, a good β-carotene bioavailability. Although some mineral-specific differences were present, purslane also displayed a higher mineral utilization efficiency. Even so, the difference with borage, and more generally with other microgreen species reported in the literature, were limited and unlikely to have a significant impact on the human diet. On the other hand, borage microgreens provided the benefit of a much higher fresh yield and a more composite and balanced phenolic profile, which may possibly increase the commercial interest for this underutilized species.

Supplementary Materials: The following is available online at https://www.mdpi.com/article/10.3390/horticulturae7080211/s1. Figure S1: Microgreens at harvest. A: *Portulaca oleracea*; B: *Borago officinalis*. Not to scale.

Author Contributions: Conceptualization, C.E.-N. and Y.R.; methodology, G.C. and C.E.-N.; software, A.P.; validation, G.C. and C.E.-N.; formal analysis, G.C., C.E.-N., G.G., A.P., and P.G.; investigation, G.C., C.E.-N., and M.C.K.; resources, G.G. and A.Z.; data curation, C.E.-N., G.G., A.P., A.Z., and P.G.; writing—original draft preparation, G.C. and C.E.-N.; writing—review and editing, G.C., C.E.-N., and Y.R.; visualization, G.C., and C.E.-N.; supervision, Y.R., A.R., M.C.K., and S.D.P.; project administration, G.C., A.R., S.D.P., and Y.R.; funding acquisition, G.C. and Y.R. All authors have read and agreed to the published version of the manuscript.

Funding: This research received no external funding.

Institutional Review Board Statement: Not applicable.

Informed Consent Statement: Not applicable.

Data Availability Statement: The data presented in this study are available in the article. Raw data are available on request from the corresponding author.

Acknowledgments: The authors would like to thank Luca Scognamiglio, Daniele Capurso, Luigi G Duri, and Francesco Cristofano for their technical help during the experiment.

Conflicts of Interest: The authors declare no conflict of interest.

References

1. Srivastava, R.; Srivastava, V.; Singh, A. Multipurpose benefits of an underexplored species purslane (*Portulaca oleracea* L.): A critical review. *Environ. Manag.* **2021**, 1–12. [CrossRef]
2. Petropoulos, S.; Karkanis, A.; Martins, N.; Ferreira, I.C. Edible halophytes of the mediterranean basin: Potential candidates for novel food products. *Trends Food Sci. Technol.* **2018**, *74*, 69–84. [CrossRef]
3. Morales, P.; Ferreira, I.C.; Carvalho, A.M.; Sánchez-Mata, M.C.; Cámara, M.; Fernández-Ruiz, V.; Pardo-de-Santayana, M.; Tardío, J. Mediterranean non-cultivated vegetables as dietary sources of compounds with antioxidant and biological activity. *LWT Food Sci. Tech.* **2014**, *55*, 389–396. [CrossRef]
4. Adlercreutz, H. Western diet and Western diseases: Some hormonal and biochemical mechanisms and associations. *Scand. J. Clin. Lab. Investig.* **1990**, *50*, 3–23. [CrossRef]
5. Guarrera, P.; Savo, V. Wild food plants used in traditional vegetable mixtures in Italy. *J. Ethnopharmacol.* **2016**, *185*, 202–234. [CrossRef]
6. Torija-Isasa, M.E.; Matallana-González, M.C. A historical perspective of wild plant foods in the mediterranean area. In *Mediterranean Wild Edible Plants*; Springer: New York, NY, USA, 2016; pp. 3–13.
7. Gonnella, M.; Charfeddine, M.; Conversa, G.; Santamaria, P. Purslane: A review of its potential for health and agricultural aspects. *Eur. J. Plant Sci. Biotech.* **2010**, *4*, 131–136.
8. Dabbou, S.; Lahbib, K.; Pandino, G.; Dabbou, S.; Lombardo, S. Evaluation of pigments, phenolic and volatile compounds, and antioxidant activity of a spontaneous population of *Portulaca oleracea* L. grown in Tunisia. *Agriculture* **2020**, *10*, 353. [CrossRef]
9. Sdouga, D.; Branca, F.; Kabtni, S.; Di Bella, M.C.; Trifi-Farah, N.; Marghali, S. Morphological traits and phenolic compounds in tunisian wild populations and cultivated varieties of *Portulaca oleracea* L. *Agronomy* **2020**, *10*, 948. [CrossRef]
10. Ochoa, J.; Niñirola, D.; Conesa, E.; Lara, L.; López-Marín, J.; Fernández, J. Predicting Purslane (*Portulaca oleracea* L.) Harvest in a Hydroponic Floating System. In *Proceedings of the V International Symposium on Seed, Transplant and Stand Establishment of Horticultural Crops 898, Murcia and Almería, Spain, 27 September–1 October 2009*; ISHS: Leuven, Belgium, 2011; pp. 205–209.
11. Alam, M.; Juraimi, A.S.; Rafii, M.; Hamid, A.A.; Aslani, F.; Hasan, M.; Zainudin, M.A.M.; Uddin, M. Evaluation of antioxidant compounds, antioxidant activities, and mineral composition of 13 collected purslane (*Portulaca oleracea* L.) accessions. *BioMed. Res. Int.* **2014**, *2014*, 296063. [CrossRef]
12. Noonan, S.; Savage, G. Oxalate content of foods and its effect on humans. *Asia Pac. J. Clin. Nutr.* **1999**, *8*, 64–74. [PubMed]
13. Corré, W.J.; Breimer, T. *Nitrate and Nitrite in Vegetables*; Pudoc: Wageningen, The Netherlands, 1979.
14. Egea-Gilabert, C.; Ruiz-Hernández, M.V.; Parra, M.Á.; Fernández, J.A. Characterization of purslane (*Portulaca oleracea* L.) accessions: Suitability as ready-to-eat product. *Sci. Hortic.* **2014**, *172*, 73–81. [CrossRef]
15. Gupta, M.; Singh, S. *Borago officinalis* linn. An important medicinal plant of mediterranean region: A review. *Int. J. Pharm. Sci. Rev. Res.* **2010**, *5*, 27–34.
16. Asadi-Samani, M.; Bahmani, M.; Rafieian-Kopaei, M. The chemical composition, botanical characteristic and biological activities of *Borago officinalis*: A review. *Asia Pac. J. Trop. Med.* **2014**, *7*, S22–S28. [CrossRef]
17. Borowy, A.; Chwil, M.; Kapłan, M. Biologically active compounds and antioxidant activity of borage (*Borago officinalis* L.) flowers and leaves. *Acta Sci. Pol. Hortorum Cultus* **2017**, *16*, 169–180. [CrossRef]
18. Abu-Qaoud, H.; Shawarb, N.; Hussen, F.; Jaradat, N.; Shtaya, M. Comparison of qualitative, quantitative analysis and antioxidant potential between wild and cultivated *Borago officinalis* leaves from palestine. *Pak. J. Pharm. Sci.* **2018**, *31*, 953–959.
19. Miceli, C.; Moncada, A.; Vetrano, F.; D'Anna, F.; Miceli, A. Suitability of *Borago officinalis* for minimal processing as fresh-cut produce. *Horticulturae* **2019**, *5*, 66. [CrossRef]
20. Urrestarazu, M.; Gallegos-Cedillo, V.M.; Ferrón-Carrillo, F.; Guil-Guerrero, J.L.; Lao, M.T.; Álvaro, J.E. Effects of the electrical conductivity of a soilless culture system on gamma linolenic acid levels in borage seed oil. *PLoS ONE* **2019**, *14*, e0207106. [CrossRef]
21. El-Nakhel, C.; Pannico, A.; Graziani, G.; Kyriacou, M.C.; Giordano, M.; Ritieni, A.; De Pascale, S.; Rouphael, Y. Variation in macronutrient content, phytochemical constitution and in vitro antioxidant capacity of green and red butterhead lettuce dictated by different developmental stages of harvest maturity. *Antioxidants* **2020**, *9*, 300. [CrossRef]
22. Caracciolo, F.; El-Nakhel, C.; Raimondo, M.; Kyriacou, M.C.; Cembalo, L.; De Pascale, S.; Rouphael, Y. Sensory attributes and consumer acceptability of 12 microgreens species. *Agronomy* **2020**, *10*, 1043. [CrossRef]
23. Pannico, A.; Graziani, G.; El-Nakhel, C.; Giordano, M.; Ritieni, A.; Kyriacou, M.C.; Rouphael, Y. Nutritional stress suppresses nitrate content and positively impacts ascorbic acid concentration and phenolic acids profile of lettuce microgreens. *Italus Hortus* **2020**, *27*, 41–52. [CrossRef]
24. Brand-Williams, W.; Cuvelier, M.-E.; Berset, C. Use of a free radical method to evaluate antioxidant activity. *LWT Food Sci. Tech.* **1995**, *28*, 25–30. [CrossRef]
25. Benzie, I.F.; Strain, J.J. The ferric reducing ability of plasma (FRAP) as a measure of "antioxidant power": The frap assay. *Anal. Biochem.* **1996**, *239*, 70–76. [CrossRef]
26. Re, R.; Pellegrini, N.; Proteggente, A.; Pannala, A.; Yang, M.; Rice-Evans, C. Antioxidant activity applying an improved abts radical cation decolorization assay. *Free. Radic. Biol. Med.* **1999**, *26*, 1231–1237. [CrossRef]
27. Lichtenthaler, H.K.; Wellburn, A.R. Determinations of total carotenoids and chlorophylls a and b of leaf extracts in different solvents. *Biochem. Soc. Trans.* **1983**, *11*, 591–592. [CrossRef]

28. Minutolo, M.; Chiaiese, P.; Di Matteo, A.; Errico, A.; Corrado, G. Accumulation of ascorbic acid in tomato cell culture: Influence of the genotype, source explant and time of in vitro cultivation. *Antioxidants* **2020**, *9*, 222. [CrossRef]

29. Kyriacou, M.C.; El-Nakhel, C.; Pannico, A.; Graziani, G.; Soteriou, G.A.; Giordano, M.; Palladino, M.; Ritieni, A.; De Pascale, S.; Rouphael, Y.; et al. Phenolic constitution, phytochemical and macronutrient content in three species of microgreens as modulated by natural fiber and synthetic substrates. *Antioxidants* **2020**, *9*, 252. [CrossRef] [PubMed]

30. Volpe, M.G.; Nazzaro, M.; Di Stasio, M.; Siano, F.; Coppola, R.; De Marco, A. Content of micronutrients, mineral and trace elements in some mediterranean spontaneous edible herbs. *Chem. Cent. J.* **2015**, *9*, 1–9. [CrossRef] [PubMed]

31. El-Nakhel, C.; Pannico, A.; Graziani, G.; Kyriacou, M.C.; Gaspari, A.; Ritieni, A.; De Pascale, S.; Rouphael, Y. Nutrient supplementation configures the bioactive profile and production characteristics of three brassica microgreens species grown in peat-based media. *Agronomy* **2021**, *11*, 346. [CrossRef]

32. Kyriacou, M.C.; El-Nakhel, C.; Graziani, G.; Pannico, A.; Soteriou, G.A.; Giordano, M.; Ritieni, A.; De Pascale, S.; Rouphael, Y. Functional quality in novel food sources: Genotypic variation in the nutritive and phytochemical composition of thirteen microgreens species. *Food Chem.* **2019**, *277*, 107–118. [CrossRef] [PubMed]

33. Paradiso, V.M.; Castellino, M.; Renna, M.; Gattullo, C.E.; Calasso, M.; Terzano, R.; Allegretta, I.; Leoni, B.; Caponio, F.; Santamaria, P. Nutritional characterization and shelf-life of packaged microgreens. *Food Funct.* **2018**, *9*, 5629–5640. [CrossRef]

34. Bulgari, R.; Baldi, A.; Ferrante, A.; Lenzi, A. Yield and quality of basil, swiss chard, and rocket microgreens grown in a hydroponic system. *N. Z. J. Crop. Hortic. Sci.* **2017**, *45*, 119–129. [CrossRef]

35. Viršilė, A.; Sirtautas, R. Light irradiance level for optimal growth and nutrient contents in borage microgreens. *Rural. Dev.* **2013**, *8*, 1241–1249.

36. Di Gioia, F.; De Bellis, P.; Mininni, C.; Santamaria, P.; Serio, F. Physicochemical, agronomical and microbiological evaluation of alternative growing media for the production of rapini (brassica rapa l.) microgreens. *J. Sci. Food Agric.* **2017**, *97*, 1212–1219. [CrossRef]

37. Kyriacou, M.C.; El-Nakhel, C.; Soteriou, G.A.; Graziani, G.; Kyratzis, A.; Antoniou, C.; Ritieni, A.; De Pascale, S.; Rouphael, Y. Preharvest nutrient deprivation reconfigures nitrate, mineral, and phytochemical content of microgreens. *Foods* **2021**, *10*, 1333. [CrossRef]

38. Chivenge, P.; Mabhaudhi, T.; Modi, A.T.; Mafongoya, P. The potential role of neglected and underutilised crop species as future crops under water scarce conditions in sub-saharan africa. *Int. J. Environ. Res. Public Heal.* **2015**, *12*, 5685–5711. [CrossRef]

39. Xiao, Z.; Codling, E.E.; Luo, Y.; Nou, X.; Lester, G.E.; Wang, Q. Microgreens of brassicaceae: Mineral composition and content of 30 varieties. *J. Food Compos. Anal.* **2016**, *49*, 87–93. [CrossRef]

40. Kyriacou, M.C.; Rouphael, Y.; Di Gioia, F.; Kyratzis, A.; Serio, F.; Renna, M.; De Pascale, S.; Santamaria, P. Micro-scale vegetable production and the rise of microgreens. *Trends Food Sci. Technol.* **2016**, *57*, 103–115. [CrossRef]

41. Giménez, A.; Martínez-Ballesta, M.d.C.; Egea-Gilabert, C.; Gómez, P.A.; Artés-Hernández, F.; Pennisi, G.; Orsini, F.; Crepaldi, A.; Fernández, J.A. Combined effect of salinity and led lights on the yield and quality of purslane (*Portulaca oleracea* L.) microgreens. *Horticulturae* **2021**, *7*, 180. [CrossRef]

42. Drewnowski, A.; Maillot, M.; Rehm, C. Reducing the sodium-potassium ratio in the us diet: A challenge for public health. *Am. J. Clin. Nutr.* **2012**, *96*, 439–444. [CrossRef] [PubMed]

43. Pannico, A.; El-Nakhel, C.; Graziani, G.; Kyriacou, M.C.; Giordano, M.; Soteriou, G.A.; Zarrelli, A.; Ritieni, A.; De Pascale, S.; Rouphael, Y. Selenium biofortification impacts the nutritive value, polyphenolic content, and bioactive constitution of variable microgreens genotypes. *Antioxidants* **2020**, *9*, 272. [CrossRef]

44. Choe, U.; Yu, L.L.; Wang, T.T. The science behind microgreens as an exciting new food for the 21st century. *J. Agric. Food Chem.* **2018**, *66*, 11519–11530. [CrossRef] [PubMed]

45. Paciolla, C.; Fortunato, S.; Dipierro, N.; Paradiso, A.; De Leonardis, S.; Mastropasqua, L.; de Pinto, M.C. Vitamin c in plants: From functions to biofortification. *Antioxidants* **2019**, *8*, 519. [CrossRef] [PubMed]

46. Di Gioia, F.; Renna, M.; Santamaria, P. Sprouts, microgreens and "baby leaf" vegetables. In *Minimally Processed Refrigerated Fruits and Vegetables*; Springer: New York, NY, USA, 2017; pp. 403–432.

47. Van den Berg, H. Effect of lutein on beta-carotene absorption and cleavage. *Int. J. Vitam. Nutr. Res.* **1998**, *68*, 360–365. [PubMed]

48. Uddin, M.; Juraimi, A.S.; Ali, M.; Ismail, M.R. Evaluation of antioxidant properties and mineral composition of purslane (*Portulaca oleracea* L.) at different growth stages. *Int. J. Mol. Sci.* **2012**, *13*, 10257–10267. [CrossRef]

49. Frutos, M.J.; Rincón-Frutos, L.; Valero-Cases, E. Rutin. In *Nonvitamin and Nonmineral Nutritional Supplements*; Elsevier: Amsterdam, The Netherlands, 2019; pp. 111–117.

50. Atanassova, M.; Bagdassarian, V. Rutin content in plant products. *J. Univ. Chem. Techn. Metall.* **2009**, *44*, 201–203.

51. Azimova, S.S.; Vinogradova, V.I. *Natural Compounds: Flavonoids*; Springer: New York, NY, USA, 2013.

52. Ganeshpurkar, A.; Saluja, A.K. The pharmacological potential of rutin. *Saudi Pharm. J.* **2017**, *25*, 149–164. [CrossRef] [PubMed]

53. Zadernowski, R.; Naczk, M.; Nowak-Polakowska, H. Phenolic acids of borage (*Borago officinalis* L.) and evening primrose (oenothera biennis l.). *J. Am. Oil Chem. Soc.* **2002**, *79*, 335–338. [CrossRef]

54. Thompson, A.M.G.; Iancu, C.V.; Neet, K.E.; Dean, J.V.; Choe, J.-y. Differences in salicylic acid glucose conjugations by ugt74f1 and ugt74f2 from arabidopsis thaliana. *Sci. Rep.* **2017**, *7*, srep46629. [CrossRef] [PubMed]

55. Calderon-Montano, J.; Burgos-Morón, E.; Pérez-Guerrero, C.; López-Lázaro, M. A review on the dietary flavonoid kaempferol. *Mini-Rev. Med. Chem.* **2011**, *11*, 298–344. [CrossRef]

 *horticulturae*

Article

Impact of Salinity on the Growth and Chemical Composition of Two Underutilized Wild Edible Greens: *Taraxacum officinale* and *Reichardia picroides*

Alexios A. Alexopoulos [1], Anna Assimakopoulou [2,*], Panagiotis Panagopoulos [1], Maria Bakea [2], Nikolina Vidalis [3], Ioannis C. Karapanos [3] and Spyridon A. Petropoulos [4,*]

[1] Laboratory of Agronomy, Department of Agriculture, University of the Peloponnese, Antikalamos, 241 00 Kalamata, Greece; a.alexopoulos@us.uop.gr (A.A.A.); panagtakis@yahoo.gr (P.P.)
[2] Laboratory of Plant Physiology, Department of Agriculture, University of the Peloponnese, Antikalamos, 241 00 Kalamata, Greece; maria.bakea@gmail.com
[3] Laboratory of Vegetable Production, Department of Crop Science, Agricultural University of Athens, Iera Odos 75, 118 55 Athens, Greece; vidalisn@aua.gr (N.V.); karapanos@aua.gr (I.C.K.)
[4] Laboratory of Vegetable Production, Department of Agriculture, Crop Production and Rural Environment, University of Thessaly, Fytokou Street, 384 46 Volos, Greece
*   Correspondence: a.assimakopoulou@us.uop.gr (A.A.); spetropoulos@uth.gr (S.A.P.); Tel.: +30-272-1045-178 (A.A.); +30-242-1093-196 (S.A.P.)

Citation: Alexopoulos, A.A.; Assimakopoulou, A.; Panagopoulos, P.; Bakea, M.; Vidalis, N.; Karapanos, I.C.; Petropoulos, S.A. Impact of Salinity on the Growth and Chemical Composition of Two Underutilized Wild Edible Greens: *Taraxacum officinale* and *Reichardia picroides*. *Horticulturae* **2021**, *7*, 160. https://doi.org/10.3390/horticulturae7070160

Academic Editors: Rosario Paolo Mauro, Carlo Nicoletto and Leo Sabatino

Received: 5 June 2021
Accepted: 19 June 2021
Published: 22 June 2021

Publisher's Note: MDPI stays neutral with regard to jurisdictional claims in published maps and institutional affiliations.

Abstract: Soil salinization is one of the major environmental factors responsible for limited crop production throughout the world. Therefore, there is urgent need to find tolerant/resistant species to exploit in commercial cultivation systems. In this context, the valorization of wild edible greens for human consumption and/or medicinal purposes is gaining more and more interest. The aim of the present work was to study the effect of salinity, e.g., electrical conductivity: 2 mS cm^{-1} (nutrient solution EC), 6 mS cm^{-1} and 10 mS cm^{-1} on plant growth and chemical composition of *Reichardia picroides* and *Taraxacum officinale* plants grown in a floating hydroponic system. The results showed that *R. picroides* is a moderately salt-tolerant species, as the majority of plant growth parameters determined were not negatively affected under the treatment of 6 mS cm^{-1}. On the other hand, the growth parameters of *T. officinale* plants were severely affected under the same conditions. Moreover, high salinity levels (EC at 10 mS cm^{-1}) impaired the growth of both species. The content of leaves in chlorophylls (a, b and total), carotenoids+xanthophylls and total soluble solids was not significantly affected by the tested EC levels in both species, whereas the titratable acidity increased under the treatment of 10 mS cm^{-1}. Moreover, *R. picroides* exhibited a more effective adaptation mechanism against saline conditions than *T. officinale*, as evidenced by the higher accumulation of osmolytes such as proline and the higher shoot K content, probably through a more efficient K/Na selectivity. In conclusion, both species were severely affected by high salinity; however, *R. picroides* showed promising results regarding its commercial cultivation under moderate salinity levels, especially in regions where resources of high-quality irrigation water are limited.

Keywords: dandelion; common brighteyes; wild edible greens; chemical composition; nutrient contents; soilless cultivation; minerals content; saline conditions

1. Introduction

Soil salinization is one of the most important environmental stressors around the globe with significant implications on crop productivity, especially in arid and semi-arid regions such as the broad Mediterranean area [1]. Among the various crops, vegetables are considered susceptible to environmental extremities which become more and more frequent due to the ongoing climate change [2–4]. The cultivation of conventional crops under these new limiting conditions is becoming difficult and less profitable for farmers due to yield losses and the increased production cost [5]. Therefore, urgent measures are needed to ensure food security, especially when considering the rapidly increasing global population and the growing demands for quality foods [6]. For this purpose, several means have been suggested during the last years including the use of cost-effective practices such as the application of biostimulants, the grafting of vegetables to tolerant rootstocks and the cropping of alternative and tolerant species, among others [7–12].

According to FAO, a significant part of world food production is obtained from only nine crops, which entails increased risk of genetic erosion due to agrobiodiversity degradation [13]. In this context, wild edible greens are a promising solution toward the sustainable increase in agrobiodiversity since they are tolerant to arduous conditions and can easily adapt to climate changes [14,15]. Most of these species are an integral part of local cuisines and are traditionally used for culinary and medicinal purposes [16–18]. Recently, the commercial cultivation of such species has gained interest both by farmers and consumers, and several studies have reported the potential of using wild edible species in sustainable cropping systems for the production of high value-added products due to increased health beneficial effects [19–22]. Considering that these species are usually collected in the wild or confronted as weeds within the fields, there is a lack of information regarding the best practice guides that should be applied to ensure high yields without compromising the quality and food safety of the final products. Therefore, several reports have suggested cultivation practices related to harvesting stage, growing period, the fertilization regimes or cropping under stress conditions and soilless cultivation systems [23–31].

Taraxacum officinale and *Reichardia picroides* are two unexploited species of the Asteraceae family with limited information regarding their requirements in agronomic practices. According to González et al. [32], who carried out an ethnobotanical survey in the Iberian peninsula, it was suggested that a cultural importance index based on the frequency and versatility of uses and *T. officinale* scored very low values. However, it is not uncommon for wild edible greens to have local interest, and their importance may vary from region to region. Recently, our team reported the soilless cultivation of both species in nutrient solution with different pH values (e.g., 4.0, 5.5 and 7.0), and the results showed that not only can these species be cultivated under unfavorable conditions, but they also can improve their bioactive properties through the increased phytochemicals content [33]. Moreover, in the earlier study of Petropoulos et al. [28], both *T. officinale* and *R. picroides* recorded a high content in phenolic compounds and tocopherols, which were significantly affected by the growing period.

The response of horticultural crops to abiotic stressors is complex and includes changes in plant physiology and morphology through the induction of secondary metabolites biosynthesis, the expression of stress-related genes and the hormonal regulation of plants [34]. The mechanisms of salt tolerance in plants have received the attention of researchers for many years focusing on the effects of soil water potential and water availability decrease on plant physiology, as well as on ion-specific impacts that limit plant growth under saline conditions, especially those caused by NaCl [35–37]. In particular, salinity may affect nutritional balance in plant tissues, since significant antagonism may be observed in the absorption and transfer of nutrients [36]. Moreover, saline conditions may variably change the pH and redox potential in nutrient solution, depending on the severity of salinity and the plant species, thus resulting in reduction of micronutrients solubility [38]. Nitrogen absorption is negatively affected under saline conditions due to the interaction and the antagonistic effects observed in Cl^- and NO_3^- and/or Na^+ and NH_4^+ [36]. However, the various plants may differ in their response to saline conditions and several species may exhibit significant tolerance under elevated salinity without significant yield reductions [39]. In addition, given that soil infertility is often associated with the presence of large amounts of salts, the identification of tolerant genotypes is considered to be a promising approach toward food security through saline agriculture [40,41].

Regarding salinity effects, sodium chloride affects the transport of ions across plasmalemma of root cells through rupturing of the cellular membranes [42], and salt tolerance in crops is based on specific physiological characteristics such as shoot- or leaf-specific ion accumulation or the production of specific osmolytes [43]. Furthermore, in order to adapt to salt stress, plants have developed various hormonal-based strategies that help to regulate plant growth through mediation of salinity stress signals [44]. Other adaptation mechanisms include the biosynthesis of bioactive compounds (e.g., phenolic compounds) and osmoprotectants such as glutames and γ-aminobutyric acid [11]. Several studies have confirmed the tolerance of wild edible greens under environmental constraints, and species such as *Reichardia picroides*, *Cichorium spinosum*, *Sonchus oleraceus* and *Urospermum picroides* have been identified as salt tolerant and should be suggested as alternative/complementary crops or for the phytoremediation of saline soils [25,45–47].

Considering the increasing interest in wild edible greens and the lack of information regarding their cultivation practices, the aim of the present study was to evaluate the effect of salinity on plant growth parameters and chemical composition of the two native to the Mediterranean basin unexploited species, namely *T. officinale* and *R. picroides*. For this purpose, plants of both species were cultivated in a floating hydroponic system under greenhouse conditions, and three electrical conductivity (EC) levels in nutrient solution were implemented, e.g., 2.0, 6.0 and 10.0 mS cm^{-1}. The results of this study increase the knowledge regarding agronomic requirements and help the domestication and commercial cultivation of these valuable species.

2. Materials and Methods

2.1. Plant Material, Experimental Treatments and Growing Conditions

The experiment was performed in the experimental greenhouse at the University of the Peloponnese (Kalamata, Messinia, Southern Greece, 37°3'22" N, 22°1'43" E). Seeds of dandelion (*Taraxacum officinale* (L.) Weber ex F.H.Wigg.-*T. officinale*, hereafter) and common brighteyes (*Reichardia picroides* (L.) Roth-*R. picroides*, hereafter) were collected in the wild (Vicinity of Kalamata, Greece) and stored at 4–7 °C [33]. Seeds of both species were sown on 28/11/2018 at a depth of 0.5–1 cm in germination containers of 19 cm $\times$ 13 cm $\times$ 5 cm filled with white peat (pH 5.5–6.5, without fertilization-base substrate, Klasmann-Deilmann GmbH, Geeste, Germany). Germination containers were placed in a walk-in growth chamber at 20 °C with 16 h photoperiod and light intensity of 55 µmol m^{-2} s^{-1} provided by fluorescent lamps. On 28/01/2019 (61 days after sowing) and when they reached the 3–4 true-leaf stage, young seedlings were transplanted to polystyrene seedling trays (cell dimension 5 $\times$ 5 $\times$ 5 cm^3) containing the same media, at a distance of 15 $\times$ 15 cm^2 (namely 44.44 plants m^2) following the method described by Alexopoulos et al. [33]. The trays were transferred in the greenhouse and placed in containers (volume of 0.25 m^3) filled with 0.2 m^3 of nutrient solution (NS). The composition of NS and its preparation process have been previously described by Alexopoulos et al. [33].

Three different treatments, namely 2.0 mS cm^{-1} (EC-2; control treatment), 6.0 mS cm^{-1} (EC-6) and 10.0 mS cm^{-1} (EC-10), were applied to plants by adding NaCl to the control NS until the desired values of EC were obtained. These levels were selected based on previous studies with wild edible greens [25,46,48]. The EC and pH values of NS were recorded on a daily basis. During the experimental period, the nutrient solution pH of all the three treatments ranged from 5.9 to 6.2, whereas the EC ranged from 2.0 to 2.2 mS cm^{-1} in treatment EC-2, from 6.0 to 6.2 mS cm^{-1} in EC-6 and from 10.0 to 10.2 mS cm^{-1} in EC-10. The temperature in the greenhouse ranged from 8.0 to 31.1 °C, whereas the temperature of nutrient solution in all the containers during the cultivation period ranged from 13 to 17 °C. The experiment was carried out in a completely randomized experimental design. Each container included 44 plants, while 10 plants were harvested for plant growth assessment and chemical analyses. For each species, four replications per treatment were implemented (24 containers in total). Harvest of plants from each species was performed when plants were adequately grown but still young and tender and before anthesis, based on the harvesting stage when collected in the wild. In *T. officinale*, the harvest was carried out on 08/03/2019 (49 days after transplanting-DAT) whereas in *R. picroides* on 22/03/2019 (63-DAT).

2.2. Growth Parameters

Growth parameters were assessed according to the methodology of Alexopoulos et al. [33]. In brief, plant leaf number, rosette diameter, and length and width of the largest leaf of the plant were measured in 10 plants from each replication and for each plant species. Then, yield parameters were recorded, e.g., the number of nonmarketable leaves (not green, dried or injured), total plant fresh weight (FW), the FW of the upper plant part and roots, and the FW of marketable leaf were assessed. The leaves obtained from the 10 plants of each replication were pooled in two batch samples, one of which was used to determine dry matter content (% DMC) and mineral composition, while the other one was stored at $-80\ °C$ for chemical analyses [33]. Similarly, roots were only used for the determination of % DMC and minerals content, since this plant part is not edible.

2.3. Leaf and Root Minerals and Nitrate Content

Minerals and nitrates content were determined according to the methodology previously described by Alexopoulos et al. [33], following the protocols of Kalra [49] for dry-ashing of samples; Boltz and Lueck [50] for P determination through the vanado-molybdo-phosphate yellow color method; the azomethin-H method for B determination [49]; atomic absorption spectrometry (SpectrAA, 240 Atomic Absorption FS; Varian, Palo Alto, CA, USA) for K, Ca, Mg, Na, Fe, Mn, Zn and Cu determination [49]; the indophenol-blue method N determination [51]; the method of Cataldo et al. [52] for the determination of N-NO_3^- in leaves; and the titration with 0.1 N silver nitrate for Cl content determination [33]. The analyses were performed in triplicate, and all the results were expressed in dry weight (DW), except for nitrates content, which was expressed in fresh weight (FW).

2.4. Chemical Composition Analyses in Leaves

Total soluble solids content (TSSC) of leaves was recorded in the juice of leaves with a portable refractometer (model HR32B, Schmidt & Haensch GmbH & Co., Berlin, Germany) after homogenized fresh samples at 20 °C [33]. Titratable acidity was assessed in aqueous extracts of homogenized samples after titration with NaOH, up to pH 8.1 [33]. The results were presented as mg of malic acid per 100 g of FW [33].

The chlorophyll content of leaves was determined using the methods described by Alexopoulos et al. [33]. In particular, one method included the use of SPAD-502 Chlorophyll Meter (Konica-Minolta Co. Ltd., Tokyo, Japan) to record SPAD index of leaves, while the other method recorded chlorophyll a, b and total chlorophyll content in acetone extracts of homogenized samples of leaves, according to Karapanos et al. [53]. The same extracts were used for carotenoids+xanthophylls quantification (absorbance at 470 nm), following the protocol of Lichtenthaler and Buschmann [54]. The results were expressed as mg per 100 g of FW.

2.4.1. Total Phenolic Compounds Content

Total phenolic compounds (TPC) were determined in methanolic extracts according to the Folin–Ciocalteu protocol [55] after slight modifications [53]. The results were expressed as mg of gallic acid equivalents (GAE) per 100 g of FW.

2.4.2. Proline Content

Free proline content was measured using the acid-ninhydrin method of Bates et al. [56]. In particular, leaf samples were extracted in 3% aqueous sulfosalicylic acid, and extracts were combined with acid ninhydrin and glacial acetic acid (1:1:1) and then incubated at 90 °C. The reaction was terminated after 1 h by putting the samples in an ice bath. The chromophore was extracted using 2 mL of toluene, and its absorbance was measured at 520 nm using a spectrophotometer (Lambda 1A, Perkin-Elmer, Waltham, MA, USA). Pure proline was used as standard, and the results were expressed as μmole of proline per g FW.

2.5. Statistical Analysis

For each plant species separately, the statistical analysis was performed with one-way ANOVA, and means were separated according to the least significant difference (LSD) test at $p \leq 0.05$. For each plant species and each treatment, four replications (n = 4) with 10 plants each were used, as described in detail in Sections 2.1–2.3. The correlations between growth parameters and minerals content in *R. picroides* and *T. officinale* were examined using the Pearson's correlation test. All statistical analyses were carried out with StatGraphics Centurion-XVI statistical package (StatPoint Technologies Inc., Warrenton, VA, USA).

3. Results and Discussion

3.1. Plant Growth

Reichardia picroides and *Taraxacum officinale* plants grown under either EC-6 or EC-10 treatment did not show any salt toxicity symptoms, i.e., local wilting or necrotic spots in the leaves. In *R. picroides*, the number of leaves per plant was the highest in EC-6 treatment (40.65 leaves per plant), followed by EC-2 and EC-10 treatments (32.28 and 24.75 leaves per plant, respectively) (Table 1). Nonmarketable leaf number per plant, leaf SPAD index values, root FW and root/shoot ratio were not affected by the studied EC treatments. On the other hand, EC-10 treatment caused a significant reduction in rosette diameter, the maximum leaf length and width, total plant FW, upper part plant FW and marketable leaves FW per plant in comparison with the EC-2 and the EC-6 treatments, whereas EC-6 treatment caused a significant reduction in leaf DMC compared to the EC-2 treatment and did not differ from the EC-10 (Table 1). In the case of *T. officinale*, the effect of EC treatments on plant growth was more profound. In particular, EC-6 and EC-10 treatments caused a significant reduction in leaf number per plant, rosette diameter, maximum leaf length and width, total plant FW, upper plant part FW, root FW and marketable leaves FW per plant compared to the EC-2 treatment. By contrast, a significant increase was observed in the case root/shoot ratio and leaf DMC for the same treatments (EC-6 and EC-10), while the number of nonmarketable leaves and the SPAD index values were not affected by the tested treatments (Table 2). According to the literature, a similar reduction in the number of leaves per plant due to high salinity was also observed in *Brassica* species [57], as was the case for *T officinalis* in our study. In addition, the reduction in leaf FW per plant in *T. officinale* with increasing salinity is in accordance with the findings of Wang and Nil [58] and El-Hendawy et al. [59], who reported that salinity mainly affects the leaf surface expansion, thus limiting leaf area and negatively influencing the development of the photosynthetically active surface area.

Table 1. Growth parameters of *R. picroides* plants grown under different nutrient solution EC (2, 6 and 10 mS cm^{-1}).

EC (mS cm^{-1})	Leaf Number Plant^{-1}	Rosette Diameter (cm)	Nonmarketable Leaf Number Plant^{-1}	SPAD Index	Maximum Leaf Length (cm)	Maximum Leaf Width (cm)
2.0	32.28 b *	35.56 b	2.18 a	49.94 a	18.41 b	2.79 b
6.0	40.65 c	28.60 b	2.60 a	54.61 a	14.63 b	2.48 b
10.0	24.75 a	17.32 a	1.93 a	52.11 a	8.97 a	1.86 a

EC (mS cm^{-1})	Total Plant FW (kg m^{-2})	Upper Plant Part FW (kg m^{-2})	Root FW (kg m^{-2})	Root/Shoot Ratio	Marketable Leaves FW (kg m^{-2})	Leaf DMC (%)
2.0	1.40 b	1.12 b	0.27 a	0.25 a	0.99 b	9.23 b
6.0	1.68 b	1.30 b	0.37 a	0.29 a	1.13 b	7.89 a
10.0	0.96 a	0.72 a	0.23 a	0.32 a	0.62 a	8.63 ab

* Means within the same column followed by the same letter do not differ significantly based on the least significant difference (LSD) at $p < 0.05$.

Table 2. Growth parameters of *T. officinale* plants grown under different nutrient solution EC (2, 6 and 10 mS cm^{-1}).

EC (mS cm^{-1})	Leaf Number Plant^{-1}	Rosette Diameter (cm)	Nonmarketable Leaf Number Plant^{-1}	SPAD Index	Maximum Leaf Length (cm)	Maximum Leaf Width (cm)
2.0	31.00 b *	37.60 b	1.80 a	42.40 a	19.15 b	6.28 b
6.0	13.96 a	18.73 a	1.78 a	42.59 a	9.62 a	3.41 a
10.0	16.09 a	19.98 a	1.62 a	39.34 a	10.41 a	3.81 a

EC (mS cm^{-1})	Total Plant FW (kg m^{-2})	Upper Plant Part FW (kg m^{-2})	Root FW (kg m^{-2})	Root/Shoot Ratio	Marketable Leaves FW (kg m^{-2})	Leaf DMC (%)
2.0	1.37 b	1.01 b	0.35 b	0.35 a	0.90 b	11.59 a
6.0	0.35 a	0.23 a	0.12 a	0.52 b	0.20 a	15.16 b
10.0	0.44 a	0.29 a	0.15 a	0.53 b	0.21 a	15.14 b

* Means within the same column followed by the same letter do not differ significantly based on the least significant difference (LSD) test at $p < 0.05$.

Regarding the other growth parameters, the response of *R. picroides* and *T. officinale* to the presence of NaCl in the nutrient solution also varied. In particular, EC-6 treatment led to a 79% reduction in the upper plant FW and to a 67% reduction in root FW in *T. officinale* plants, whereas in *R. picroides* EC-6 treatment increased both the upper plant part and the root FW (Tables 1 and 2). It has been reported that salt stress may lead to a considerable decrease in the FW of leaves, upper plant part and roots of various plants resulting in stunted growth habit [60–65]. Plant growth restriction occurred because of the accumulation of specific ions that affect plant metabolism and physiology or/and due to the adverse water relations which have an impact on water and nutrients uptake [36,66]. However, there is considerable variation in salinity tolerance among plant species that may belong to the same family or even to the same genus. In the Asteraceae family, in which several important leafy crops and numerous wild edible herbs (including the tested species) belong, there are also included moderately salt-sensitive species (e.g., lettuce-*Lactuca sativa* [67]), moderately to highly resistant ones (e.g., wild chicory-*Cichorium intybus* [68] and spiny chicory-*Cichorium spinosum* [25]) and halophytes (e.g., sea fennel-*Crithmum maritimum* [41,69]). Differences in salt tolerance based on the plant growth restriction have been also reported between genotypes of the same plant species [70]. In addition, the higher root/shoot ratio in salt-treated plants of *T. officinale* may indicate its sensitivity to saline conditions. Pérez-Alfocea et al. [71] suggested a greater proportion of assimilates for the root compared to assimilates for the shoot in salt-treated tomato plants, leading to a greater reduction in the growth of the aboveground plant part compared to the roots. Moreover, stress hormones are involved in the plant defense mechanism and could be involved in mediating salinity stress signals and in controlling the balance between growth and stress responses [44,72,73]. Moreover, the increased DMC (by up to 31%) in combination with the significant decrease in *T. officinale* growth could be attributed to the high concentrations of NaCl in the nutrient solution resulting in hyperosmotic conditions which hinder water and nutrients uptake [66,74]. On the contrary, the values for *R. picroides* growth parameters such as the root/shoot ratio and the upper plant weight indicate moderate tolerance of the species which retains its ability to uptake water and nutrients from the nutrient solution up to EC-6.

Available data on the salinity tolerance of the tested species are scarce in the literature. However, in agreement with our results, high salt content in soil (>0.7%) significantly reduced plant growth in *Taraxacum erythropodium*, whereas at salt content below 0.7% the declining trend weakened [45]. On the other hand, there is evidence that *R. picroides* is resistant to high salinity levels, as its natural habitats include saline sand dunes in the coastal areas of the Mediterranean [75], while other reports indicate the effectiveness of the species to withstand saline irrigation water of 8 dS m^{-1} without significantly compromising plant growth [46].

3.2. Leaf and Root Minerals Concentrations

3.2.1. Total Leaf Nitrogen Content

In *R. picroides*, total leaf N was higher in the EC-10 treatment than in EC-6 and EC-2 treatments, whereas in *T. officinale*, the lowest N content was observed in EC-10 treatment (Table 3). Moreover, leaf N content did not significantly differ between EC-2 and EC-6 treatments in both plant species. Our results for *T. officinale* are in accordance with those of Pessarakli and Tucker [76] who reported that nitrogen concentration in tomato leaves was not significantly affected at relatively low salt concentrations, but at 140 and 200 mM NaCl, it was reduced by approximately 33% compared to plants grown under nonsaline conditions. Similarly, Camalle et al. [77] suggested that high salinity may lead to nitrogen deficiency since Na and Cl exhibit antagonistic effects to nitrate uptake. On the other hand, the fact that the total leaf N in *R. picroides* plants for the EC-10 was the highest could be attributed to the efficient defense mechanism that allowed plants to retain root functionality and nutrients uptake, as well as to the variable effects of salinity on the activities of N metabolizing enzymes which may depend on the species and numerous soil/nutrient solution parameters [78].

Table 3. Leaf nutrient concentrations of *R. picroides* and *T. officinale* plants grown under different nutrient solution EC (2, 6 and 10 mS cm^{-1}).

Leaf	N	P	K	Ca	Mg	Na	Cl	Fe	Mn	Zn	Cu	B	K/Na	Ca/Na
				% leaf DW				mg kg^{-1} leaf DW						
EC (mS cm^{-1})						*Reichardia picroides*								
2.0	4.25 a *	0.77 a	7.04 c	0.90 b	0.20 b	0.55 a	0.82 a	60.0 a	42.9 a	37.5 a	4.8 a	143.2 b	12.7 b	1.63 b
6.0	4.37 a	1.04 b	4.49 b	0.72 a	0.16 a	3.68 b	1.09 a	57.2 a	46.2 a	65.7 b	3.5 a	138.2 ab	1.23 a	0.19 a
10.0	5.12 b	1.29 c	3.66 a	0.63 a	0.16 a	4.02 c	1.79 b	55.6 a	71.6 b	75.4 c	4.5 a	129.1 a	0.91 a	0.16 a
						Taraxacum officinale								
2.0	4.77 b	1.16 a	5.48 b	0.92 b	0.33 c	0.05 a	0.53 a	77.2 ab	30.7 a	39.1 a	6.8 a	46.8 a	137.4 b	23.18 b
6.0	4.81 b	1.37 b	3.02 a	0.75 a	0.26 a	2.00 b	0.99 b	68.5 a	23.7 a	45.8 ab	6.1 a	35.0 a	1.51 a	0.38 a
10.0	4.17 a	1.69 c	2.99 a	0.83 ab	0.29 b	3.84 c	1.58 c	90.0 b	31.9 a	50.3 b	5.7 a	38.8 a	0.78 a	0.22 a

* Means within the same column followed by the same letter do not differ significantly based on the least significant difference (LSD) test at $p < 0.05$.

3.2.2. Leaf and Root Phosphorus Content

Leaf and root P content of *R. picroides* and leaf P content of *T. officinale* increased with increasing NaCl concentration in the nutrient solution (Tables 3 and 4), whereas root P content of *T. officinale* was the lowest in treatment EC-6 (Table 4). Based on the recommended dietary allowances (RDA) for P (700 mg per day for adults) [79], the consumption of 100 g FW of *R. picroides* leaves can cover up to 36.5% of RDI when plants are grown under highest salinity (10 mS cm^{-1}), whereas 100 g of fresh *T. officinale* leaves can cover only 15.9% of RDI (plants grown at EC-10 treatment). In contrast to the findings of the present work, salinity decreased the concentration of P in tomato plant tissues [80], whereas other studies indicated that salinity either increased or had no effect on P uptake [81]. Moreover, plant tissue is also important, since according to Villora et al. [82], increased salinity resulted to P accumulation in zucchini leaves, while in fruit P accumulation differed among the different parts (pulp, skin and whole fruit). Finally, P availability is also a key factor, and Tang et al. [83] suggested that salinity affected differently maize plants depending on the available P amount, while P deficiency improved tolerance to salinity through the selective absorption of K and Na.

Table 4. Root nutrient concentrations at harvest date of *R. picroides* and *T. officinale* plants grown under different nutrient solution EC (2, 6 and 10 mS cm^{-1}).

Root	P	K	Ca	Mg	Na	Fe	Mn	Zn	Cu	B
		% leaf DW				mg kg^{-1} leaf DW				
EC (dS m^{-1})				*Reichardia picroides*						
2.0	1.04 a *	5.98 b	0.37 a	0.17 b	0.19 a	221.8 a	60.3 a	84.2 b	15.1 a	14.1 a
6.0	1.46 b	5.35 a	0.33 a	0.15 ab	1.02 b	385.9 b	46.2 a	98.8 c	16.7 a	15.4 a
10.0	1.67 c	5.00 a	0.41 a	0.14 a	1.22 b	257.8 a	55.5 a	62.8 a	14.9 a	17.4 a
				Taraxacum officinale						
2.0	1.47 b	5.27 b	0.46 a	0.18 a	0.11 a	162.1 a	16.3 a	50.8 a	15.5 b	16.7 a
6.0	1.18 a	4.01 a	0.43 a	0.17 a	0.71 b	158.5 a	16.7 a	44.6 a	10.7 a	21.7 b
10.0	1.28 ab	3.62 a	0.51 a	0.17 a	1.20 c	157.3 a	20.0 a	44.6 a	9.4 a	15.6 a

* Means within the same column followed by the same letter do not differ significantly based on the least significant difference (LSD) test at $p < 0.05$.

3.2.3. Leaf and Root Potassium Content

In *R. picroides* plants, leaf K content decreased gradually with increasing salinity (Table 3), whereas K content in roots of the same species as well as K content in both tissues (leaves and roots) of *T. officinale* were significantly decreased at both salinity treatments compared to the control (Tables 3 and 4). Moreover, the reduction of K concentration in the leaves of both species was higher (36–45% in EC-6 and 46–48% in EC-10 treatments, compared to the control) in relation to that in the roots (10–24% in EC-6 and 16–31% in EC-10 treatments, compared to the control). This finding is in accordance with that of Pérez-Alfocea et al. [71], who reported decreased K levels in all tissues of tomato plants grown under salt stress, although the lowest relative reduction in K concentrations was found in the roots. Our results also indicate competition effects between Na$^+$ and K$^+$ ions which most likely share the same transport system at the root surface [84], an effect of Na$^+$ on the K$^+$ transport into the xylem, or indirect inhibition of the uptake process, i.e., through the H+-ATPase activity [85]. The role of K homeostasis in salt-tolerance mechanisms of salinized plants is highly recognized [86], due to the fact that high NaCl uptake competes with the uptake of other nutrient ions, especially K, resulting in growth and yield reduction of various crops [87–90]. Contrary to our results, Semiz et al. [91] did not find any effect of salinity on K and Mg concentrations in pepper leaves, whereas Assimakopoulou et al. [80] have reported increased leaf K, Ca and Mg concentrations in salt-treated tomato plants. According to Shahid et al. [92], the salinity-tolerance mechanisms are still highly controversial and are influenced by growth conditions, growth medium (soil or soilless culture), stress duration and plant genotype, among others. Regarding the nutritional value of edible leaves, the RDA values for potassium are 3400 and 2600 mg per day for adult males and females, respectively [93]. Considering the suggested values, the consumption of 100 g of fresh from either *R. picroides* or *T. officinale* plants grown at EC-2 treatment were the most nutritious, since they could cover up to approximately 19.0% and 25.0% of RDI for male and female adults, respectively.

3.2.4. Leaf and Root Calcium and Magnesium Content

In both species, leaf Ca and Mg contents were the highest in the EC-2 treatment (Table 3). On the other hand, no significant effects of salinity treatments were observed for root Ca content in both species (Table 4). According to Yu et al. [44], this finding could be due to plant adaptation to salinity stress through a flexible system of hormone regulation and/or through signaling via glycosyl inositol phosphorylceramide (GIPC) sphingolipids in the plasma membrane, which allow the sensing of Na$^+$ in the apoplastic space and increasing of Ca^{2+} influx channels in plants. The exploration of potential Na+ receptors has undoubtedly provided new opportunities to the understanding of salt-stress perception by plants [94]. Similarly to our study, Bolarin et al. [95] reported slight changes in root Ca and K of tomato plants grown under saline conditions. The retention of root Ca content under saline conditions at levels similar to control could also induce the retention of K, since the presence of Ca seems to be necessary for K-Na selectivity and for the retention of the required K content in plant cells [62,96]. Saline conditions reduced the nutritional value of edible leaves in the case of *T. officinale*, whereas increasing salinity increased Ca and Mg content in fresh leaves of *R. picroides*. According to the literature, the RDI values for Ca 1000 mg per day for male adults and between 1000 and 1200 mg per day for female adults [97]. Therefore, the consumption of 100 g of fresh edible leaves of *R. picroides* plants grown under high salinity can cover 12% and 10.0% of RDA values for male and female adults. In the case of *T. officinale*, leaves collected from plants grown under the EC-2 treatment can provide only 8% and 7% of daily requirements of male and female adults, respectively. Regarding the Mg, the allowance intake (AI) refers to 400–420 mg per day for male adults and to 310–320 mg per day for female adults [79]. Considering the results of our study, 100 g of fresh *R. picroides* leaves (grown at EC-10 treatment) can cover 10.4% and 13.7% of male and female adults, respectively, whereas *T. officinale* leaves (grown at EC-2 treatment) can cover only 4.3% and 5.8% of AI of male and female adults, respectively.

3.2.5. Leaf and Root Sodium and Leaf Chlorine Content

Na and Cl content increased with increasing salinity in both species, although no significant differences were observed between EC-2 and EC-6 treatments in the case of leaf Cl content of *R. picroides*, as well as between EC-6 and EC-10 treatments in root Cl content of the same species (Tables 3 and 4). Moreover, *R. picroides* plants under both salinity levels presented increased leaf Na^+ concentration by 6.6 and 7.3 times compared to the control; however, the relevant leaf Na^+ increase in *T. officinale* was much higher (by 41 and 79 times, respectively) (Table 3). Taking into consideration that the leaf Na content in *R. picroides* was equal to 3.7 g kg^{-1} DW and the growth was unaffected in EC-6 treatment, whereas in *T. officinale*, the relevant leaf Na content was 2.0 g kg^{-1} DW with severe effects on plant growth, and it could be suggested that *T. officinale* cannot tolerate Na accumulation in the leaves. According to the literature, the regulation of root-to-leaf Na and Cl transport is important for increased tolerance under saline conditions since it may affect photosynthetic activity [98], while the presence of ion specific transporters in cell membranes is pivotal in plant defense system against salinity [99].

Regarding the nutritional parameters of edible leaves, high intake of Na and Cl is associated with high blood pressure; therefore, AI values of 1500 mg per day and 3100 mg per day for male and female adults have been set for Na and Cl, respectively [93,100]. Based on that, high salinity (EC-10) results to final products that may significantly contribute to the overall daily intake of Na; therefore, excessive consumption should be avoided. In particular, the consumption of 100 g of fresh leaves accounts to 23.1% of Na and 5.0% of Cl of AI values in the case of *T. officinale* and 38.7% of Na and 7.7% Cl of AI in the case of *R. picroides*. This indicates that consumption of high amounts of the latter species should be avoided when plants are grown under high salinity.

3.2.6. Leaf K/Na and Ca/Na Ratios

The leaf K/Na and Ca/Na ratios of *R. picroides* and *T. officinale* decreased significantly in EC-6 and EC-10 treatments compared to control without significant differences with each other (Table 3). The K/Na ratio, which is widely used as a salinity tolerance predictor in many plant species, was found to be 10 times lower in EC-6 treatment and 14 times lower in EC-10 treatment compared to control in *R. picroides*, while in the case of *T. officinale*, the reduction was even higher (by 91 and 176 times in EC-6 EC-10 treatments, respectively). Similarly, a significant reduction was also observed in leaf Ca/Na ratio, namely 8 times lower in EC-6 and 10 times lower in EC-10 treatment compared to the control in the case of *R. picroides* and 62 times lower in EC-6 and 108 times in EC-10 treatment compared to the control in the case of *T. officinale*. These differences in K/Na and Ca/Na in the EC-2 treatment between the species are due to the very low content of Na in *T. officinale*, which consequently results in considerably higher K/Na and Ca/Na values. These findings indicate the differences in salt tolerance between the studied species, since the high K^+/Na^+ ratio in cytosol is associated with high salinity tolerance [101], which was the case for *R. picroides* in our study. Moreover, the present results are in agreement with previous reports [3,70], while Pérez-Alfocea et al. [71] also suggested that high values of K/Na and Ca/Na ratios indicate an equilibrium of nutrients more similar to the nonsalinized plants.

3.2.7. Leaf and Root Micronutrients Content

In the case of *R. picroides*, leaf Fe and Cu as well as root Mn, Cu and B contents were not significantly affected by salinity treatments (Tables 3 and 4). On the other hand, the highest content of leaf Mn, Zn and B was the highest for the highest salinity level (EC-10), while Fe and Zn content in roots was the highest at moderate salinity levels (EC-6) (Table 4). Regarding *T. officinale* plants, Mn, Cu and B content in leaves as well as leaf Fe, Mn and Zn content were not affected by salinity treatments, while increasing trends with increasing salinity were observed in the case of leaf Fe and Zn content (Tables 3 and 4). Finally, Cu content in roots was significantly reduced under saline conditions compared to the control treatment, whereas the highest content of B was observed in the EC-6 treatment (Table 4). According to Bingham et al. [102], no significant effects of salinity on B content were observed in wheat plants, while Hasana et al. [103] reported a varied response for different micronutrients content under salinity stress in maize plants. Contrary to our results, a reduced Mn uptake with salinity had been reported in corn [104,105], while Shibli et al. [106] mentioned that leaf Fe, B, Zn, Mn and Cu content decreased with elevated salinity. Regarding the Fe uptake in plants under salinity, inconsistent results are reported as salinity increased or decreased leaf Fe content in red lettuce [107] and tomato plants [107], respectively.

Comparing leaf Zn content between the two plant species under the same salinity level, leaf Zn concentration in *R. picroides* under EC-6 and EC-10 was increased by 75% and 101%, respectively, whereas in *T. officinale* the increase was only 17% and 29% compared to the control (Table 3). According to Rahman et al. [104] salinity increased Zn content in corn shoots, while the increase in zinc content via foliar spraying alleviated salinity stress effects in pak choi plants through the decrease in oxidative damage [108]. The importance of increased Zn concentration on plants' adaptation to salinity stress could also be related to the role of zinc in auxin biosynthesis, as phytohormones under salinity stress play a crucial role in modulating plant physiological responses [73]. In addition, Zn is required for scavenging of reactive oxygen species (ROS) that are produced under salinity stress [92,109].

Regarding the intake of the tested micronutrients on a daily basis, different thresholds have been set. In particular, RDI values of 8 mg per day (male and female adults) have been suggested for Fe, 8–11 mg per day for Zn (female and male adults, respectively), and 900 mg per day for Cu (male and female adults) [97]. In the case of Mn, an AI value of 2.3 mg per day (male and female adults) has been suggested, while for B the tolerable upper intake level (UL) has been set to 20 mg per day (male and female adults) [97]. The results of our study show that the consumption of 100 g of fresh leaves of both species does not significantly contribute to the overall daily intake for most of the micronutrients, especially in the case of B, Zn and Cu. However, Fe intake accounts for 13.0% in the case of *R. picroides* (EC-10 treament) and for 6.9% of RDI in the case of *T. officinale* (EC-2 treatment), while the intake of Mn is even higher (16.0% and 28.7% of AI values for *R. picroides* and *T. officinale* plants grown at EC-10 treatment, respectively).

3.2.8. Proline Content

Regardless of salt treatment, *R. picroides* and *T. officinale* plants accumulated higher amounts of proline compared to control, although no significant differences were observed between the control and the EC-6 treatment (Figure 1). Given that proline is an important osmolyte for the osmotic adjustment under salinity stress conditions [110], the salt tolerance of *R. picroides* based on the recorded growth results could be attributed to the induction of proline biosynthesis, especially at the highest salinity level (EC-10), where an increase by 127.7% was observed. Moreover, *R. picroides* contained higher amounts of proline compared to *T. officinale* in all the studied treatments, which also indicates the better adaptation of the species under saline conditions compared to *T. officinale*. On the other hand, the increase in proline content in *T. officinale* was 1089.5 times higher than the control treatment, a finding which probably indicates that accumulation of proline indicates a symptom of salt injury instead of salt tolerance of the species [111,112]. Moreover, the increased biosynthetic rate of proline in *T. officinale* highlights the high energy expenditure for the alleviation of salt stress and the deficit of energy for biomass production, as indicated by the limited growth of plants under saline conditions [112,113]. Moreover, the increased leaf K concentration (7.04% DW) determined in *R. picroides* compared to the leaf K (5.48% DW) of *T. officinale* (Table 3) could be related to the higher proline biosynthesis in *R. picroides*, since Chou et al. [114] mentioned that proline accumulation depends on the availability of potassium. As mentioned before, in both species, proline content in EC-6 treatment did not differ from the control; however, the increase in *T. officinale* was more profound than that in *R. picroides*, a finding which is in agreement with the results of Wu et al. [45], who reported that *Taraxacum erythropodium* plants exhibited a rapid response to the increased salinity by increasing leaf proline content. Similar results were reported by Slabbert and Krüger [115] for *Amaranthus* sp., Xue et al. [113] for *Brassica napus* and Saleh [116] for *Vigna radiata* plants.

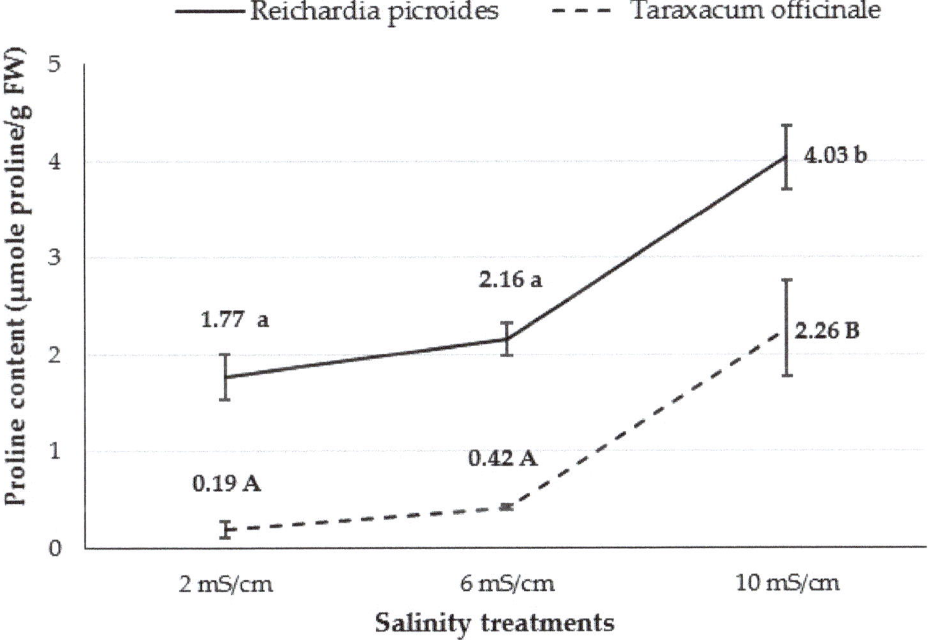

Figure 1. Leaf proline content (μmole proline/g FW) of *R. picroides* and *T. officinale* plants grown under different nutrient solution electrical conductivities (2, 6 and 10 mS cm^{-1}). Different lowercase and capital Latin letters indicate significant differences between salinity treatments for *Recihardia picroides* and *Taraxacum officinale* plants, respectively, according to the least significant difference (LSD) test ($p < 0.05$).

3.3. Chemical Composition

The studied EC treatments had no effect on chlorophylls (a, b and total), carotenoids+ xanthophylls of leaves in both species (Table 5). In general, the effect of salinity on the content and function of photosynthetic pigments in green vegetables is related to the salt tolerance of the species, the severity of the stress, the plant growth stage, the duration and the method of stress application [117]. In agreement to our results, the photosynthetic pigments content in *Cichorium spinosum* leaves was not affected by 20 and 40 mM NaCl [118], whereas total chlorophyll content of *Taraxacum erythropodium* decreased when soil salt content exceeded 0.4% [45]. The fact that the chlorophyll content was not affected even when both plants were grown under high salinity (EC-10) indicates that despite the reduced plant growth, the harvested leaves retained their greenness and thus their visual quality [119,120]. Total soluble solids content (TSSC) of leaves was also not affected by salinity in both species, whereas titratable acidity (TA) increased with increasing salinity, particularly in EC-10 treatment (Table 5). It is well established that mild salt stress (salinity eustress) may favor both TSSC and TA in fruit vegetables, apart from other flavor and taste characteristics [121]. However, in leafy greens, salinity has been proven either beneficial (e.g., in lettuce, in combination with elevated CO_2 [122]) or detrimental (e.g., spiny chicory [25]) in regards to sugar accumulation. On the other hand, the observed elevated titratable acidity could be partly attributed to the enhanced biosynthesis of organic acids in plants grown under EC-10 in order to counteract the excess of cations in relation to anions [123] or to concentration effects due to the increasing DMC with increasing salinity (see Table 1).

Table 5. Chlorophyll (a, b and total), carotenoids+xanthophylls, total phenolics, total soluble solids content (TSSC), titratable acidity (TA) and leaf nitrate content of *Reichardia picroides* and *Taraxacum officinale* grown under different nutrient solution EC (2, 6 and 10 mS cm^{-1}).

Nutrient Solution EC	Chlorophyll a	Chlorophyll b	Total Chlorophyll	Carotenoids+ Xanthophylls	Total Phenolics	TSSC	TA	Nitrate Content
EC (dS m^{-1})	mg/100 g leaf FW				mg GAE/100 g leaf FW	^{0} Brix	g malic acid/100 g leaf FW	mg/kg leaf FW
				Reichardia picroides				
2	41.0 a *	18.7 a	59.6 a	7.08 a	113.6 b	5.00 a	0.16 a	5509.8 ab
6	41.6 a	19.0 a	60.6 a	7.21 a	100.9 b	4.58 a	0.17 a	6877.6 b
10	41.4 a	18.3 a	59.7 a	7.01 a	62.2 a	4.73 a	0.22 b	4586.9 a
				Taraxacum officinale				
2	77.7 a	38.2 a	115.9 a	12.3 a	79.6 a	5.50 a	0.16 a	2410.5 b
6	82.4 a	39.6 a	122.0 a	13.2 a	91.6 a	6.10 a	0.20 ab	885.7 a
10	87.0 a	43.0 a	130.0 a	13.5 a	95.8 a	5.85 a	0.22 b	597.7 a

* Means within the same column followed by the same letter do not differ significantly based on the least significant difference (LSD) test at $p < 0.05$.

Regarding total phenolics content (TPC), contrasting effects of salinity treatments were observed in the studied species. In particular, TPC was not significantly affected in the case of *T. officinale*, whereas a significant decrease (by 45.2%) was observed for the EC-10 treatment in *R. picroides* plants. According to the literature, abiotic stress factors such as salinity may induce the biosynthesis of phenolic compounds content in various species, including wild and commercially cultivated leafy vegetables [124–126]. However, this is not always the case, and decreasing trends of TPC content have also been recorded under saline conditions, as for example in romaine lettuce following a long-term mild salt stress (5 mM NaCl) [127], or in green and red baby lettuce under 10 mM NaCl [107]. Moreover, salinity was not beneficial for biosynthesis of phenolic compounds, as shown by Kim et al. [127], in green and red baby lettuce under 10 mM NaCl, or in spiny chicory under 8 dS m^{-1} [25].

The leaf nitrate content in *R. picroides* plants was the highest for the EC-6 treatment without being significantly different from the control treatment, while the lowest content was recorded for the EC-10 treatment (Table 5). On the other hand, nitrate content was significantly reduced when plants were subjected to EC-6 or EC-10 treatments. The observed trends for nitrates content follow the pattern of organic acids content, since according to Gent [128], organic acids and nitrates content are inversely related. Moreover, according to Bonasia et al. [129] and Cantabella et al. [130], the decrease in nitrates content could be related to the competitive effects of Cl$^-$ on NO$_3^-$, which may inhibit nitrates accumulation in plant tissues under saline conditions. As nitrates are considered an important antinutrient factor in leafy vegetables [131], the reduction of leaf nitrates content under the EC-10 in *R. picroides* or EC-6 in *T. officinale* is of high importance for the commercial cultivation of these species in saline areas, due to the compensation of yield loss by the production of a safer produce. Moreover, the decrease in nitrates was more profound in the case of *T. officinale* under the EC-6 and EC-10 treatments compared to control (reduced by 63% and 75%, respectively), while in *R. picroides* the decrease for the EC-10 treatment was 17% (Table 5). These findings are in accordance with those reported by Kafkafi et al. [132], who suggested that in tomato and melon plants, salt-tolerant genotypes exhibited higher nitrate influx rates than the more sensitive ones. Moreover, the dramatic decrease in plant growth in *T. officinale* under saline treatments could be related to the severe limitations in water uptake as that species showed significant growth decrease in combination with significant increased DMC. On the contrary, plant growth of *R. picroides* was unaffected and was followed by lower DMC under the EC-6 treatment (Tables 1 and 2). These findings are in accordance with those of Abdelgadir et al. [133], who suggested that the inhibition of nitrates absorption by tomato plants was more strongly related to the reduced water uptake than to Cl$^-$ antagonism from salt stress. For the same reason, the uptake of nitrates by *R. picroides* was not hindered by salinity up to EC-6 as the water uptake and growth of the species was found to be unaffected despite the high presence of NaCl in nutrient solution.

3.4. Correlations

Most of plant growth parameters of *T. officinale* (i.e., the total plant FW, the upper plant part FW, the root FW, the marketable leaf FW, the leaf number per plant, the rosette diameter, the maximum leaf length and the maximum leaf width) were found to be significantly positively correlated with the leaf nitrate concentration, root P, leaf and root K, leaf Ca, leaf Mg, leaf K/Na and Ca/Na ratios, root Zn and root Cu but significantly negatively correlated with leaf and root Na concentrations (see Supplementary Material Table S1). In particular, the correlation coefficients between the rosette diameter, maximum leaf length and maximum leaf width of *T. officinale* with leaf Na concentration were r = −0.81, r = −0.80 and r = −0.78, respectively, whereas the relevant correlation coefficients with leaf K were r = 0.99, r = 0.99 and r = 0.98. By contrast, the determined growth parameters of *R. picroides* were not significantly correlated with the majority of minerals content. Significant correlations were detected mainly between the rosette diameter (and the maximum leaf length) of the species with the leaf N (r = −0.79), P (r = −0.85), K (r = 0.70), Na (r = −0.68), Mn (r = −0.74), Zn (r = −0.77) and B (r = 0.78), root K (r = 0.64) and root Na (r = −0.79), as well as with proline (r = −0.88) and malic acid contents (r = −0.81). Significant correlations between plant growth parameters and ion contents under salinity have been indicated by several researchers [71,110,134], while Bosiacka et al. [47] reported that the strongest correlations were found between soil salinity and the leaf Na, Mn, Ca, Fe, K and Zn content of three *Taraxacum* microspecies.

4. Conclusions

The wild edible greens *R. picroides* and *T. officinale* tested in the present study responded differently to salinity treatments indicating different tolerance mechanisms. In particular, plant growth of *R. picroides* was negatively affected only when grown under nutrient solution with EC values equal to 10 mS cm^{-1} (EC-10 treatment), whereas *T. officinale* was more sensitive and plant growth rapidly decreased when EC increased at 6 mS cm^{-1}. The leaf and root Na and Cl concentration changes under salinity could partially explain the aforementioned salt-tolerance differentiation between the two species as the more salt-tolerant *R. picroides* accumulated more Cl and Na in the leaves as compared to the sensitive *T. officinale*. Therefore, the higher salt tolerance of *R. picroides* could be due to its ability to develop a better adaptation mechanism of water uptake, to effectively accumulate osmolytes such as proline and to keep high shoot K probably through a more efficient K/Na selectivity, in combination with an increased Zn uptake ability under salinity stress. Moreover, the studied species differed in the contribution of secondary metabolites such as phenolic compounds to the overall antioxidant mechanism, since it seems that in *R. picroides*, phenolic compounds have an important role in plant defense against abiotic stressors, whereas in *T. officinale*, no such effect was observed. In conclusion, the response of *R. picroides* to moderate and high salinity (EC-6 and EC-10) is of great importance for its commercial exploitation under saline soils or in regions where irrigation water is of low quality. However, the ability of the species to adapt to saline conditions that are unsuitable for most leafy greens as well as the relevant adaptation mechanisms should be further studied.

Supplementary Materials: The following are available online at https://www.mdpi.com/article/10.3390/horticulturae7070160/s1, Table S1: Correlation coefficients of plant growth parameters and chemical composition of *Reichardia picroides* and *Taraxacum officinale*.

Author Contributions: Conceptualization, A.A.A.; methodology, A.A.A., A.A., P.P., M.B., N.V. and I.C.K.; software, A.A., P.P., M.B. and N.V.; validation, I.C.K. and A.A.; formal analysis, A.A.A. and I.C.K.; investigation, A.A.A., A.A., P.P., M.B. and N.V.; resources, A.A.A.; data curation, A.A., S.A.P. and I.C.K.; writing—original draft preparation, A.A.A., A.A., S.A.P. and I.C.K.; writing—review and editing, A.A.A., S.A.P. and I.C.K.; visualization, A.A.A.; supervision, A.A.A.; project administration, A.A.A., S.A.P. and I.C.K. All authors have read and agreed to the published version of the manuscript.

Funding: This research received no external funding.

Institutional Review Board Statement: Not applicable.

Informed Consent Statement: Not applicable.

Data Availability Statement: The study did not report any data.

Acknowledgments: The authors thank George Georgiopoulos for providing seeds and Anastasios Kotsiras for his technical advice for nutrient solution preparation.

Conflicts of Interest: The authors declare no conflict of interest.

References

1. Raza, A.; Razzaq, A.; Mehmood, S.S.; Zou, X.; Zhang, X.; Lv, Y.; Xu, J. Impact of climate change on crops adaptation and strategies to tackle its outcome: A review. *Plants* **2019**, *8*, 34. [CrossRef]
2. Dong, J.; Gruda, N.; Li, X.; Tang, Y.; Zhang, P.; Duan, Z. Sustainable vegetable production under changing climate: The impact of elevated CO_2 on yield of vegetables and the interactions with environments—A review. *J. Clean. Prod.* **2020**, *253*, 119920. [CrossRef]
3. Paranychianakis, N.V.; Chartzoulakis, K.S. Irrigation of Mediterranean crops with saline water: From physiology to management practices. *Agric. Ecosyst. Environ.* **2005**, *106*, 171–187. [CrossRef]
4. Mahajan, S.; Tuteja, N. Cold, salinity and drought stresses: An overview. *Arch. Biochem. Biophys.* **2005**, *444*, 139–158. [CrossRef]
5. Slama, I.; Abdelly, C.; Bouchereau, A.; Flowers, T.; Savouré, A. Diversity, distribution and roles of osmoprotective compounds accumulated in halophytes under abiotic stress. *Ann. Bot.* **2015**, *115*, 433–447. [CrossRef] [PubMed]
6. Iriti, M.; Vitalini, S. Sustainable crop protection, global climate change, food security and safety—Plant immunity at the crossroads. *Vaccines* **2020**, *8*, 42. [CrossRef]
7. Petropoulos, S.A. Practical applications of plant biostimulants in greenhouse vegetable crop production. *Agronomy* **2020**, *10*, 1569. [CrossRef]
8. Shahrajabian, M.H.; Chaski, C.; Polyzos, N.; Petropoulos, S.A. Biostimulants Application: A Low Input Cropping Management Tool for Sustainable Farming of Vegetables. *Biomolecules* **2021**, *11*, 698. [CrossRef]
9. Voutsela, S.; Yarsi, G.; Petropoulos, S.A.; Khan, E.M. The effect of grafting of five different rootstocks on plant growth and yield of tomato plants cultivated outdoors and indoors under salinity stress. *Afr. J. Agric. Res.* **2012**, *7*, 5553–5557. [CrossRef]
10. Petropoulos, S.A.; Khah, E.M.; Passam, H.C. Evaluation of rootstocks for watermelon grafting with reference to plant development, yield and fruit quality. *Int. J. Plant Prod.* **2012**, *6*, 481–492.
11. Petropoulos, S.A.; Karkanis, A.; Martins, N.; Ferreira, I.C.F.R. Edible halophytes of the Mediterranean basin: Potential candidates for novel food products. *Trends Food Sci. Technol.* **2018**, *74*, 69–84. [CrossRef]
12. Shahrajabian, M.H.; Chaski, C.; Polyzos, N.; Tzortzakis, N.; Petropoulos, S.A. Sustainable Agriculture Systems in Vegetable Production Using Chitin and Chitosan as Plant Biostimulants. *Biomolecules* **2021**, *11*, 819. [CrossRef]

13. FAO. The State of the World's Biodiversity for Food and Agriculture. In *The State of the World's Biodiversity for Food and Agriculture*; Pilling, D., Bélanger, J., Eds.; FAO Commission on Genetic Resources for Food and Agriculture Assessments: Rome, Italy, 2019; p. 572, ISBN 9789251312704.
14. Corrêa, R.C.G.; Gioia, F.D.; Ferreira, I.C.F.R.; Petropoulos, S.A. Wild greens used in the Mediterranean diet. In *The Mediterranean Diet: An Evidence-based Approach*; Preedy, V., Watson, R., Eds.; Academic Press: London, UK, 2020; pp. 209–228, ISBN 9788578110796.
15. Chatzopoulou, E.; Carocho, M.; Di Gioia, F.; Petropoulos, S.A. The beneficial health effects of vegetables and wild edible greens: The case of the mediterranean diet and its sustainability. *Appl. Sci.* **2020**, *10*, 9144. [CrossRef]
16. Capurso, C.; Vendemiale, G. The Mediterranean Diet Reduces the Risk and Mortality of the Prostate Cancer: A Narrative Review. *Front. Nutr.* **2017**, *4*, 38. [CrossRef] [PubMed]
17. Ceccanti, C.; Landi, M.; Benvenuti, S.; Pardossi, A.; Guidi, L. Mediterranean wild edible plants: Weeds or "new functional crops"? *Molecules* **2018**, *23*, 2299. [CrossRef] [PubMed]
18. Guarrera, P.M.; Savo, V. Wild food plants used in traditional vegetable mixtures in Italy. *J. Ethnopharmacol.* **2016**, *185*, 202–234. [CrossRef] [PubMed]
19. Benvenuti, S.; Maggini, R.; Pardossi, A. Agronomic, Nutraceutical, and Organoleptic Performances of Wild Herbs of Ethnobotanical Tradition. *Int. J. Veg. Sci.* **2017**, *23*, 270–281. [CrossRef]
20. Johns, T.; Powell, B.; Maundu, P.; Eyzaguirre, P.B. Agricultural biodiversity as a link between traditional food systems and contemporary development, social integrity and ecological health. *J. Sci. Food Agric.* **2013**, *93*, 3433–3442. [CrossRef]
21. Morales, P.; Ferreira, I.C.F.R.; Carvalho, A.M.; Sánchez-Mata, M.C.; Cámara, M.; Fernández-Ruiz, V.; Pardo-de-Santayana, M.; Tardío, J. Mediterranean non-cultivated vegetables as dietary sources of compounds with antioxidant and biological activity. *LWT-Food Sci. Technol.* **2014**, *55*, 389–396. [CrossRef]
22. Zimmerer, K.S.; Vanek, S.J. Toward the integrated framework analysis of linkages among agrobiodiversity, livelihood diversification, ecological systems, and sustainability amid global change. *Land* **2016**, *5*, 10. [CrossRef]
23. Conversa, G.; Elia, A. Effect of seed age, stratification, and soaking on germination of wild asparagus (*Asparagus acutifolius* L.). *Sci. Hortic.* **2009**, *119*, 241–245. [CrossRef]
24. Cros, V.; Martínez-Sánchez, J.J.; Franco, J.A. Good yields of common purslane with a high fatty acid content can be obtained in a peat-based floating system. *Horttechnology* **2007**, *17*, 14–20. [CrossRef]
25. Petropoulos, S.; Levizou, E.; Ntatsi, G.; Fernandes, Â.; Petrotos, K.; Akoumianakis, K.; Barros, L.; Ferreira, I. Salinity effect on nutritional value, chemical composition and bioactive compounds content of *Cichorium spinosum* L. *Food Chem.* **2017**, *214*, 129–136. [CrossRef]
26. Petropoulos, S.; Fernandes, Â.; Karkanis, A.; Ntatsi, G.; Barros, L.; Ferreira, I. Successive harvesting affects yield, chemical composition and antioxidant activity of *Cichorium spinosum* L. *Food Chem.* **2017**, *237*, 83–90. [CrossRef]
27. Petropoulos, S.; Fernandes, Â.; Karkanis, A.; Antoniadis, V.; Barros, L.; Ferreira, I. Nutrient solution composition and growing season affect yield and chemical composition of *Cichorium spinosum* plants. *Sci. Hortic.* **2018**, *231*, 97–107. [CrossRef]
28. Petropoulos, S.A.; Fernandes, Â.; Tzortzakis, N.; Sokovic, M.; Ciric, A.; Barros, L.; Ferreira, I.C.F.R. Bioactive compounds content and antimicrobial activities of wild edible Asteraceae species of the Mediterranean flora under commercial cultivation conditions. *Food Res. Int.* **2019**, *119*. [CrossRef] [PubMed]
29. Finimundy, T.C.; Karkanis, A.; Fernandes, Â.; Petropoulos, S.A.; Calhelha, R.; Petrović, J.; Soković, M.; Rosa, E.; Barros, L.; Ferreira, I.C.F.R. Bioactive properties of *Sanguisorba minor* L. cultivated in central Greece under different fertilization regimes. *Food Chem.* **2020**, *327*. [CrossRef] [PubMed]
30. Petropoulos, S.A.; Fernandes, Â.; Dias, M.I.; Pereira, C.; Calhelha, R.C.; Ivanov, M.; Sokovic, M.D.; Ferreira, I.C.F.R.; Barros, L. The Effect of Nitrogen Fertigation and Harvesting Time on Plant Growth and Chemical Composition of *Centaurea raphanina* subsp. *mixta* (DC.) Runemark. *Molecules* **2020**, *25*, 1–21.
31. Petropoulos, S.; Karkanis, A.; Fernandes, Â.; Barros, L.; Ferreira, I.C.F.R.; Ntatsi, G.; Petrotos, K.; Lykas, C.; Khah, E. Chemical composition and yield of six genotypes of common purslane (*Portulaca oleracea* L.): An alternative source of omega-3 fatty acids. *Plant Foods Hum. Nutr.* **2015**, *70*, 420–426. [CrossRef]
32. González, J.A.; García-Barriuso, M.; Amich, F. The consumption of wild and semi-domesticated edible plants in the Arribes del Duero (Salamanca-Zamora, Spain): An analysis of traditional knowledge. *Genet. Resour. Crop. Evol.* **2011**, *58*, 991–1006. [CrossRef]
33. Alexopoulos, A.A.; Marandos, E.; Assimakopoulou, A.; Vidalis, N.; Petropoulos, S.A.; Karapanos, I.C. Effect of Nutrient Solution pH on the Growth, Yield and Quality of *Taraxacum officinale* and *Reichardia picroides* in a Floating Hydroponic System. *Agronomy* **2021**, *11*, 1118. [CrossRef]
34. Rao, N.S.; Shivashankara, K.S.; Laxman, R.H. *Abiotic Stress Physiology of Horticultural Crops*; Rao, N.S., Shivashankara, K.S., Laxman, R.H., Eds.; Springer India: New Delhi, India, 2016; ISBN 9788132227236.
35. Hasegawa, P.M.; Bressan, R.A.; Zhu, J.-K.; Bohnert, H.J. Plant cellular and molecular responses to high salinity. *Annu. Rev. Plant Physiol. Plant Mol. Biol.* **2000**, *51*, 463–499. [CrossRef]
36. Giordano, M.; Petropoulos, S.A. Response and Defence Mechanisms of Vegetable Crops against Drought, Heat and Salinity Stress. *Agriculture* **2021**, *11*, 463. [CrossRef]
37. Giordano, M.; Petropoulos, S.A.; Cirillo, C. Biochemical, Physiological, and Molecular Aspects of Ornamental Plants Adaptation to Deficit Irrigation. *Horticulturae* **2021**, *7*, 107. [CrossRef]

38. Parihar, P.; Singh, S.; Singh, R.; Singh, V.P.; Prasad, S.M. Effect of salinity stress on plants and its tolerance strategies: A review. *Environ. Sci. Pollut. Res.* **2015**, *22*, 4056–4075. [CrossRef] [PubMed]

39. Parida, A.K.; Das, A.B. Salt tolerance and salinity effects on plants: A review. *Ecotoxicol. Environ. Saf.* **2005**, *60*, 324–349. [CrossRef] [PubMed]

40. Cheeseman, J. Food Security in the Face of Salinity, Drought, Climate Change, and Population Growth. In *Halophytes for Food Security in Dry Lands*; Khan, M.A., Ozturk, M., Gul, B., Ahmed, M.Z., Eds.; Academic Press: Cambridge, MA, USA, 2016; pp. 111–123, ISBN 9780128018545.

41. Correa, R.C.G.; Di Gioia, F.; Ferreira, I.; SA, P. Halophytes for Future Horticulture: The Case of Small-Scale Farming in the Mediterranean Basin. In *Halophytes for Future Horticulture: From Molecules to Ecosystems towards Biosaline Agriculture*; Grigore, M.-N., Ed.; Springer Nature Switzerland AG: Cham, Switzerland, 2020; pp. 1–28, ISBN 9783030178543.

42. Alleva, K.; Niemietz, C.M.; Sutka, M.; Maurel, C.; Parisi, M.; Tyerman, S.D.; Amodeo, G. Plasma membrane of *Beta vulgaris* storage root shows high water channel activity regulated by cytoplasmic pH and a dual range of calcium concentrations. *J. Exp. Bot.* **2006**, *57*, 609–621. [CrossRef]

43. Flowers, T.J. Improving crop salt tolerance. *J. Exp. Bot.* **2004**, *55*, 307–319. [CrossRef] [PubMed]

44. Yu, Z.; Duan, X.; Luo, L.; Dai, S.; Ding, Z.; Xia, G. How Plant Hormones Mediate Salt Stress Responses. *Trends Plant Sci.* **2020**, *25*, 1117–1130. [CrossRef]

45. Wu, Z.; Xue, Z.; Li, H.; Zhang, X.; Wang, X.; Lu, X. Cultivation of dandelion (*Taraxacum erythropodium*) on coastal saline land based on the control of salinity and fertilizer. *Folia Hortic.* **2019**, 277–284. [CrossRef]

46. Salonikioti, A.; Petropoulos, S.; Antoniadis, V.; Levizou, E.; Alexopoulos, A. Wild Edible Species with Phytoremediation Properties. *Procedia Environ. Sci.* **2015**, *29*, 98–99. [CrossRef]

47. Bosiacka, B.; Mysliwy, M.; Bosiacki, M. Habitat conditions strongly affect macroand microelement concentrations in *Taraxacum* microspecies growing on coastal meadows along a soil salinity gradient. *PeerJ* **2020**, *8*. [CrossRef]

48. Petropoulos, S.A.; Fernandes, Â.; Dias, M.I.; Pereira, C.; Calhelha, R.C.; Chrysargyris, A.; Tzortzakis, N.; Ivanov, M.; Sokovic, M.D.; Barros, L.; et al. Chemical composition and plant growth of *Centaurea raphanina* subsp. *mixta* plants cultivated under saline conditions. *Molecules* **2020**, *25*, 2204. [CrossRef]

49. Kalra, Y. *Handbook of Reference Methods for Plant. Analysis*; CRC Press: New York, NY, USA, 1998.

50. Boltz, D.F.; Lueck, C.H. *Colorimetric Determination of Non-Metals*; Interscience Publishers: New York, NY, USA, 1958.

51. Allen, S. *Chemical Analysis of Ecological Materials*, 2nd ed.; Blackwell Scientific Publications: Oxford and London, UK, 1989.

52. Cataldo, D.A.; Maroon, M.; Schrader, L.E.; Youngs, V.L. Rapid colorimetric determination of nitrate in plant tissue by nitration of salicylic acid. *Commun. Soil Sci. Plant Anal.* **1975**, *6*, 71–80. [CrossRef]

53. Karapanos, I.; Papandreou, A.; Skouloudi, M.; Makrogianni, D.; Fernández, J.A.; Rosa, E.; Ntatsi, G.; Bebeli, P.J.; Savvas, D. Cowpea fresh pods—a new legume for the market: Assessment of their quality and dietary characteristics of 37 cowpea accessions grown in southern Europe. *J. Sci. Food Agric.* **2017**, *97*, 4343–4352. [CrossRef]

54. Lichtenthaler, H.K.; Buschmann, C. Chlorophylls and Carotenoids: Measurement and Characterization by UV-VIS Spectroscopy. *Curr. Protoc. Food Anal. Chem.* **2001**, *1*, F4.3.1–F4.3.8. [CrossRef]

55. Singleton, V.; Rossi, J.J. Colorimetry of Total Phenolics with Phosphomolybdic-Phosphotungstic Acid Reagents. *Am. J. Enol. Vitic.* **1965**, *16*, 144–158.

56. Bates, L.S. Rapid determination of free proline for water-stress studies. *Plant Soil* **1973**, *39*, 205–207. [CrossRef]

57. Uddin, M.N.; Tariqul Islam, M.; Karim, M.A. Salinity tolerance of three mustard/rapeseed cultivars. *J. Bangladesh Agril. Univ.* **2005**, *3*, 203–208.

58. Wang, Y.; Nii, N. Changes in chlorophyll, ribulose bisphosphate carboxylase-oxygenase, glycine betaine content, photosynthesis and transpiration in *Amaranthus tricolor* leaves during salt stress. *J. Hortic. Sci. Biotechnol.* **2000**, *75*, 623–627. [CrossRef]

59. El-Hendawy, S.E.; Hu, Y.; Schmidhalter, U. Growth, ion content, gas exchange, and water relations of wheat genotypes differing in salt tolerances. *Aust. J. Agric. Res.* **2005**, *56*, 123–134. [CrossRef]

60. Chartzoulakis, K.; Klapaki, G. Response of two greenhouse pepper hybrids to NaCl salinity during different growth stages. *Sci. Hortic.* **2000**, *86*, 247–260. [CrossRef]

61. Rangaiah, D. V High temperature and salt stress response in French bean (*Phaseolus vulgaris*). *Aust. J. Crop. Sci.* **2008**, *2*, 40–48.

62. Turhan, A.; Eniz, V.; Kuşçu, H. Genotypic variation in the response of tomato to salinity. *Afr. J. Biotechnol.* **2009**, *8*, 1062–1068. [CrossRef]

63. Ziaf, K.; Amjad, M.; Pervez, M.A.; Iqbal, Q.; Rajwana, I.A.; Ayyub, M. Evaluation of different growth and physiological traits as indices of salt tolerance in hot pepper (*Capsicum annuum* L.). *Pak. J. Bot.* **2009**, *41*, 1797–1809.

64. Fageria, N.K.; Gheyi, H.R.; Moreira, A. Nutrient bioavailability in salt affected soils. *J. Plant Nutr.* **2011**, *34*, 945–962. [CrossRef]

65. Akladious, S.A.; Mohamed, H.I. Ameliorative effects of calcium nitrate and humic acid on the growth, yield component and biochemical attribute of pepper (*Capsicum annuum*) plants grown under salt stress. *Sci. Hortic.* **2018**, *236*, 244–250. [CrossRef]

66. Ryu, H.; Cho, Y.G. Plant hormones in salt stress tolerance. *J. Plant Biol.* **2015**, *58*, 147–155. [CrossRef]

67. Shannon, M.C.; Grieve, C.M. Tolerance of vegetable crops to salinity. *HortScience* **1999**, *78*, 5–38. [CrossRef]

68. Sergio, L.; de Paola, A.; Cantore, V.; Pieralice, M.; Cascarano, N.A.; Bianco, V.V.; Di Venere, D. Effect of salt stress on growth parameters, enzymatic antioxidant system, and lipid peroxidation in wild chicory (*Cichorium intybus* L.). *Acta Physiol. Plant* **2012**, *34*, 2349–2358. [CrossRef]

69. Ksouri, R.; Ksouri, W.M.; Jallali, I.; Debez, A.; Magné, C.; Hiroko, I.; Abdelly, C. Medicinal halophytes: Potent source of health promoting biomolecules with medical, nutraceutical and food applications. *Crit. Rev. Biotechnol.* **2012**, *32*, 289–326. [CrossRef] [PubMed]

70. Assimakopoulou, A.; Salmas, I.; Roussos, P.A.; Nifakos, K.; Kalogeropoulos, P.; Kostelenos, G. Salt tolerance evaluation of nine indigenous Greek olive cultivars. *J. Plant Nutr.* **2017**, *40*, 1099–1110. [CrossRef]

71. Pérez-Alfocea, F.; Balibrea, M.E.; Santa Cruz, A.; Estañ, M.T. Agronomical and physiological characterization of salinity tolerance in a commercial tomato hybrid. *Plant Soil* **1996**, *180*, 251–257. [CrossRef]

72. Zhou, J.; Li, Z.; Xiao, G.; Zhai, M.; Pan, X.; Huang, R.; Zhang, H. CYP71D8L is a key regulator involved in growth and stress responses by mediating gibberellin homeostasis in rice. *J. Exp. Bot.* **2020**, *71*, 1160–1170. [CrossRef] [PubMed]

73. Iqbal, N.; Umar, S.; Khan, N.A.; Khan, M.I.R. A new perspective of phytohormones in salinity tolerance: Regulation of proline metabolism. *Environ. Exp. Bot.* **2014**, *100*, 34–42. [CrossRef]

74. Mohamed, H.I.; Akladious, S.A.; Ashry, N.A. Evaluation of Water Stress Tolerance of Soybean Using Physiological Parameters and Retrotransposon-Based Markers. *Gesunde Pflanz.* **2018**, *70*, 205–215. [CrossRef]

75. Maggini, R.; Benvenuti, S.; Leoni, F.; Pardossi, A. Terracrepolo (*Reichardia picroides* (L.) Roth.): Wild food or new horticultural crop? *Sci. Hortic.* **2018**, *240*, 224–231. [CrossRef]

76. Pessarakli, M.; Tucker, T.C. Dry Matter Yield and Nitrogen-15 Uptake by Tomatoes under Sodium Chloride Stress. *Soil Sci. Soc. Am. J.* **1988**, *52*, 698–700. [CrossRef]

77. Camalle, M.; Standing, D.; Jitan, M.; Muhaisen, R.; Bader, N.; Bsoul, M.; Ventura, Y.; Soltabayeva, A.; Sagi, M. Effect of salinity and nitrogen sources on the leaf quality, biomass, and metabolic responses of two ecotypes of *Portulaca oleracea*. *Agronomy* **2020**, *10*, 656. [CrossRef]

78. Ashraf, M.; Shahzad, S.M.; Imtiaz, M.; Rizwan, M.S. Salinity effects on nitrogen metabolism in plants–focusing on the activities of nitrogen metabolizing enzymes: A review. *J. Plant Nutr.* **2018**, *41*, 1065–1081. [CrossRef]

79. Pizzorno, L. *203-Osteoporosis*, 5th ed.; Pizzorno, J.E., Murray, M.T., Eds.; Churchill Livingstone: St. Louis, MO, USA, 2020; pp. 1633–1658.e17. ISBN 978-0-323-52342-4.

80. Assimakopoulou, A.; Nifakos, K.; Salmas, I.; Kalogeropoulos, P. Growth, Ion Uptake, and Yield Responses of Three Indigenous Small-Sized Greek Tomato (*Lycopersicon esculentum* L.) Cultivars and Four Hybrids of Cherry Tomato under NaCl Salinity Stress. *Commun. Soil Sci. Plant Anal.* **2015**, *46*, 2357–2377. [CrossRef]

81. Grattan, S.R.; Grieve, C.M. Salinity-mineral nutrient relations in horticultural crops. *Sci. Hortic.* **1998**, *78*, 127–157. [CrossRef]

82. Villora, G.; Moreno, D.A.; Pulgar, G.; Romero, L. Salinity affects phosphorus uptake and partitioning in zucchini. *Commun. Soil Sci. Plant Anal.* **2000**, *31*, 501–507. [CrossRef]

83. Tang, H.; Niu, L.; Wei, J.; Chen, X.; Chen, Y. Phosphorus limitation improved salt tolerance in maize through tissue mass density increase, osmolytes accumulation, and Na$^+$ uptake inhibition. *Front. Plant Sci.* **2019**, *10*. [CrossRef] [PubMed]

84. Rus, A.; Lee, B.H.; Muñoz-Mayor, A.; Sharkhuu, A.; Miura, K.; Zhu, J.K.; Bressan, R.A.; Hasegawa, P.M. AtHKT1 facilitates Na+ homeostasis and K$^+$ nutrition in planta. *Plant Physiol.* **2004**, *136*, 2500–2511. [CrossRef] [PubMed]

85. Mansour, M.; Van Hasselt, P.; Kuiper, P. NaCl effects on root plasma membrane ATPase of salt tolerant wheat. *Biol. Plant.* **2000**, *43*, 61–66. [CrossRef]

86. Zhang, J.L.; Flowers, T.J.; Wang, S.M. Mechanisms of sodium uptake by roots of higher plants. *Plant Soil* **2010**, *326*, 45–60. [CrossRef]

87. Shibli, R.A.; Sawwan, J.; Swaidat, I.; Tahat, M. Increased phosphorus mitigates the adverse effects of salinity in tissue culture. *Commun. Soil Sci. Plant Anal.* **2001**, *32*, 429–440. [CrossRef]

88. Essa, T.A. Effect of Salinity Stress on Growth and Nutrient Composition of Three Soybean (*Glycine max* L. Merrill) Cultivars. *J. Agron. Crop. Sci.* **2002**, *188*, 86–93. [CrossRef]

89. Weisany, W.; Sohrabi, Y.; Heidari, G.; Siosemardeh, A.; Badakhshan, H. Effects of Zinc Application on Growth, Absorption and Distribution of Mineral Nutrients under Salinity Stress in Soybean (*Glycine max* L.). *J. Plant Nutr.* **2014**, *37*, 2255–2269. [CrossRef]

90. Assimakopoulou, A.; Salmas, I.; Nifakos, K.; Kalogeropoulos, P. Effect of salt stress on Three Green Bean (*Phaseolus vulgaris* L.) Cultivars. *Not. Bot. Horti Agrobot. Cluj-Napoca* **2015**, *43*, 113–118. [CrossRef]

91. Semiz, G.D.; Suarez, D.L.; Ünlükara, A.; Yurtseven, E. Interactive Effects of Salinity and N on Pepper (*Capsicum annuum* L.) Yield, Water Use Efficiency and Root Zone and Drainage Salinity. *J. Plant Nutr.* **2014**, *37*, 595–610. [CrossRef]

92. Shahid, M.A.; Sarkhosh, A.; Khan, N.; Balal, R.M.; Ali, S.; Rossi, L.; Gómez, C.; Mattson, N.; Nasim, W.; Garcia-Sanchez, F. Insights into the physiological and biochemical impacts of salt stress on plant growth and development. *Agronomy* **2020**, *10*, 938. [CrossRef]

93. Stallings, V.A.; Harrison, M.; Oria, M. *Dietary Reference Intakes for Sodium and Potassium*; Stallings, V.A., Harrison, M., Oria, M., Eds.; The National Academies Press: Washington, DC, USA, 2019.

94. Jiang, Z.; Zhou, X.; Tao, M.; Yuan, F.; Liu, L.; Wu, F.; Wu, X.; Xiang, Y.; Niu, Y.; Liu, F.; et al. Plant cell-surface GIPC sphingolipids sense salt to trigger Ca^{2+} influx. *Nature* **2019**, *572*, 341–346. [CrossRef] [PubMed]

95. Bolarín, M.C.; Fernández, F.G.; Cruz, V.; Cuartero, J. Salinity Tolerance in Four Wild Tomato Species using Vegetative Yield-Salinity Response Curves. *J. Am. Soc. Hortic. Sci.* **2019**, *116*, 286–290. [CrossRef]

96. Subbarao, G.V.; Johansen, C.; Jana, M.K.; Kumar Rao, J.V.D.K. Effects of the Sodium/Calcium Ratio in Modifying Salinity Response of Pigeonpea (*Cajanus cajan*). *J. Plant Physiol.* **1990**, *136*, 439–443. [CrossRef]

97. Hunt, C.D.; Johnson, L.A.K. Calcium requirements: New estimations for men and women by cross-sectional statistical analyses of calcium balance data from metabolic studies. *Am. J. Clin. Nutr.* **2007**, *86*, 1054–1063. [CrossRef]

98. Onodera, M.; Nakajima, T.; Nanzyo, M.; Takahashi, T.; Xu, D.; Homma, K.; Kokubun, M. Regulation of root-to-leaf Na and Cl transport and its association with photosynthetic activity in salt-tolerant soybean genotypes. *Plant Prod. Sci.* **2019**, *22*, 262–274. [CrossRef]

99. Cai, Z.Q.; Gao, Q. Comparative physiological and biochemical mechanisms of salt tolerance in five contrasting highland quinoa cultivars. *BMC Plant Biol.* **2020**, *20*, 1–15. [CrossRef]

100. EFSA. Dietary Reference Values for chloride. *EFSA J.* **2019**, *17*, e05779.

101. Almeida, D.M.; Margarida Oliveira, M.; Saibo, N.J.M. Regulation of Na+ and K+ homeostasis in plants: Towards improved salt stress tolerance in crop plants. *Genet. Mol. Biol.* **2017**, *40*, 326–345. [CrossRef] [PubMed]

102. Bingham, F.T.; Strong, J.E.; Rhoades, J.D.; Keren, R. Effects of salinity and varying boron concentrations on boron uptake and growth of wheat. *Plant Soil* **1987**, *351*, 345–351. [CrossRef]

103. Hasana, R.; Miyake, H. Salinity Stress Alters Nutrient Uptake and Causes the Damage of Root and Leaf Anatomy in Maize. *KnE Life Sci.* **2017**, *3*, 219. [CrossRef]

104. Rahman, S.; Vance, G.F.; Munn, L.C. Salinity Induced Effects on the Nutrient Status of Soil, Corn Leaves and Kernels. *Commun. Soil Sci. Plant Anal.* **1993**, *24*, 2251–2269. [CrossRef]

105. Izzo, R.; Navari-Izzo, F.; Quartacci, M.F. Growth and mineral absorption in maize seedlings as affected by increasing NaCl concentrations. *J. Plant Nutr.* **1991**, *14*, 687–699. [CrossRef]

106. Shibli, R.A.; Kushad, M.; Yousef, G.G.; Lila, M.A. Physiological and biochemical responses of tomato microshoots to induced salinity stress with associated ethylene accumulation. *Plant Growth Regul.* **2007**, *51*, 159–169. [CrossRef]

107. Neocleous, D.; Koukounaras, A.; Siomos, A.S.; Vasilakakis, M. Assessing the Salinity Effects on Mineral Composition and Nutritional Quality of Green and Red "Baby" Lettuce. *J. Food Qual.* **2014**, *37*, 1–8. [CrossRef]

108. Fatemi, H.; Carvajal, M.; Rios, J.J. Foliar application of Zn alleviates salt stress symptoms of pak choi plants by activating water relations and glucosinolate synthesis. *Agronomy* **2020**, *10*, 1528. [CrossRef]

109. Cakmak, I. Tansley Review No. 111 Possible roles of zinc in protecting plant cells from damage by reactive oxygen species. *New Phytol.* **2000**, *146*, 185–205. [CrossRef]

110. Ashraf, M.; Foolad, M.R. Roles of glycine betaine and proline in improving plant abiotic stress resistance. *Environ. Exp. Bot.* **2007**, *59*, 206–216. [CrossRef]

111. Dar, M.I.; Naikoo, M.I.; Rehman, F.; Naushin, F.; Khan, F.A. Proline Accumulation in Plants: Roles in Stress Tolerance and Plant Development BT-Osmolytes and Plants Acclimation to Changing Environment: Emerging Omics Technologies. In *Osmolytes and Plants Acclimation to Changing Environment: Emerging Omics Technologies*; Iqbal, N., Nazar, R., A. Khan, N., Eds.; Springer India: New Delhi, India, 2016; pp. 155–166, ISBN 978-81-322-2616-1.

112. Hayat, S.; Hayat, Q.; Alyemeni, M.N.; Wani, A.S.; Pichtel, J.; Ahmad, A. Role of proline under changing environments: A review. *Plant Signal. Behav.* **2012**, *7*, 1456–1466. [CrossRef]

113. Xue, X.; Liu, A.; Hua, X. Proline accumulation and transcriptional regulation of proline biothesynthesis and degradation in *Brassica napus. BMB Rep.* **2009**, *42*, 28–34. [CrossRef]

114. Chou, I.T.; Chen, C.T.; Kao, C.H. Characteristics of the induction of the accumulation of proline by abscisic acid and isobutyric acid in detached rice leaves. *Plant Cell Physiol.* **1991**, *32*, 269–272. [CrossRef]

115. Slabbert, M.M.; Krüger, G.H.J. Antioxidant enzyme activity, proline accumulation, leaf area and cell membrane stability in water stressed *Amaranthus* leaves. *S. Afr. J. Bot.* **2014**, *95*, 123–128. [CrossRef]

116. Saleh, A.A.H. Amelioration of Chilling Injuries in Mung Bean (*Vigna radiata* L.) Seedlings by Paclobutrazol, Abscisic Acid and Hydrogen Peroxide. *Am. J. Plant Physiol.* **2007**, *2*, 318–332. [CrossRef]

117. Xu, C.; Mou, B. Responses of spinach to salinity and nutrient deficiency in growth, physiology, and nutritional value. *J. Am. Soc. Hortic. Sci.* **2016**, *141*, 12–21. [CrossRef]

118. Chatzigianni, M.; Alkhaled, B.; Livieratos, I.; Stamatakis, A.; Ntatsi, G.; Savvas, D. Impact of nitrogen source and supply level on growth, yield and nutritional value of two contrasting ecotypes of *Cichorium spinosum* L. grown hydroponically. *J. Sci. Food Agric.* **2017**, *98*, 1615–1624. [CrossRef]

119. Colonna, E.; Rouphael, Y.; Barbieri, G.; De Pascale, S. Nutritional quality of ten leafy vegetables harvested at two light intensities. *Food Chem.* **2016**, *199*, 702–710. [CrossRef]

120. Singh, H.; Dunn, B.; Payton, M.; Brandenberger, L. Fertilizer and cultivar selection of lettuce, Basil, and Swiss chard for hydroponic production. *Horttechnology* **2019**, *29*, 50–56. [CrossRef]

121. Rouphael, Y.; Petropoulos, S.A.; Cardarelli, M.; Colla, G. Salinity as eustressor for enhancing quality of vegetables. *Sci. Hortic.* **2018**, *234*, 361–369. [CrossRef]

122. Pérez-López, U.; Miranda-Apodaca, J.; Lacuesta, M.; Mena-Petite, A.; Muñoz-Rueda, A. Growth and nutritional quality improvement in two differently pigmented lettuce cultivars grown under elevated CO_2 and/or salinity. *Sci. Hortic.* **2015**, *195*, 56–66. [CrossRef]

123. Petersen, K.K.; Willumsen, J.; Kaack, K. Composition and taste of tomatoes as affected by increased salinity and different salinity sources. *J. Hortic. Sci. Biotechnol.* **1998**, *73*, 205–215. [CrossRef]

124. Zhao, J.; Davis, L.C.; Verpoorte, R. Elicitor signal transduction leading to production of plant secondary metabolites. *Biotechnol. Adv.* **2005**, *23*, 283–333. [CrossRef]
125. Klados, E.; Tzortzakis, N. Effects of substrate and salinity in hydroponically grown *Cichorium spinosum*. *J. Soil Sci. Plant Nutr.* **2014**, *14*, 211–222. [CrossRef]
126. Garrido, Y.; Tudela, J.A.; Marín, A.; Mestre, T.; Martínez, V.; Gil, M.I. Physiological, phytochemical and structural changes of multi-leaf lettuce caused by salt stress. *J. Sci. Food Agric.* **2014**, *94*, 1592–1599. [CrossRef]
127. Kim, H.J.; Fonseca, J.M.; Choi, J.H.; Kubota, C.; Dae, Y.K. Salt in irrigation water affects the nutritional and visual properties of romaine lettuce (*Lactuca sativa* L.). *J. Agric. Food Chem.* **2008**, *56*, 3772–3776. [CrossRef]
128. Gent, M.P.N. Composition of hydroponic lettuce: Effect of time of day, plant size, and season. *J. Sci. Food Agric.* **2012**, *92*, 542–550. [CrossRef]
129. Bonasia, A.; Lazzizera, C.; Elia, A.; Conversa, G. Nutritional, biophysical and physiological characteristics of wild rocket genotypes as affected by soilless cultivation system, salinity level of nutrient solution and growing period. *Front. Plant Sci.* **2017**, *8*, 1–15. [CrossRef]
130. Cantabella, D.; Piqueras, A.; Acosta-Motos, J.R.; Bernal-Vicente, A.; Hernández, J.A.; Díaz-Vivancos, P. Salt-tolerance mechanisms induced in *Stevia rebaudiana* Bertoni: Effects on mineral nutrition, antioxidative metabolism and steviol glycoside content. *Plant Physiol. Biochem.* **2017**, *115*, 484–496. [CrossRef]
131. Santamaria, P. Nitrate in vegetables: Toxicity, content, intake and EC regulation. *J. Sci. Food Agric.* **2006**, *86*, 10–17. [CrossRef]
132. Kafkafi, U.; Yaeesh Siddiqi, M.; Ritchie, R.J.; Glass, A.D.M.; Ruth, T.J. Reduction of nitrate (^{13}no$_3$) influx and nitrogen (^{13}n) translocation by tomato and melon varieties after short exposure to calcium and potassium chloride salts. *J. Plant Nutr.* **1992**, *15*, 959–975. [CrossRef]
133. Abdelgadir, E.M.; Oka, M.; Fujiyama, H. Characteristics of nitrate uptake by plants under salinity. *J. Plant Nutr.* **2005**, *28*, 33–46. [CrossRef]
134. Rameeh, V.; Rezai, A.; Saeidi, G. Study of salinity tolerance in rapeseed. *Commun. Soil Sci. Plant Anal.* **2004**, *35*, 2849–2866. [CrossRef]

horticulturae

Article

The Impact of Salt Stress on Plant Growth, Mineral Composition, and Antioxidant Activity in *Tetragonia decumbens* Mill.: An Underutilized Edible Halophyte in South Africa

Avela Sogoni, Muhali Olaide Jimoh *, Learnmore Kambizi and Charles Petrus Laubscher *

Department of Horticultural Sciences, Faculty of Applied Sciences, Cape Peninsula University of Technology, Bellville, Cape Town 7535, South Africa; sogoniavis@gmail.com (A.S.); kambizil@cput.ac.za (L.K.)
* Correspondence: Jimohm@cput.ac.za (M.O.J.); LaubscherC@cput.ac.za (C.P.L.)

Abstract: Climate change, expanding soil salinization, and the developing shortages of freshwater have negatively affected crop production around the world. Seawater and salinized lands represent potentially cultivable areas for edible salt-tolerant plants. In the present study, the effect of salinity stress on plant growth, mineral composition (macro-and micro-nutrients), and antioxidant activity in dune spinach (*Tetragonia decumbens*) were evaluated. The treatments consisted of three salt concentrations, 50, 100, and 200 mM, produced by adding NaCl to the nutrient solution. The control treatment had no NaCl but was sustained and irrigated by the nutrient solution. Results revealed a significant increase in total yield, branch production, and ferric reducing antioxidant power in plants irrigated with nutrient solution incorporated with 50 mM NaCl. Conversely, an increased level of salinity (200 mM) caused a decrease in chlorophyll content (SPAD), while the phenolic content, as well as nitrogen, phosphorus, and sodium, increased. The results of this study indicate that there is potential for brackish water cultivation of dune spinach for consumption, especially in provinces experiencing the adverse effect of drought and salinity, where seawater or underground saline water could be diluted and used as irrigation water in the production of this vegetable.

Keywords: dune spinach; NaCl; functional food; salt tolerance; underexploited vegetable

Citation: Sogoni, A.; Jimoh, M.O.; Kambizi, L.; Laubscher, C.P. The Impact of Salt Stress on Plant Growth, Mineral Composition, and Antioxidant Activity in *Tetragonia decumbens* Mill.: An Underutilized Edible Halophyte in South Africa. *Horticulturae* **2021**, *7*, 140. https://doi.org/10.3390/horticulturae7060140

Academic Editors: Rosario Paolo Mauro, Carlo Nicoletto and Leo Sabatino

Received: 5 May 2021
Accepted: 24 May 2021
Published: 8 June 2021

Publisher's Note: MDPI stays neutral with regard to jurisdictional claims in published maps and institutional affiliations.

1. Introduction

Global agriculture feeds over 7 billion people and alarmingly, this number is expected to increase by a further 50% by 2050 [1]. The global need for food production has never been greater, especially in developing nations where an increase of 90% of the population growth anticipates that food insecurity will become a greater problem [2,3]. To meet the additional food demand, the world development report has estimated that crop production should increase between 70% and 100% by 2050 [4]. However, increasing crop production has led to a loss of soil fertility and the phenomena of salinization and desertification, which makes soils unsuitable for cultivation [5]. This is caused by the accumulation of soluble salts in the root zone. These salts restrict the absorption of water by the plant roots, which leads to osmotic stress and thus nutritional imbalance due to the high concentration of toxic salts in plant cells [6,7]. In addition, the accumulation of toxic ions inhibits physiological processes such as photosynthesis, respiration, and nitrogen fixation [8]. These effects can result in reduced leaf area, plant biomass production, and yield [9,10]. Moreover, soil salinity and drought stress are known factors to induce oxidative stress in plants through the production of superoxide radicals by the process of the Mehler reaction [11]. These free radicals initiate the chain of reactions that produce more harmful oxygen radicals [12]. These reactive oxygen species (ROS) are continuously generated during normal metabolic processes in mitochondria, peroxisomes, and cytoplasm, which disturb normal metabolism through oxidative damage of lipids, proteins, and nucleic acids when produced in excess [13,14].

To overcome salt-mediated oxidative stress, plants detoxify ROS by up-regulating antioxidative enzymes, which includes the superoxide dismutase (SOD) found in various cell compartments [15]. This enzyme catalyzes a conversion from two O_2 radicals to H_2O_2 and O_2 [16]. In alternative ways, several antioxidant enzymes can also eliminate the H_2O_2, such as catalases (CAT) and peroxidases (POX), by converting it to water [17]. During this process, the antioxidant capacity of some species increases when exposed to salinity stress to eliminate or reduce the ROS. Thus, research on oxidative stress is imperative due to the usefulness of these antioxidants against free radicals that predispose humans to sickness and diseases. The author of [18] stated that these antioxidants exert a large spectrum of biological and physiological functions on human health, such as anti-allergic, anti-atherogenic, anti-inflammatory, and anti-microbial activities.

The catastrophic effect of salinity and drought on crop yield call for a creative, sustainable, and sufficient crop production method, given the rising population and increasing demand for plant-based food [19,20]. With this in mind, numerous researchers have pointed out the use of salt-tolerant plant species with possible commercial value as a proposed upfront strategy for saline lands [21,22]. This led to a worldwide interest in edible salt-tolerant plant species in addressing the challenges of food and nutritional deficiency. Currently, underutilized edible halophytes are slowly becoming a viable alternative to popular crops in regions experiencing the adverse effect of drought and salinity [23]. Moreover, edible halophytes have been reported to be rich in nutrients and bioactive compounds [24], which are considered as important mediators of various health effects [25]. The medicinal value of edible halophytes has been documented and proven for prophylaxis against various chronic diseases that afflict modern societies [26].

Tetragonia decumbens commonly known as 'dune spinach' or 'duinespinasie' (Afrikaans) is an edible halophyte belonging to the Aizoaceae family and is largely distributed along the coastal regions from southern Namibia to the Eastern Cape [27,28]. It is an endemic sprawling perennial shrub with branches (runners) that can grow up to 1 m long [29]. The leaves and soft stems have a salty taste and can be used like spinach, served raw in green salads, or cooked with other vegetables. They can also be fermented, pickled, and used in stews and soups and are particularly tasty in a stir fry. However, the leaves and soft stems are foraged rather than cultivated and are known only by a small group of local chefs and food enthusiasts [30]. Thus, there is a need for agronomical studies to support its domestication and ensure its sustainable use. The cultivation of this native halophyte for food production in South Africa could be a climate change adaptation strategy, as freshwater continues to become scarce and rain becomes more sporadic particularly in sub-Saharan Africa [31,32]. Moreover, it has also been stated that South Africa is approaching physical water scarcity by 2025, and its agricultural sector has been directly hampered by the recent drought [33]. Hence, it is of utmost importance to cultivate crops that are adapted to harsh conditions within the framework of saline agriculture [34].

This study was therefore undertaken to evaluate the effect of salt stress on plant growth, mineral composition, and antioxidant activity in dune spinach, to lay a potential growing protocol for the use of brackish water or saline soil. Moreover, the dearth of literature on the nutritional value of this halophyte under saline conditions is a contributing factor to its underutilization and consumption among coastal households. Hence, data from this study are expected to serve as a template for future researchers, households, and potential farmers, who may want to exploit this plant for diet diversity and as pharmaceutical precursors.

2. Materials and Methods

2.1. Experimental Location

The experiment was conducted in the greenhouse of the Department of Horticultural Sciences at the Cape Peninsula University of Technology (CPUT), Bellville campus, Cape Town, South Africa, located at 33°55′56″ S, 18°38′25″ E. The greenhouse was equipped with environmental control with temperatures set to range from 21 to 26 °C during the day and 12–18 °C at night, with relative humidity averages of 60%. The average daily photosynthetic photon flux density (PPFD) was 420 μmol/m^2/s and the maximum was 1020 μmol/m^2/s.

2.2. Plant Preparation, Irrigation and Treatments

Softwood cuttings of *T. decumbens* were harvested on 1 August 2019 from a selected plant population growing along the coast at the Granger Bay campus of CPUT located at 33°53'58.2'' S, 18°24'41.4'' E. Only cuttings taken using homogeneous methods, i.e., stem cuttings with about two-thirds of leaves removed, $\pm$15 cm long with a stem thickness of approx. 8 mm were used for the experiment. One hundred cuttings were made to ensure the minimum number of 60 rooted cuttings required for the experiment was available. The cuttings were then soaked in 0.1% Sporekil™ for precaution against fungal infection and, thereafter, were dipped in a rooting hormone (Dynaroot™ No. 1 with active ingredient 0.1% I.B.A) for two seconds. The cuttings were then placed in trays containing washed and sterilized coarse river sand and peat of equal volume. The trays were then placed in the main greenhouse on heated propagation beds. Once rooted, 80 *T. decumbens* cuttings of uniform size were individually transplanted in 12.5 cm plastic pots containing a mixture of commercial peat and sand (V:V) and placed in a greenhouse to acclimatize. Only cuttings showing the strongest growth were selected and left to grow for two weeks. During this period, rooted cuttings were irrigated with a nutrient solution three times a week. The nutrient solution was formed by adding NUTRIFEED™ (manufactured by STARKE AYRES Pty. Ltd. Hartebeesfontein Farm, Bredell Rd, Kaalfontein, Kempton Park, Gauteng, South Africa, 1619) to municipal water at 10 g per 5L. The nutrient solution contained the following ingredients: N (65 mg/kg), P (27 mg/kg), K (130 mg/kg), Ca (70 mg/kg), Cu (20 mg/kg), Fe (1500 mg/kg), Mo (10 mg/kg), Mg (22 mg/kg), Mn (240 mg/kg), S (75 mg/kg), B (240 mg/kg), and Zn (240 mg/kg). After 14 days of growth, the established cuttings were watered with clean water for 5 days to wash off any salt residue and, thereafter, were organized into 4 treatments each containing 15 replicates. Salt concentrations were set up on three treatments by adding increasing concentrations of NaCl in the nutrient solution (50, 100, and 200 mM). A total of 300 mL of the nutrient solution was prepared for each plant with and/or without NaCl. The plants were then watered every three days. The control treatment was sustained and irrigated only by the nutritive solutions. In all of the treatments, the pH was maintained at 6.0. Ten weeks after salt treatments, all plants were harvested, and various postharvest measurements were made.

2.3. Determination of Plant Growth

2.3.1. Plant Weight

The weight of the plants was measured using a standard laboratory scale (RADWAG® Model PS 750.R2) before planting out to ensure homogeneity within the samples. Postharvest, shoots, stems, and roots were separated, and the fresh/wet weights of the individual samples were recorded. The plant material was then oven-dried at 55 °C in a LABTECH™ model LDO 150F (Daihan Labtech India. Pty. Ltd. 3269 Ranjit Nagar, New Delhi, India) to a constant weight and recorded. The difference between the fresh and dry weight was compared with the amount of water held within the plants' tissues [35,36].

2.3.2. Shoot Length and Branch Number

The shoot length and branch number were used as a variable to determine new growth. Shoot length was measured every two weeks with a metal tape measure from the substrate level to the tip of the tallest shoot, while branch number was counted [37].

2.4. Mineral Analysis

To determine the mineral composition of each set of replicates in the experiment, three plants (shoots/leaves) were randomly selected from each treatment at the end of the experiment. The vegetative material was then removed, labelled, and sent to Bemlab

Laboratory, located at 16 van der Berg Crescent, Gant's Centre, Strand, Cape Town for mineral analysis. The methodology to determine macronutrients (N, K, P, Ca, Mg, and Na) and micronutrients (Cu, Zn, Mn, Fe, Al, and B) was conducted by ashing 1 g ground sample of plant material in a porcelain crucible at 500 °C overnight. This was followed by dissolving the ash in 5 mL of HCI and placing it in an oven at 50 °C for 30 min. Thirty-five milliliters of deionized water was then added and the extract filtered through Whatman No. 1 filter paper. Nutrient concentrations in plant extracts were determined using an inductively coupled plasma (ICP) emission spectrophotometer (IRIS/AP HR DUO Thermo Electron Corporation, Franklin, MA, USA) [38,39].

2.5. Chlorophyll Readings

The chlorophyll content was measured every two weeks using a Soil Plant Analysis Development (SPAD-502) meter supplied by Konica Minolta. The readings of two fully formed leaves were taken from each plant, and the figures were averaged out by the SPAD-502 meter to produce a final number. The readings were taken between 11 a.m. and midday from week 4 to 10 of the experiment [40].

2.6. The Antioxidant Analysis

2.6.1. Sample Preparation

Harvested shoot materials were immediately dried in a fan-drying laboratory oven (Oxidative Stress Research Centre, Faculty of Health and Wellness Sciences at CPUT, Bellville) at 40 °C for 7–14 days. The dried plants were ground into a fine powder using a Junkel and Kunkel model A 10 mill. Shoot material was extracted by mixing 100 mg of the dried powdered material with 25 mL of 70% (*v/v*) ethanol (EtOH) (Merck, Modderfontein, South Africa) for 1 h. It was centrifuged at 4000 rpm for 5 min, and the supernatants were used for all analyses.

2.6.2. Determination of Antioxidant Capacity and Content

Antioxidant activity and accumulation of secondary metabolites within the leaves were assessed using assays for total polyphenols, ABTS, and ferric reducing antioxidant power (FRAP).

2.6.3. Polyphenol Assay

The total polyphenols assay (Folin assay) was performed as described by [41]. Folin and Ciocalteu's phenol reagent (2 N, Sigma, Gauteng, South Africa) were diluted 10 times with distilled water, and a 7.5% sodium carbonate (Sigma-Aldrich, Gauteng, South Africa) solution was prepared. In a 96-well plate, 25 µL of the crude extract was mixed with 125 µL of Folin and Ciocalteu's phenol reagent and 100 µL of sodium carbonate. The plate was incubated for 2 h at room temperature. The absorbance was then measured at 765 nm in a Multiskan Spectrum plate reader (Thermo Electron Corporation, USA). The samples' polyphenol values were calculated using a gallic acid (Sigma-Aldrich, Gauteng, South Africa) standard curve with concentration varying between 0 and 500 mg/L. The results were expressed as mg gallic acid equivalents (GAE) per g dry weight (mg GAE/g DW).

2.6.4. ABTS Assay

The ABTS assay was performed following the method of [42]. The stock solutions included a 7 mM ABTS and 140 mM potassium–peroxodisulphate ($K_2S_2O_8$) (Merck, Modderfontein, South Africa) solution. The working solution was then prepared by adding 88 µL of $K_2S_2O_8$ to 5 mL of ABTS solution. The two solutions were mixed well and allowed to react for 24 h at room temperature in the dark. Trolox (6-Hydrox-2,5,7,8-tetramethylchroman-2-20 carboxylic acid) was used as the standard with concentrations ranging between 0 and 500 µM. Crude sample extracts (25 µL) were allowed to react with 300 µL of ABTS in the dark at room temperature for 30 min before the absorbance was read at 734 nm at 25 °C in a plate reader. The results were expressed as μM/Trolox equivalent per g dry weight (μM TE/g DW).

2.6.5. Ferric Reducing Antioxidant Power (FRAP) Assay

The FRAP assay was performed using the method of [43]. The FRAP reagent was prepared by mixing 30 mL of acetate buffer (0.3 M, pH 3.6) (Merck, Modderfontein, South Africa) with 3 mL of 2,4,6- tripyridyl-s-triazine (10 mM in 0.1 M hydrochloric acid) (Sigma-Aldrich, Gauteng, South Africa), 3 mL of iron (III) chloride hexahydrate ($FeCl_3 \cdot 6H_2O$) (Sigma-Aldrich, Gauteng, South Africa) and 6 mL of distilled water. In a 96-well plate, 10 μL of the crude sample extract was mixed with 300 μL of the FRAP reagent and incubated for 30 min at room temperature. The absorbance was then measured at 593 nm in a Multiskan Spectrum plate reader (Thermo Electron Corporation, USA). The samples' FRAP values were calculated using an L-Ascorbic acid (Sigma-Aldrich, Gauteng, South Africa) standard curve with concentrations varying between 0 and 1000 μM. The results were expressed as μM ascorbic acid equivalents (AAE) per g dry weight (μM AAE/g DW) [41,44].

2.7. Statistical Analysis

For minerals, three samples were analyzed for each treatment, while all the assays were carried out in triplicate. The results were expressed as mean values and standard error (SE) and analyzed using one-way analysis of variance (ANOVA) followed by Fisher's least significant test at $p \leq 0.05$ significance level. This analysis was carried out using the STATISTICA version 13.5.0.17 program [45].

3. Results

3.1. Effects of Salt Stress on Plant Growth

3.1.1. Shoot Length and Lateral Branch Number

The results showed that *T. decumbens* growth response to NaCl was variable (Table 1). The shoot length and lateral branch number were significantly affected by salinity concentrations at $p \leq 0.05$. The control had the highest shoot length, and this was significantly higher than 100 mM NaCl and 200 NaCl mM concentrations, respectively, but did not differ significantly from the 50 mM NaCl concentration. However, this was not the case with the lateral branch number, where the 50 mM NaCl concentration had the highest number of branches compared to the control.

Table 1. Effects of salt stress on growth parameters of *T. decumbens*.

Treatments	SL (cm)	BN (n)	FWS (g)	DWS (g)	FWSR (g)	DWSR (g)	TFW (g)	TDW (g)
Control	101.1 ± 3.8 [a]	3.4 ± 0.2 [c]	136.8 ± 6.7 [c]	23.1 ± 1.2 [c]	61.4 ± 5.4 [b]	14.5 ± 1 [b]	198.3 ± 8.3 [c]	37.6 ± 1.1 [c]
50 mM	96.5 ± 3.1 [ab]	5.4 ± 0.4 [a]	210.8 ± 10.4 [a]	32.1 ± 1.3 [a]	89.3 ± 6.7 [a]	18.6 ± 1 [a]	300.2 ± 12.8 [a]	50.6 ± 1.6 [a]
100 mM	90.7 ± 2.2 [b]	4.7 ± 0.2 [ab]	186.4 ± 6.8 [b]	26.9 ± 0.7 [b]	78.0 ± 3.8 [a]	17.3 ± 0.8 [ab]	264.4 ± 6.6 [b]	44.1 ± 0.9 [b]
200 mM	73.7 ± 3.1 [c]	3.8 ± 0.3 [bc]	119.7 ± 6.3 [c]	15.6 ± 0.8 [d]	79.5 ± 6.2 [a]	15.3 ± 1.1 [b]	199.2 ± 7.8 [c]	30.9 ± 1.2 [d]
F-statistic	14.6 *	7.6 *	29.8 *	45 *	4.2 *	3.2 *	30 *	45 *

Note. SL: shoot length; BN: branch number; FWS: fresh weight of shoots; DWS: dry weight of shoots; FWSR: fresh weight of stem and roots; DWSR: dry weight of stem and roots; TFW: total fresh weight; TDW: total dry weight. The values (mean ± SE) followed by dissimilar letters in each column are significantly different at $p \leq 0.05$ (*).

3.1.2. Fresh and Dry Weight of Shoots

Both fresh and dry weights of shoots significantly differed between treatments (Table 1). The highest fresh weight was obtained at 50 mM NaCl concentration. This was significantly higher than all NaCl concentrations, including the control. The lowest fresh weight was obtained at 200 mM NaCl concentration; however, this was not significantly different from the control. As for the dry weight, the highest mean value was again obtained at 50 mM NaCl concentration; this was significantly higher than all other treatments, including the control. The lowest dry weight was obtained at 200 mM NaCl concentration, and this was significantly lower compared to the other treatments, including the control.

3.1.3. Fresh and Dry Weight of Stem and Roots

NaCl concentrations positively influenced the fresh and dry weights of the stem and roots. The highest fresh weight of the stem and roots was obtained at 50 mM NaCl concentration. This was significantly higher than the control but did not differ significantly to 100 mM and 200 mM NaCl concentrations, respectively. The highest dry weight of the stem and roots was again recorded at 50 mM concentration. This was significantly higher than the control and 200 mM NaCl but did not differ significantly from the 100 mM NaCl concentration.

3.1.4. Total Fresh and Dry Weight

Experimental results also showed that NaCl concentrations significantly ($p \leq 0.05$) affected the fresh and dry weight of dune spinach (Table 1). Plants exposed to different salt concentrations had fresh weights that were higher than the control. The highest measurement of fresh weight was obtained at 50 mM NaCl concentration, and this was significantly higher than all treatments, including the control. Although plants exposed to the 200 mM NaCl concentration had a higher fresh weight than the control, they did not differ significantly from each other. Conversely, this was not the case in total dry weight, where the 200 mM NaCl concentration recorded the lowest dry weight compared to all treatments, including the control. The highest total dry weight was again obtained at 50 mM NaCl concentration, and this was significantly higher than all other treatments, including the control.

3.2. Effect of Salt Stress on the Mineral Content of Dried Leaves of T. decumbens

3.2.1. Macronutrients

Salinity stress significantly increased macronutrients (N, P, and Na) in the leaves of dune spinach. The highest N, P, and Na were all obtained from the highest salt concentration (200 mM). Conversely, the accumulation of K, Ca, and Mg significantly lowered in comparison to the control (Table 2).

Table 2. Effect of salt stress on the concentration of macronutrients in the leaves of *T. decumbens*.

Treatments	N (g/kg)	P (g/kg)	K (g/kg)	Ca (g/kg)	Mg (g/kg)	Na (g/kg)
Control	23.5 ± 0.4 [a]	4.7 ± 0.4 [a]	44.1 ± 1.1 [a]	6.9 ± 0.7 [a]	6.9 ± 0.5 [a]	6.9 ± 0.4 [b]
50 mM	20.8 ± 0.1 [b]	3.3 ± 0.3 [a]	26.8 ± 1.1 [b]	3.6 ± 0.2 [b]	3.3 ± 0.1 [b]	35.8 ± 1.3 [a]
100 mM	20.9 ± 0.3 [b]	3.4 ± 0.4 [a]	22.6 ± 1 [c]	3.7 ± 0.5 [b]	3.2 ± 0.2 [b]	54.7 ± 1.3 [c]
200 mM	24.4 ± 0.3 [a]	4.8 ± 0.4 [a]	21.9 ± 0.7 [c]	2.5 ± 0.1 [b]	2.8 ± 0.1 [b]	59 ± 0.9 [c]
F-statistic	14.9 *	1.9 ns	81.3 *	18.6 *	18.6 *	58.8 *

Values (mean $\pm$ SE) followed by dissimilar letters in each column are significantly different at $p \leq 0.05$ (*); ns = not significant.

3.2.2. Micronutrients

Salt stress positively influenced the micronutrients (Mn, Fe, and Cu) in the leaves of dune spinach. The highest Mn, Fe, and Cu were all obtained from the moderate salt concentration (100 mM). However, the highest mean values in Mn and Cu were not significantly ($p \leq 0.05$) different from all other treatments, including the control. However, the opposite was true for Fe. Conversely, salt stress negatively influenced Zn and B accumulation in the leaves. The control had the highest mean values in both Zn and B, and these were significantly different from all salt treatments (Table 3).

Table 3. Effect of salt stress on the concentration of micronutrients in the leaves of *T. decumbens*.

Treatments	Mn (mg/kg)	Fe (mg/kg)	Cu (mg/kg)	Zn (mg/kg)	B (mg/kg)
Control	74.3 ± 1.3 [a]	66.4 ± 2.2 [b]	1.7 ± 0.3 [a]	56.7 ± 1.8 [a]	35.1 ± 1.5 [a]
50 mM	67.5 ± 1.9 [a]	61.3 ± 0.83 [bc]	1.7 ±0.2 [a]	35.4 ± 0.5 [b]	20.7 ± 0.7 [b]
100 mM	81.3 ± 1.9 [a]	89.2 ± 2.7 [a]	4.2 ±1.9 [a]	51.6 ± 1.4 [a]	24.7 ± 2.4 [b]
200 mM	64.4 ± 1.2 [a]	57.9 ± 2.2 [c]	2.3 ± 0.5 [a]	36 ± 0.3 [b]	22.1 ± 0.9 [b]
F-statistic	0.4 ns	44.9 *	1.2 ns	17.9 *	5.2 *

Values (mean ± SE) followed by dissimilar letters in a column are significantly different at $p \leq 0.05$ (*); ns = not significant.

3.3. Effect of Salt Stress on Chlorophyll Content

As shown in Figure 1, the total chlorophyll contents were negatively affected by 50 mM, 100 mM, and 200 mM salinity concentrations during the fourth week of growth. However, plants exposed to lower salinity (50 mM NaCl) had the highest SPAD-502 values, which was significantly higher than the other treatments, including the control. During the sixth week, salinity concentrations positively affected the chlorophyll values and were significantly different from one another at $p \leq 0.05$. The highest SPAD-502 values were obtained at 200 mM NaCl concentration followed by 100 mM, 50 mM, and the control. During the eighth week, chlorophyll values were negatively affected by salinity as all treatments, including the control, had lower chlorophyll values when compared to week 6. However, a higher concentration (200 mM NaCl) had the highest SPAD-502 value, but it was not significantly different from the other treatments except for the control at $p \leq 0.05$. During the 10th week, salinity stress further reduced the chlorophyll values of all treatments except the control. The control had the highest mean value, which was significantly higher than all salt treatments.

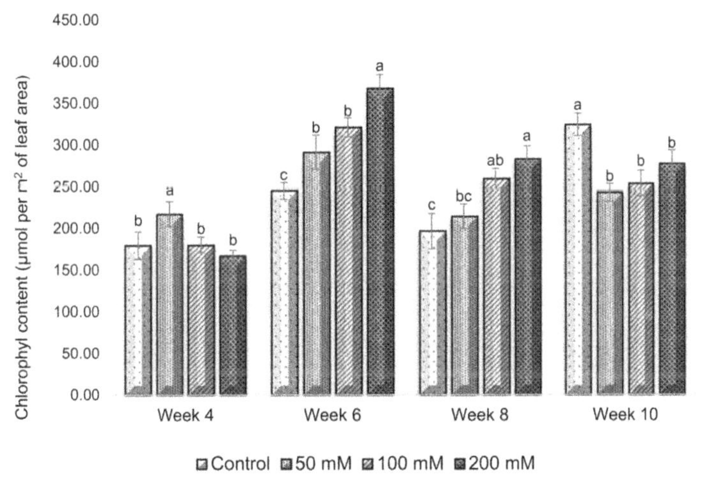

Figure 1. The effect of NaCl concentrations on the chlorophyll readings of *T. decumbens* leaves. a–c indicate significant differences in mean values measured with Fisher's least significant difference. Bars with different letters in the same week are significantly different at $p \leq 0.05$.

3.4. Effects of Salt Stress on Phenolic Content and Antioxidant Capacity

3.4.1. Polyphenol Content

The polyphenol content in the leaves of *T. decumbens* varied significantly at $p \leq 0.05$ when different NaCl concentrations were compared with each other and with the control (Table 4). Plants exposed to the higher NaCl concentration (200 mM) had the highest polyphenol content (2.6 GAE/g DW) compared to all treatments, including the control. This was significantly different from the control and the 50 mM NaCl concentration but did not differ significantly from the moderate NaCl concentration (100 mM).

Table 4. Effect of salt stress on the phenolic content and antioxidant capacity of *T. decumbens* leaves.

Treatments	Total Polyphenols (mg GAE/g DW^{-1})	ABTS (µM TE/g DW^{-1})	FRAP (µM AAE/g DW^{-1})
Control	1.3 ± 0.2 [b]	82.5 ± 8 [a]	10.9 ± 0.1 [b]
50 mM	1.6 ± 0.3 [b]	78.4 ± 6 [a]	14.3 ± 0.4 [a]
100 mM	1.7 ± 0.4 [ab]	70.4 ± 4 [a]	12.1 ± 1.1 [ab]
200 mM	2.6 ± 0.2 [a]	77.8 ± 4.4 [a]	11.8 ± 0.7 [b]
F-statistic	3.3 *	0.41 ns	3.6 *

Values (mean $\pm$ SE) followed by dissimilar letters in a column are significantly different at $p \leq 0.05$ (*); ns = not significant.

3.4.2. ABTS Capacity

Salt stress had a negative influence on ABTS capacity in the leaves of *T. decumbens*. The control had the highest ABTS capacity; however, this was not significantly different from all the treatments (Table 4).

3.4.3. FRAP Capacity

The total FRAP capacity in the leaves of *T. decumbens* was significantly influenced by the NaCl concentrations at $p \leq 0.05$. The lower NaCl concentration (50 mM) had the highest FRAP capacity (14 µM AAE/g DW) compared to the other treatments, including the control. This was significantly higher than the control and 200 mM NaCl concentration but did not differ significantly from the 100 mM NaCl concentration (Table 4).

4. Discussion

There is extensive literature on the reduction of growth caused by salt stress in many plants, which are facilitated by homeostatic transport of Na$^+$ across intra- and inter-cellular cell boundaries predominated by NaCl [46,47]. Nevertheless, the effect of salt stress varies among plants. In the present study, increasing NaCl concentrations led to a significant decrease in plant height. Height reduction as a result of salinity stress has been reported in several plant species and has been mainly associated with the osmotic stress and ion toxicity that causes a reduction in plant growth [48–50]. However, when comparing the number of branches among the treatments, all plants irrigated with NaCl had more branches compared to the control. This might be caused by the natural adaptation of the species to saline environments, which enhances the ability of the species to remediate saline soil and stabilize the coastal dunes. The increase in branch numbers resulted in a higher total fresh weight. These findings agree with the findings of [51], where the halophyte *Ammophila arenaria* showed increased plant biomass in lower to moderate soil salinity. However, the results contradict those reported by [52], where the authors observed that after 12 weeks of cultivating some halophytes, such as *Inula crithmoides* L., *Plantago crassifolia* Forssk. and *Medicago marina* L., the plants whose irrigation water had not been spiked with salt showed better productivity and growth rates. The ability of dune spinach to withstand these varying salt concentrations could be attributed to osmotic, ion, and tissue tolerance. At high salt concentrations, the growth of dune spinach reduced drastically. This has been reported in numerous studies conducted on halophytes, where increasing salinity negatively affected plant growth performance, causing a reduction in biomass, leaf number, and plant height [47,53,54]. Reference [55] also reported that longer salt exposure in the root zone restricts the flow of water and nutrients into the plant, causing direct injury to plant cells through the accumulation of toxic ions causing a decline in plant growth.

Salt stress has been reported as one of the major environmental factors affecting the nutritional value of many edible plants. In the present study, salt stress increased the uptake of N, P, and Na in the leaves, while K, Ca, and Mg were reduced drastically. The reduction of these elements may be directly linked to excessive Na^+ absorption by the roots as reported by [56]. However, sufficient K, Ca, and Mg are required to meet basic metabolic processes such as intracellular K homeostasis, which is essential for optimal functioning of the photosynthetic machinery and maintenance of stomatal opening [57]. These results suggest that dune spinach can transport K, Ca, and Mg to new shoots and leaves under salt stress and maintain a suitable ratio needed for normal metabolism; hence, the chlorophyll content was not affected for 8 weeks. This could be attributed to the water use efficiency and carbon fixing capacity of this species, which uses Crassulacean acid metabolism (CAM) to adapt to harsh conditions. Our results agree with those conducted by [58,59] on *Chenopodium quinoa* (genotype A7) and *Cichorium spinosum* in saline conditions, respectively, where it was reported that higher transport of K and Ca into new shoots and leaves contributed to mitigating ion toxicity in leaf cells.

Moreover, salinity stress also increased the Mn, Fe, and Cu contents in the leaves, while Zn and B were negatively affected. Similar findings were reported on the edible halophyte *Salicornia ramosissima* by [60]. These results indicate that salt stress caused Zn and B deficiency in the leaves of dune spinach, but since they are required in small quantities, visual symptoms of nutrient deficiency did not occur.

It has been reported in the literature that salinity stress damages nutrition and promotes senescence mechanisms in plants, thereby causing a reduction of chlorophyll content in the leaves [49,50]. However, the extent of reduction depends on the salt tolerance of the plant species [61]. Reference [62] reported that salt-tolerant species, such as *Thellungiella halophila*, indicated more or unchanged chlorophyll content when exposed to 0–500 mM NaCl, while salt-sensitive species (glycophyte), such as *Arabidopsis thaliana*, had lower chlorophyll content. In the present study, chlorophyll content was used as a biochemical marker to screen the salt tolerance of dune spinach. SPAD values (chlorophyll content) varied among treatments during the growing weeks, with salt treatments having higher SPAD values on week 6 and 8 when compared to the control (Figure 1). The findings of this study are in agreement with the results obtained by [63] in M-81E sweet sorghum (salt-tolerant genotype), where the chlorophyll content was not affected by 50 mM NaCl. In another study conducted by [64], spinach cultivar raccoon treated with saline irrigation water maintained SPAD chlorophyll levels but had a reduced photosynthetic rate, stomatal conductance, and transpiration rate.

Under salinity stress, the balance between reactive oxygen species production and activities of an antioxidative enzyme determines whether oxidative damage will occur [65]. To reduce the effects of oxidative stress, plants accumulate metabolites, such as phenolic compounds, which act as reducing agents, hydrogen donors, and singlet oxygen quenchers [49,66]. Moreover, phenolic compounds are of great interest due to the relevant role they play in the taste and flavor of food products, as well as their health-promoting properties [67,68]. In the present study, the total phenolic content was significantly increased by salinity levels, with more prominent content in plants irrigated with the highest NaCl concentration (200 mM). These findings validate that of [69], where an increase in the total phenolic content, antioxidant activity, and cyanidin-3- glucoside content was found in Khamdoisaket and KDML 105 Thai rice cultivars subjected to salinity stress. A similar trend was also reported by [70] on the effects of salinity on biochemical characteristics of the stock plant (*Matthiola incana* L.), where the phenolic content in severe salt-stressed plants of both cultivars was higher than the control. Reference [71] stated that phenolic compounds may be affected by salinity, but this critically depends on the salt sensitivity of a considered species. The results of the study prove that dune spinach can grow under severe salinity concentrations and could be considered as an alternative source of nutritional antioxidant in areas with higher and problematic saline soils.

Contrary to the increase in phenolic content, the antioxidant activity (ABTS) in the leaves of *T. decumbens* exposed to various salinity concentrations showed a much weaker antioxidant capacity compared to the control. This contradicts the findings of [16] on edible flowers, where three antioxidant assays (FRAP, DPPH, and ABTS) increased with the application of salinity, with a more pronounced impact at a salinity of 100 mM NaCl. Furthermore, these results also substantiate that of [72] on a traditional Chinese herb, where a much weaker antioxidant capacity was found with increasing NaCl concentrations at 50 mM and 100 mM, respectively, compared to the control. Conversely, the FRAP capacity was positively influenced by salt stress with the strongest antioxidant capacity obtained at 50 mM NaCl. A similar trend was reported earlier by [16] in edible flowers, where the application of salinity enhanced antioxidant activities.

5. Conclusions

The results of this study showed that there is a significant advantage of using nutrient solution incorporated with NaCl to increase yield, mineral composition, and antioxidant activity in dune spinach. Plants irrigated with nutrient solution incorporated with 50 mM NaCl revealed a significant increase in growth parameters and FRAP capacity. As the plant grew beyond week 6, an increased level of salinity caused a decrease in chlorophyll content (SPAD), while the phenolic content, as well as nitrogen, phosphorus, and sodium, increased. Based on this study, it is evident that there is potential for brackish water cultivation of dune spinach for consumption as a leafy vegetable and as a natural source of nutritional antioxidants. This could be a water-saving option in provinces experiencing the adverse effect of drought and salinity, where seawater or underground saline water could be diluted and used as irrigation water in the production of this vegetable.

Author Contributions: A.S.: investigation, data collection, documentation and analysis, methodology, and writing the original draft of the manuscript. M.O.J.: referencing and statistical software, critical review of the article with technical input, final approval of the article, and manuscript preparation. L.K.: validation of experimental design, supervision, and critical revision of the article. C.P.L.: conceptualization, supervision, funding acquisition, and critical revision of the article. All authors have read and agreed to the published version of the manuscript.

Funding: This research was funded by the National Research Foundation of South Africa (Grant numbers: 117032 and 121087).

Institutional Review Board Statement: Not applicable.

Informed Consent Statement: Not applicable.

Data Availability Statement: Data is contained within the article.

Acknowledgments: The authors would like to thank Xatyiswa Salukazana, Sihle Ngxabi, Lizeka Gana and Nomnqophiso Zantanta for their valuable assistance in data collection and experimental setup. To Deborah Erasmus and Phumlani Roto (Technicians in Horticultural Department) thank you for your assistance in procuring the needed equipment for this research.

Conflicts of Interest: The authors declare no conflict of interest.

References

1. Molotoks, A.; Smith, P.; Dawson, T.P. Impacts of land use, population, and climate change on global food security. *Food Energy Secur.* **2021**, *10*, e261. [CrossRef]
2. Gidamis, A.B.; Chove, B.E. Biotechnology and Biosafety: Exploring the Debate and Public Perception in Developing Countries. *J. Knowl. Glob.* **2009**, *2*, 45–62.
3. Ogunniyi, A.I.; Mavrotas, G.; Olagunju, K.O.; Fadare, O.; Adedoyin, R. Governance quality, remittances and their implications for food and nutrition security in Sub-Saharan Africa. *World Dev.* **2020**, *127*, 104752. [CrossRef]
4. World Bank. *World Development Report 2008: Agriculture for Development*; World Bank: Washington, DC, USA, 2007.
5. Acquaah, G. *Principles of Plant Genetics and Breeding*; Blackwell Publishing: Oxford, UK, 2007.

6. Hussain, M.; Park, H.W.; Farooq, M.; Jabran, K.; Lee, D.J. Morphological and physiological basis of salt resistance in different rice genotypes. *Int. J. Agric. Biol.* **2013**, *15*, 113–118.

7. Shrivastava, P.; Kumar, R. Soil salinity: A serious environmental issue and plant growth promoting bacteria as one of the tools for its alleviation. *Saudi J. Biol. Sci.* **2015**, *22*, 123–131. [CrossRef]

8. Farooq, M.; Hussain, M.; Wakeel, A.; Siddique, K.H.M. Salt stress in maize: Effects, resistance mechanisms, and management. A review. *Agron. Sustain. Dev.* **2015**, *35*, 461–481. [CrossRef]

9. Atieno, J.; Li, Y.; Langridge, P.; Dowling, K.; Brien, C.; Berger, B.; Varshney, R.K.; Sutton, T. Exploring genetic variation for salinity tolerance in chickpea using image-based phenotyping. *Sci. Rep.* **2017**, *7*, 1–11. [CrossRef]

10. Huang, Y.; Guan, C.; Liu, Y.; Chen, B.; Yuan, S.; Cui, X.; Zhang, Y.; Yang, F. Enhanced growth performance and salinity tolerance in transgenic switchgrass via overexpressing vacuolar Na+ (K+)/H+ antiporter gene (PvNHX1). *Front. Plant Sci.* **2017**, *8*, 1–13. [CrossRef]

11. García-Caparrós, P.; De Filippis, L.; Gul, A.; Hasanuzzaman, M.; Ozturk, M.; Altay, V.; Lao, M.T. Oxidative Stress and Antioxidant Metabolism under Adverse Environmental Conditions: A Review. *Bot. Rev.* **2020**, 1–46. [CrossRef]

12. Chao, Y.Y.; Hsu, Y.T.; Kao, C.H. Involvement of glutathione in heat shock- and hydrogen peroxide-induced cadmium tolerance of rice (*Oryza sativa* L.) seedlings. *Plant Soil* **2009**, *318*, 37–45. [CrossRef]

13. Hernández, J.A.; Ferrer, M.A.; Jiménez, A.; Barceló, A.R.; Sevilla, F. Antioxidant systems and O_2-/H_2O_2 production in the apoplast of pea leaves. Its relation with salt-induced necrotic lesions in minor veins. *Plant Physiol.* **2001**, *127*, 817–831. [CrossRef]

14. Ahmad, P.; Jaleel, C.A.; Sharma, S. Antioxidant defence system, lipid peroxidation, proline-metabolizing enzymes, and biochemical activities in two Morus alba genotypes subjected to NaCl stress. *Russ. J. Plant Physiol.* **2010**, *57*, 509–517. [CrossRef]

15. Rasheed, R.; Ashraf, M.A.; Iqbal, M.; Hussain, I.; Akbar, A.; Farooq, U.; Shad, M.I. Major Constraints for Global Rice Production: Changing Climate, Abiotic and Biotic Stresses. In *Rice Research for Quality Improvement: Genomics and Genetic Engineering*; Springer: Singapore, 2020; pp. 15–45.

16. Chrysargyris, A.; Tzionis, A.; Xylia, P.; Tzortzakis, N. Effects of salinity on tagetes growth, physiology, and shelf life of edible flowers stored in passive modified atmosphere packaging or treated with ethanol. *Front. Plant Sci.* **2018**, *871*, 1–13. [CrossRef]

17. Chrysargyris, A.; Michailidi, E.; Tzortzakis, N. Physiological and biochemical responses of Lavandula angustifolia to salinity under mineral foliar application. *Front. Plant Sci.* **2018**, *9*, 489. [CrossRef]

18. Durazzo, A. Study approach of antioxidant properties in foods: Update and considerations. *Foods* **2017**, *6*, 17. [CrossRef]

19. Jimoh, M.O.; Afolayan, A.J.; Lewu, F.B. Nutrients and antinutrient constituents of *Amaranthus caudatus* L. Cultivated on different soils. *Saudi J. Biol. Sci.* **2020**, *27*, 3570–3580. [CrossRef]

20. Salami, S.O.; Afolayan, A.J. Suitability of *Roselle—Hibiscus sabdariffa* L. as raw material for soft drink production. *J. Food Qual.* **2020**, *2020*, 8864142. [CrossRef]

21. Debez, A.; Saadaoui, D.; Slama, I.; Huchzermeyer, B.; Abdelly, C. Responses of Batis maritima plants challenged with up to two-fold seawater NaCl salinity. *J. Plant Nutr. Soil Sci.* **2010**, *173*, 291–299. [CrossRef]

22. Ventura, Y.; Sagi, M. Halophyte crop cultivation: The case for salicornia and sarcocornia. *Environ. Exp. Bot.* **2013**, *92*, 144–153. [CrossRef]

23. Bueno, M.; Cordovilla, M.P. Ecophysiology and Uses of Halophytes in Diverse Habitats. In *Handbook of Halophytes*; Springer International Publishing: Cham, Switzerland, 2020; pp. 1–25.

24. Ruiz-Rodríguez, B.M.; Morales, P.; Fernández-Ruiz, V.; Sánchez-Mata, M.C.; Cámara, M.; Díez-Marqués, C.; Pardo-de-Santayana, M.; Molina, M.; Tardío, J. Valorization of wild strawberry-tree fruits (*Arbutus unedo* L.) through nutritional assessment and natural production data. *Food Res. Int.* **2011**, *44*, 1244–1253. [CrossRef]

25. Trichopoulou, A.; Vasilopoulou, E.; Hollman, P.; Chamalides, C.; Foufa, E.; Kaloudis, T.; Kromhout, D.; Miskaki, P.; Petrochilou, I.; Poulima, E.; et al. Nutritional composition and flavonoid content of edible wild greens and green pies: A potentially rich source of antioxidant nutrients in the Mediterranean diet. *Food Chem.* **2000**, *70*, 319–323. [CrossRef]

26. Ksouri, R.; Megdiche, W.; Debez, A.; Falleh, H.; Grignon, C.; Abdelly, C. Salinity effects on polyphenol content and antioxidant activities in leaves of the halophyte Cakile maritima. *Plant Physiol. Biochem.* **2007**, *45*, 244–249. [CrossRef]

27. Bittrich, V. AIZOACEAE. *Bothalia* **1990**, *20*, 217–219. [CrossRef]

28. Klak, C.; Hanáček, P.; Bruyns, P.V. Out of southern Africa: Origin, biogeography and age of the Aizooideae (Aizoaceae). *Mol. Phylogenet. Evol.* **2017**, *109*, 203–216. [CrossRef]

29. Forrester, J. Tetragonia decumbens. Plantz Africa. South African National Biodiversity Institute. 2004, pp. 1–3. Available online: http://opus.sanbi.org/bitstream/20.500.12143/3865/1/Tetragoniadecumbens_PlantzAfrica.pdf (accessed on 28 May 2021).

30. Tembo-Phiri, C. *Edible Fynbos Plants: A Soil Types and Irrigation Regime Investigation on Tetragonia decumbens and Mesembryanthemum Crystallinum*; Stellenbosch University: Stellenbosch, South Africa, 2019.

31. Bartels, D.; Sunkar, R. Drought and Salt Tolerance in Plants. *CRC Crit. Rev. Plant Sci.* **2005**, *24*, 23–58. [CrossRef]

32. Glenn, E.P.; Brown, J.J.; Blumwald, E. Salt Tolerance and Crop Potential of Halophytes. *CRC Crit. Rev. Plant Sci.* **1999**, *18*, 227–255. [CrossRef]

33. World Wide Fund. *Scenarios for the Future of Water in South Africa*; World Wide Fund For Nature: Cape Town, South Africa, 2017.

34. Panta, S.; Flowers, T.; Lane, P.; Doyle, R.; Haros, G.; Shabala, S. Halophyte agriculture: Success stories. *Environ. Exp. Bot.* **2014**, *107*, 71–83. [CrossRef]

35. Unuofin, J.O.; Otunola, G.A.; Afolayan, A.J. Nutritional evaluation of *Kedrostis africana* (L.) Cogn: An edible wild plant of South Africa. *Asian Pac. J. Trop. Biomed.* **2017**, *7*, 443–449. [CrossRef]

36. Olatunji, T.L.; Afolayan, A.J. Comparison of nutritional, antioxidant vitamins and capsaicin contents in *Capsicum annuum* and *C. frutescens. Int. J. Veg. Sci.* **2019**, *18*, 1–18. [CrossRef]

37. Jimoh, M.O.; Afolayan, A.J.; Lewu, F.B. Heavy metal uptake and growth characteristics of *Amaranthus caudatus* L. under five different soils in a controlled environment. *Not. Bot. Horti Agrobot.* **2020**, *48*, 417–425. [CrossRef]

38. Hoenig, M.; Baeten, H.; Vanhentenrijk, S.; Vassileva, E.; Quevauviller, P. Critical discussion on the need for an efficient mineralization procedure for the analysis of plant material by atomic spectrometric methods. *Anal. Chim. Acta* **1998**, *358*, 85–94. [CrossRef]

39. Idris, O.A.; Wintola, O.A.; Afolayan, A.J. Comparison of the Proximate Composition, Vitamins Analysis of the Essential Oil of the Root and Leaf of *Rumex crispus* L. *Plants* **2019**, *8*, 51. [CrossRef]

40. Nkcukankcuka, M.; Jimoh, M.O.; Griesel, G.; Laubscher, C.P. Growth characteristics, chlorophyll content and nutrients uptake in *Tetragonia decumbens* Mill. cultivated under different fertigation regimes in hydroponics. *Crop Pasture Sci.* **2021**, *72*, 1–12. [CrossRef]

41. Ainsworth, E.A.; Gillespie, K.M. Estimation of total phenolic content and other oxidation substrates in plant tissues using Folin-Ciocalteu reagent. *Nat. Protoc.* **2007**, *2*, 875–877. [CrossRef]

42. Jimoh, M.O.; Afolayan, A.J.; Lewu, F.B. Antioxidant and phytochemical activities of *Amaranthus caudatus* L. harvested from different soils at various growth stages. *Sci. Rep.* **2019**, *9*, 12965. [CrossRef]

43. Benzie, I.F.F.; Strain, J.J. Ferric reducing/antioxidant power assay: Direct measure of total antioxidant activity of biological fluids and modified version for simultaneous measurement of total antioxidant power and ascorbic acid concentration. *Methods Enzymol.* **1999**, *299*, 15–27. [CrossRef]

44. Jimoh, M.A.; Idris, O.A.; Jimoh, M.O. Cytotoxicity, Phytochemical, Antiparasitic Screening, and Antioxidant Activities of Mucuna pruriens (Fabaceae). *Plants* **2020**, *9*, 1249. [CrossRef]

45. Faber, R.J.; Laubscher, C.P.; Rautenbach, F.; Jimoh, M.O. Variabilities in alkaloid concentration of *Sceletium tortuosum* (L.) N.E. Br in response to different soilless growing media and fertigation regimes in hydroponics. *Heliyon* **2020**, *6*, e05479. [CrossRef]

46. Hasegawa, P.M. Sodium (Na+) homeostasis and salt tolerance of plants. *Environ. Exp. Bot.* **2013**, *92*, 19–31. [CrossRef]

47. Rahneshan, Z.; Nasibi, F.; Moghadam, A.A. Effects of salinity stress on some growth, physiological, biochemical parameters and nutrients in two pistachio (*Pistacia vera* L.) rootstocks. *J. Plant Interact.* **2018**, *13*, 73–82. [CrossRef]

48. Fageria, N.K.; Stone, L.F.; Santos, A.B. Dos Breeding for salinity tolerance. In *Plant Breeding for Abiotic Stress Tolerance*; Springer: Berlin/Heidelberg, Germany, 2012; pp. 103–122. ISBN 9783642305535.

49. Munns, R.; Tester, M. Mechanisms of salinity tolerance. *Annu. Rev. Plant Biol.* **2008**, *59*, 651–681. [CrossRef] [PubMed]

50. Petretto, G.L.; Urgeghe, P.P.; Massa, D.; Melito, S. Effect of salinity (NaCl)on plant growth, nutrient content, and glucosinolate hydrolysis products trends in rocket genotypes. *Plant Physiol. Biochem.* **2019**, *141*, 30–39. [CrossRef] [PubMed]

51. van Puijenbroek, M.E.B.; Teichmann, C.; Meijdam, N.; Oliveras, I.; Berendse, F.; Limpens, J. Does salt stress constrain spatial distribution of dune building grasses Ammophila arenaria and Elytrichia juncea on the beach? *Ecol. Evol.* **2017**, *7*, 7290–7303. [CrossRef] [PubMed]

52. Grigore, M.N.; Villanueva, M.; Boscaiu, M.; Vicente, O. Do Halophytes Really Require Salts for Their Growth and Development? An Experimental Approach. *Not. Sci. Biol.* **2012**, *4*, 23–29. [CrossRef]

53. Guo, J.; Li, Y.; Han, G.; Song, J.; Wang, B. NaCl markedly improved the reproductive capacity of the euhalophyte Suaeda salsa. *Funct. Plant Biol.* **2018**, *45*, 350–361. [CrossRef]

54. Orlovsky, N.; Japakova, U.; Zhang, H.; Volis, S. Effect of salinity on seed germination, growth and ion content in dimorphic seeds of *Salicornia europaea* L. (Chenopodiaceae). *Plant Divers.* **2016**, *38*, 183–189. [CrossRef]

55. Tian, F.; Hou, M.; Qiu, Y.; Zhang, T.; Yuan, Y. Salinity stress effects on transpiration and plant growth under different salinity soil levels based on thermal infrared remote (TIR) technique. *Geoderma* **2020**, *357*, 113961. [CrossRef]

56. Benito, B.; Haro, R.; Amtmann, A.; Cuin, T.A.; Dreyer, I. The twins K+ and Na+ in plants. *J. Plant Physiol.* **2014**, *171*, 723–731. [CrossRef]

57. Shabala, S.; Pottosin, I. Regulation of potassium transport in plants under hostile conditions: Implications for abiotic and biotic stress tolerance. *Physiol. Plant.* **2014**, *151*, 257–279. [CrossRef]

58. Parvez, S.; Abbas, G.; Shahid, M.; Amjad, M.; Hussain, M.; Asad, S.A.; Imran, M.; Naeem, M.A. Effect of salinity on physiological, biochemical and photostabilizing attributes of two genotypes of quinoa (*Chenopodium quinoa* Willd.) exposed to arsenic stress. *Ecotoxicol. Environ. Saf.* **2020**, *187*, 109814. [CrossRef]

59. Petropoulos, S.A.; Levizou, E.; Ntatsi, G.; Fernandes, Â.; Petrotos, K.; Akoumianakis, K.; Barros, L.; Ferreira, I.C.F.R. Salinity effect on nutritional value, chemical composition and bioactive compounds content of *Cichorium spinosum* L. *Food Chem.* **2017**, *214*, 129–136. [CrossRef]

60. Lima, A.R.; Castañeda-Loaiza, V.; Salazar, M.; Nunes, C.; Quintas, C.; Gama, F.; Pestana, M.; Correia, P.J.; Santos, T.; Varela, J.; et al. Influence of cultivation salinity in the nutritional composition, antioxidant capacity and microbial quality of Salicornia ramosissima commercially produced in soilless systems. *Food Chem.* **2020**, *333*, 127525. [CrossRef] [PubMed]

61. Gong, D.H.; Wang, G.Z.; Si, W.T.; Zhou, Y.; Liu, Z.; Jia, J. Effects of Salt Stress on Photosynthetic Pigments and Activity of Ribulose-1,5-bisphosphate Carboxylase/Oxygenase in Kalidium foliatum. *Russ. J. Plant Physiol.* **2018**, *65*, 98–103. [CrossRef]

62. Stepien, P.; Johnson, G.N. Contrasting responses of photosynthesis to salt stress in the glycophyte arabidopsis and the halophyte thellungiella: Role of the plastid terminal oxidase as an alternative electron sink. *Plant Physiol.* **2009**, *149*, 1154–1165. [CrossRef]
63. Sui, N.; Yang, Z.; Liu, M.; Wang, B. Identification and transcriptomic profiling of genes involved in increasing sugar content during salt stress in sweet sorghum leaves. *BMC Genom.* **2015**, *16*, 1–18. [CrossRef]
64. Ferreira, J.; Cornacchione, M.; Liu, X.; Suarez, D. Nutrient Composition, Forage Parameters, and Antioxidant Capacity of Alfalfa (*Medicago sativa*, L.) in Response to Saline Irrigation Water. *Agriculture* **2015**, *5*, 577–597. [CrossRef]
65. Møller, I.M.; Jensen, P.E.; Hansson, A. Oxidative modifications to cellular components in plants. *Annu. Rev. Plant Biol.* **2007**, *58*, 459–481. [CrossRef]
66. Hasegawa, P.M.; Bressan, R.A.; Zhu, J.-K.; Bohnert, H.J. Plant cellular and molecular responses to high salinity. *Annu. Rev. Plant Physiol. Plant Mol. Biol.* **2000**, *51*, 463–499. [CrossRef]
67. Cheynier, V. Phenolic compounds: From plants to foods. *Phytochem. Rev.* **2012**, *11*, 153–177. [CrossRef]
68. Gil, A.M.; Duarte, I.F.; Godejohann, M.; Braumann, U.; Maraschin, M.; Spraul, M. Characterization of the aromatic composition of some liquid foods by nuclear magnetic resonance spectrometry and liquid chromatography with nuclear magnetic resonance and mass spectrometric detection. *Anal. Chim. Acta* **2003**, *488*, 35–51. [CrossRef]
69. Daiponmak, W.; Theerakulpisut, P.; Thanonkao, P.; Vanavichit, A.; Prathepha, P. Changes of anthocyanin cyanidin-3-glucoside content and antioxidant activity in Thai rice varieties under salinity stress. *ScienceAsia* **2010**, *36*, 286–291. [CrossRef]
70. Jafari, S.; Hashemi Garmdareh, S.E. Effects of salinity on morpho-physiological, and biochemical characteristics of stock plant (*Matthiola incana* L.). *Sci. Hortic.* **2019**, *257*, 108731. [CrossRef]
71. Kim, H.J.; Fonseca, J.M.; Choi, J.H.; Kubota, C.; Dae, Y.K. Salt in irrigation water affects the nutritional and visual properties of romaine lettuce (*Lactuca sativa* L.). *J. Agric. Food Chem.* **2008**, *56*, 3772–3776. [CrossRef] [PubMed]
72. Zhou, Y.; Tang, N.; Huang, L.; Zhao, Y.; Tang, X.; Wang, K. Effects of salt stress on plant growth, antioxidant capacity, glandular trichome density, and volatile exudates of schizonepeta tenuifolia briq. *Int. J. Mol. Sci.* **2018**, *19*, 252. [CrossRef] [PubMed]

horticulturae

Article

Genetic Diversity and Population Differentiation of *Pinus koraiensis* in China

Xiang Li [1], Minghui Zhao [1], Yujin Xu [1], Yan Li [1], Mulualem Tigabu [2] and Xiyang Zhao [1,*]

[1] State Key Laboratory of Tree Genetics and Breeding, School of Forestry, Northeast Forestry University, Harbin 150040, China; lx2016bjfu@163.com (X.L.); zhaominghui66@163.com (M.Z.); xuyjphd@163.com (Y.X.); Ly2019nefu@163.com (Y.L.)

[2] Southern Swedish Forest Research Centre, Swedish University of Agricultural Sciences, SE-230 53 Alnarp, Sweden; mulualem.tigabu@slu.se

* Correspondence: zhaoxyphd@163.com; Tel.: +86-0451-8219-2225

Abstract: *Pinus koraiensis* is a well-known precious tree species in East Asia with high economic, ornamental and ecological value. More than fifty percent of the *P. koraiensis* forests in the world are distributed in northeast China, a region with abundant germplasm resources. However, these natural *P. koraiensis* sources are in danger of genetic erosion caused by continuous climate changes, natural disturbances such as wildfire and frequent human activity. Little work has been conducted on the population genetic structure and genetic differentiation of *P. koraiensis* in China because of the lack of genetic information. In this study, 480 *P. koraiensis* individuals from 16 natural populations were sampled and genotyped. Fifteen polymorphic expressed sequence tag-simple sequence repeat (EST-SSR) markers were used to evaluate genetic diversity, population structure and differentiation in *P. koraiensis*. Analysis of molecular variance (AMOVA) of the EST-SSR marker data showed that 33% of the total genetic variation was among populations and 67% was within populations. A high level of genetic diversity was found across the *P. koraiensis* populations, and the highest levels of genetic diversity were found in HH, ZH, LS and TL populations. Moreover, pairwise Fst values revealed significant genetic differentiation among populations (mean Fst = 0.177). According to the results of the STRUCTURE and Neighbor-joining (NJ) tree analyses and principal component analysis (PCA), the studied geographical populations cluster into two genetic clusters: cluster 1 from Xiaoxinganling Mountains and cluster 2 from Changbaishan Mountains. These results are consistent with the geographical distributions of the populations. The results provide new genetic information for future genome-wide association studies (GWAS), marker-assisted selection (MAS) and genomic selection (GS) in natural *P. koraiensis* breeding programs and can aid the development of conservation and management strategies for this valuable conifer species.

Keywords: *Pinus koraiensis*; EST-SSRs; genetic diversity; population structure; population differentiation; gene flow

Citation: Li, X.; Zhao, M.; Xu, Y.; Li, Y.; Tigabu, M.; Zhao, X. Genetic Diversity and Population Differentiation of *Pinus koraiensis* in China. *Horticulturae* **2021**, *7*, 104. https://doi.org/10.3390/horticulturae7050104

Academic Editors: Rosario Paolo Mauro, Carlo Nicoletto and Leo Sabatino

Received: 6 April 2021
Accepted: 6 May 2021
Published: 9 May 2021

Publisher's Note: MDPI stays neutral with regard to jurisdictional claims in published maps and institutional affiliations.

1. Introduction

Pinus koraiensis (Sieb. et Zucc), commonly known as Korean pine, is a perennial evergreen tree in the Pinaceae family with five needles per fascicle [1–3]. It is an ancient and valued forest tree in East Asia, and natural forests of this species have undergone long-term succession and are described as tertiary forest [4]. Compared with other *Pinus* species, *P. koraiensis* is long-lived and is a dominant species in mixed conifer and broadleaved forest [5,6]. Currently, *P. koraiensis* is distributed mainly in cool-temperate regions in northeast China, the Russian Far East, the Korea peninsula (note that information is not available from North Korea due to limited access) and Honshu, Japan. It typically occurs in mild regions with more than 70% humidity and at altitudes from 600 m to 1500 m [7,8]. However, in China, it only grows from the Changbai Mountains to Xiaoxinganling Mountains in northeast China, mainly on slopes and rolling hills and in river valleys [9]. Nearly half of the germplasm resources of *P. koraiensis* in the world are found in Xiaoxinganling Mountains in Yichun city, China, where the largest and most undisturbed primeval forest remaining in Asia and a natural climax community of *P. koraiensis* exists [10].

P. koraiensis has high economic, ornamental and ecological values in East Asia. Timber of *P. koraiensis* is widely used for architecture, bridges, furniture and ships because of the light, soft, fine structure and straight texture of the wood and its strong corrosion resistance [11]. Furthermore, it produces edible nuts that are nutritious and distinctly flavored, containing abundant unsaturated fatty acids, vitamins and minerals [12]. It also has high medicinal value, able to lower cholesterol levels and allay ultraviolet injury and tiredness [13]. Natural *P. koraiensis* forest absorbs large amounts of carbon dioxide and contributes to climate change regulation [14]. Therefore, it is a prominent conifer tree species of great value for the maintenance and protection of the environment in East Asia.

Genetic improvement of *P. koraiensis* began in the 1960s, which then developed slowly due to a lack of systematic breeding strategies and objectives [8]. In the early stages of selective breeding, large numbers of superior trees or natural populations were selected from natural forest to establish primary seed orchards, mainly through phenotype selection [15–17]. Earlier studies have mainly focused on propagation technology [18], provenance division [19], progeny determination [20] and selection of improved varieties [21], while studies of molecular plant breeding, including studies of genetic diversity, genomic selection and construction of genetic maps, are lacking [22]. Existing natural forests of this species have great significance for the conservation of breeding materials, the development of gene resources and the study of population genetic diversity [23]. However, in the past few decades, with the increasing demand for wood and cones of *P. koraiensis* as well as increasing wildfire, the area of natural *P. koraiensis* forest has decreased extensively [24]. Thus, to protect existing natural forests under the background of illegal logging and unpredictable biotic stress, such as white pine blister rust diseases, the collection and evaluation of germplasm resources of *P. koraiensis* are urgently needed.

Genetic diversity and population structure are key parameters of population genetics research. Analyses of genetic variation among and within populations can guide the formulation of conservation strategies. The use of molecular markers identified from whole-genome, chloroplast genome and transcriptome analysis is a primary method of revealing genetic diversity and population structure. Many DNA molecular markers are codominant and highly polymorphic, and many have been identified in the genome and transcriptome, unlike morphological and biochemical markers [25–27]. Simple sequence repeats (SSRs) are considered powerful and advantageous molecular tools due to their low cost, easy detection by polymerase chain reaction (PCR), high polymorphism, and codominance. Thus, they can be used for genetic diversity analysis, genome-wide association analysis, core collections and genetic linkage map construction in many plants and animals [28–30]. Furthermore, multiple EST-SSR markers can easily be developed from microsatellite loci of public transcriptome data. At present, there are few reports of analyses of genetic diversity in *P. koraiensis* based on DNA molecular markers; studies to date have employed random amplified polymorphic DNA (RAPD) analysis [31], single primer amplification reaction (SPAR) [32], intersimple sequence repeat (ISSR) analysis [33,34] and expressed sequence tag-simple sequence repeat (EST-SSR) analysis [35]. All these studies have identified high levels of genetic diversity in *P. koraiensis*, with the greatest levels of genetic differentiation occurring within populations. However, those previous studies focused on a limited number of populations, few molecular markers and population size. Thus, a systematic and comprehensive population genetic study, involving widespread germplasm collection and abundant polymorphic markers developed from high-throughput sequencing, is necessary to study the genetic relationships and diversity of *P. koraiensis* populations.

In this study, germplasm resources from 480 individuals of 16 natural populations of *P. koraiensis* were collected within the species' main distribution area in northeastern China, and analyzed for genetic diversity using 15 EST-SSRs. This study is the first comprehensive study evaluating the genetic diversity and population structure of *P. koraiensis* in China using large samples and wider distribution as well as a sufficient number of molecular markers. The aims of the study were to (1) investigate genetic variation using polymorphic EST-SSRs, (2) evaluate the genetic diversity and structure of natural populations, (3) conduct a comprehensive, range-wide genetic diversity study of *P. koraiensis* in China, and (4) propose a protection conservation strategy. The hypothesis of the study was that high genetic diversity could be detected within populations and significant genetic differentiation could exist among populations due to restricted natural distribution of the species and low to moderate degree of gene flow between populations. Thus, the results will provide insights into the conservation of this species and lay a foundation for further studies of marker-assisted selection (MAS) and genomic selection (GS) in *P. koraiensis* for genetic improvement.

2. Results

2.1. Genetic Diversity at Different Loci among Populations

The genetic diversity analysis was performed on 480 individuals from 16 natural *P. koraiensis* populations using 15 EST-SSRs markers (Table 1). The allele size ranged from 151 bp at locus NEPK-65 to 301 bp at loci NEPK-168 and NEPK-184. In total, 155 alleles across all 15 loci were detected in the sampled individuals; the number of alleles per locus ranged from 4 (NEPK-67) to 21 (NEPK-145), with a mean value of 10.33. There were 58 private alleles, accounting for 37.42% of the alleles. The number of effective alleles (Ne) ranged from 1.170 at locus NEPK-40 to 6.605 at locus NEPK-145, with an average of 2.514 per locus. The observed (Ho) and expected (He) heterozygosity ranged from 0.008 to 0.984 and from 0.145 to 0.849, respectively, with mean values of 0.374 and 0.521, respectively. The polymorphic information content (PIC) varied from 0.142 (NEPK-40) to 0.833 (NEPK-145), with a mean value of 0.461. Four loci exhibited high polymorphism (PIC > 0.5) and 8 loci exhibited moderate polymorphism (0.2 < PIC < 0.5). In addition, across the 480 samples, all of the loci conformed to Hardy–Weinberg equilibrium. F-statistics were calculated to detect genetic subdivision and revealed moderate inbreeding and the mean value of Fst was 0.347, indicating moderate genetic variation. Regarding gene flow, the number of effective migrants (Nm) value ranged from 0.080 to 17.691 among populations, with an average of 2.667.

Table 1. Characteristics of the 15 polymorphic EST-SSR markers used in this study.

Locus	Allele Size Range (bp)	Na	Ne	I	Ho	He	PIC	HWE	NRA	Fis	Fit	Fst	Nm
NEPK-218	196–230	12	2.080	0.949	0.056	0.519	0.443	***	6	0.647	0.893	0.695	0.109
NEPK-40	196–238	13	1.170	0.399	0.063	0.145	0.142	***	3	0.327	0.593	0.396	0.382
NEPK-32	206–224	5	2.431	0.979	0.618	0.589	0.503	***	1	−0.236	−0.053	0.148	1.438
NEPK-53	197–218	5	1.902	0.697	0.008	0.474	0.367	***	3	0.923	0.981	0.757	0.080
NEPK-65	151–259	12	2.257	1.068	0.195	0.557	0.488	***	5	0.259	0.649	0.526	0.225
NEPK-71	215–239	7	2.013	0.772	0.635	0.503	0.392	***	3	−0.275	−0.254	0.016	15.141
NEPK-117	205–217	9	3.995	1.601	0.984	0.750	0.714	***	2	−0.594	−0.290	0.191	1.058
NEPK-72	222–234	7	1.649	0.637	0.043	0.394	0.324	***	4	0.346	0.894	0.838	0.048
NEPK-67	226–232	4	2.802	1.067	0.479	0.643	0.567	***	1	0.034	0.259	0.233	0.822
NEPK-38	229–255	15	1.211	0.498	0.096	0.174	0.172	***	6	0.396	0.477	0.134	1.622
NEPK-145	251–289	21	6.605	2.243	0.637	0.849	0.833	***	6	0.229	0.309	0.104	2.154
NEPK-168	241–301	18	3.550	1.602	0.279	0.718	0.674	***	9	0.265	0.602	0.459	0.295
NEPK-181	275–289	10	2.039	0.884	0.165	0.509	0.426	***	5	0.375	0.667	0.467	0.285
NEPK-213	276–298	12	1.846	1.039	0.421	0.458	0.435	***	4	−0.151	0.109	0.226	0.856
NEPK-184	291–301	5	2.152	0.856	0.937	0.535	0.430	***	0	−0.743	−0.719	0.014	17.691
Average		10.33	2.514	1.019	0.374	0.521	0.461		3.867	0.12	0.341	0.347	2.667

Note: Na, number of different alleles; Ne, number of effective alleles; I, Shannon's diversity index; Ho, observed heterozygosity; He, expected heterozygosity; PIC, polymorphic information content; HWE, deviation from Hardy–Weinberg equilibrium (*** $p < 0.001$); NRA, number private allele; Fis, inbreeding coefficient; Fit, over inbreeding coefficient; Fst, genetic differentiation index; Nm, number of effective migrants.

2.2. Genetic Diversity within Pinus koraiensis Populations

The levels of genetic diversity in the 16 populations are shown in Table 2. Across the sampled populations, the number of different alleles (Na) varied from 2.667 (HL) to 4.467 (TL), with a mean value of 3.271, and the number of effective alleles (Ne) ranged from 1.586 (Jiaohe) to 2.257 (Linjiang), with a mean value of 1.870. The populations with the highest levels of genetic diversity were Heihe (Ne = 1.939, Ho = 0.340 and He = 0.439), Zhanhe (Ne = 2.009, Ho = 0.356 and He = 0.413), Liangshui (Ne = 1.914, Ho = 0.470 and He = 0.370) and Tieli (Ne = 2.222, Ho = 0.373 and He = 0.414), whereas those with the lowest levels were Jiaohe (Ne = 1.586, Ho = 0.293 and He = 0.275) and Helong (Ne = 1.663, Ho = 0.390 and He = 0.310). The Zhanhe population had not only the highest genetic diversity but also the largest number of private alleles, identifying it as a unique natural *P. koraiensis* population. The F value ranged from -0.235 to 0.325 among the populations, with a mean value of 0.02, indicating that there existed a deficiency of heterozygosity in the natural *P. koraiensis* populations.

Table 2. Genetic diversity estimates for the 16 *P. koraiensis* populations based on 15 EST-SRRs markers.

Population	Size	Na	Ne	I	Ho	He	uHe	F (Null)	NRA
Liangzihe	30	3.267	1.713	0.639	0.415	0.374	0.388	−0.054	7
Helong	30	2.667	1.663	0.529	0.390	0.310	0.321	−0.195	2
Lushuihe	30	3.133	1.829	0.628	0.394	0.350	0.359	0.015	1
Heihe	30	3.133	1.939	0.746	0.340	0.439	0.449	0.325	3
Liangshui	10	3.333	1.914	0.664	0.470	0.370	0.380	−0.235	6
Zhanhe	30	4.133	2.009	0.782	0.356	0.413	0.422	0.156	12
Tieli	30	4.467	2.222	0.806	0.373	0.414	0.422	0.178	8
Hegang	30	3.533	1.702	0.634	0.401	0.354	0.361	−0.088	7
Linjiang	28	3.667	2.257	0.860	0.470	0.488	0.502	0.018	4
Jiaohe	30	3.067	1.586	0.497	0.293	0.275	0.280	0.127	1
Hulin	29	2.733	1.877	0.523	0.324	0.294	0.300	−0.025	0
Boli	30	3.400	1.970	0.666	0.328	0.361	0.367	0.116	2
Muling	30	3.000	1.840	0.525	0.335	0.285	0.290	−0.013	3
Maoershan	30	3.000	1.784	0.539	0.368	0.293	0.300	−0.186	2
Fangzheng	30	3.000	1.711	0.531	0.341	0.286	0.292	0.033	1
Wangqing	30	2.800	1.913	0.579	0.327	0.328	0.337	0.148	2
Mean		3.271	1.870	0.634	0.370	0.352	0.361	0.020	3.813

Note: Na, number of different alleles; Ne, number of effective alleles; I, Shannon's diversity index; Ho, observed heterozygosity; He, expected heterozygosity; uHe, unbiased expected heterozygosity; F (null), null allele frequencies; NRA, number private allele.

2.3. Genetic Variation among Pinus koraiensis Populations

To evaluate the genetic variation among the collected samples, AMOVA was performed, and Fst among natural populations, genetic clusters and geographical regions were calculated; the results are shown in Table 3. The AMOVA results indicate that 67% of the total genetic variation existed within populations, indicating high genetic diversity within populations. AMOVA of the two genetic clusters identified by the STRUCTURE analysis indicated that 63.79% of the total variation was attributable to differences within populations, and the overall Fst was 0.362 (Fst > 0.25), indicating high genetic differentiation between the two clusters. In addition, the AMOVA of two groups classified according to geographical location indicated low genetic variation among populations within each group (2.77%). All of these results indicated high genetic differentiation within populations and groups.

Table 3. Analysis of molecular variance (AMOVA) results for 16 populations of *Pinus koraiensis* in China.

Type	Source of Variation	d.f.	Sum of Squares	Variance Component	Percentage of Variation	Fixation Index
Variance partition [a]	Among populations	15	2191.019	4.559	33.00	
	Within populations	464	4311.567	9.292	67.00	FST = 0.516
	Total	479	6502.585	13.851		
Variance partition [b]	Among groups	1	115.322	0.266	26.67	FST = 0.362
	Among populations within groups	14	88.917	0.095	9.55	FSC = 0.130
	Within populations	942	600.429	0.637	63.79	FCT = 0.267
	Total	957	804.668	0.999		
Variance partition [c]	Among groups	1	23.753	0.024	2.77	FST = 0.264
	Among populations within groups	14	180.487	0.205	23.64	FSC = 0.243
	Within populations	942	600.429	0.637	73.59	FCT = 0.028
	Total	957	804.668	0.866		

Note: [a] The analysis included all collected populations as one hierarchical group. [b] The analysis included two geographical groups (G1 and G2). [c] The analysis included two genetic clusters (Clusters 1 and Clusters 2).

The Nei's genetic distance and pairwise Fst values are shown in Table 4. Fst was considered the main genetic parameter for evaluating genetic differentiation among populations. In this study, the pairwise Fst values ranged from 0.014 to 0.348, and most of the *P. koraiensis* population pairs exhibited high values (Fst > 0.15), indicating high levels of genetic diversity. The greatest level of differentiation was observed between populations Helong and Liangshui, and the lowest was observed between Jiaohe and Hulin. The highest genetic distance was observed between populations Helong and Liangshui (0.813), consistent with the pairwise Fst values and indicating pronounced differentiation between these two populations. The relative migration network among the 16 *P. koraiensis* populations was constructed using relative migration rate with the divMigrate function in R software. Analysis of gene flow between populations suggested a biased geographic distribution, and gene flow was not uniform among all populations (Figure 1). A high degree of gene flow was observed among three populations located near one another (Muling, Maoershan and Fangzheng), consistent with the principal coordinate analysis and dendrogram analysis. In addition, one genetically isolated population (Boli) displayed high levels of gene flow with the three nearby populations Muling, Maoershan and Fangzheng. Moreover, a moderate level of gene flow was found among three admixed populations, and two genetically distinct populations (Zhanhe and Wangqing) exhibited distant segregation from the other populations.

Table 4. Pairwise genetic differentiation index values (below the diagonal) and Nei's genetic distance values (above the diagonal). **** indicates the diagonal division of the pairwise genetic differentiation index values and Nei's genetic distance values.

	P1	P2	P3	P4	P5	P6	P7	P8	P9	P10	P11	P12	P13	P14	P15	P16
P1	****	0.008	0.002	0.383	0.737	0.452	0.498	0.550	0.426	0.418	0.397	0.418	0.437	0.432	0.422	0.471
P2	0.029	****	0.007	0.443	0.813	0.403	0.520	0.619	0.458	0.395	0.388	0.406	0.439	0.466	0.434	0.423
P3	0.017	0.021	****	0.410	0.778	0.434	0.475	0.543	0.425	0.421	0.400	0.422	0.436	0.440	0.432	0.425
P4	0.188	0.233	0.208	****	0.048	0.154	0.106	0.167	0.104	0.368	0.337	0.285	0.258	0.244	0.245	0.612
P5	0.298	0.348	0.322	0.054	****	0.222	0.172	0.216	0.145	0.411	0.385	0.342	0.368	0.338	0.351	0.769
P6	0.202	0.203	0.202	0.091	0.124	****	0.093	0.207	0.143	0.317	0.297	0.260	0.255	0.266	0.234	0.430
P7	0.221	0.245	0.220	0.070	0.104	0.051	****	0.022	0.041	0.427	0.414	0.312	0.315	0.309	0.299	0.439

Table 4. *Cont.*

	P1	P2	P3	P4	P5	P6	P7	P8	P9	P10	P11	P12	P13	P14	P15	P16
P8	0.239	0.271	0.239	0.098	0.124	0.104	0.025	****	0.059	0.540	0.520	0.402	0.432	0.403	0.410	0.498
P9	0.183	0.213	0.193	0.058	0.092	0.071	0.036	0.051	****	0.2387	0.225	0.170	0.271	0.260	0.264	0.474
P10	0.219	0.240	0.224	0.212	0.228	0.169	0.218	0.257	0.140	****	0.005	0.025	0.248	0.258	0.262	0.533
P11	0.211	0.233	0.218	0.202	0.222	0.164	0.213	0.251	0.134	0.014	****	0.035	0.232	0.237	0.245	0.591
P12	0.210	0.236	0.217	0.151	0.183	0.134	0.160	0.195	0.099	0.030	0.038	****	0.146	0.152	0.155	0.457
P13	0.229	0.258	0.232	0.161	0.206	0.142	0.171	0.216	0.147	0.156	0.147	0.097	****	0.002	0.001	0.593
P14	0.228	0.260	0.237	0.154	0.188	0.145	0.165	0.203	0.143	0.158	0.151	0.102	0.015	****	0.001	0.591
P15	0.229	0.247	0.237	0.158	0.196	0.134	0.163	0.208	0.145	0.155	0.155	0.097	0.014	0.014	****	0.590
P16	0.230	0.244	0.228	0.275	0.347	0.196	0.217	0.241	0.229	0.286	0.306	0.238	0.300	0.299	0.318	****

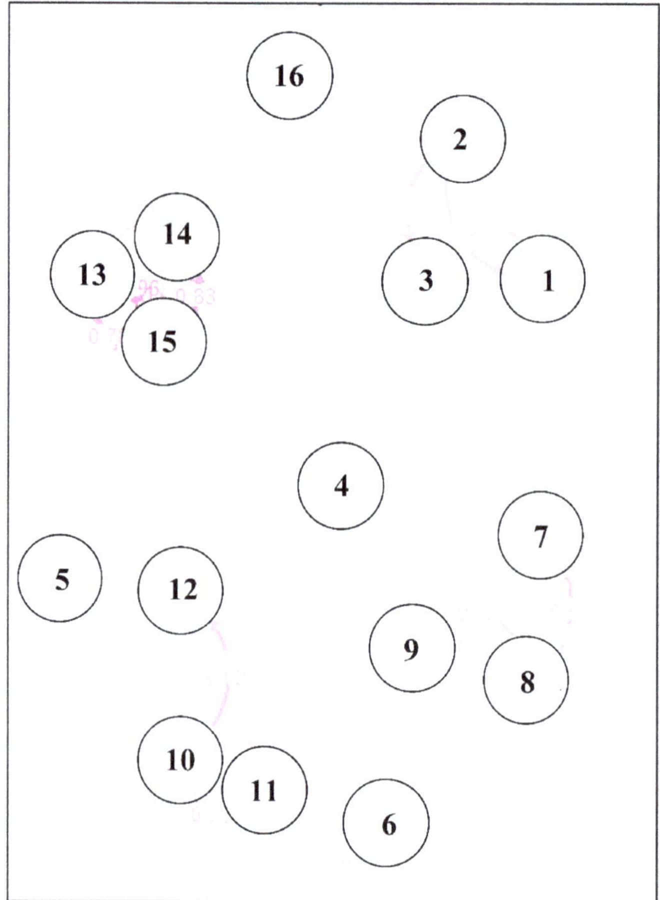

Figure 1. The relative migration network of 16 populations generated via divMigrates. The width of the line and the number shown next to the arrows indicate the migration rate.

2.4. Population Structure

The population structure analysis of the 16 natural *P. koraiensis* populations was performed based on a Bayesian approach using STRUCTURE software. The number of clusters within the range of 1 to 10 was evaluated for 10 repetitions in each run. In the structure plot (Figure 2), the maximum delta K value appeared at K = 2, with an obvious peak apparent at this value; this value was considered the optimal genetic cluster number for all EST-SSR markers (Figure 2B,C). The 480 sampled individuals of *P. koraiensis* were divided into two genetic groups (Group 1 and Group 2) at K = 2: Group 1 comprised 149 individuals from 5 populations (Heihe, Liangshui, Zhanhe, Tieli and Hegang), and Group 2 comprised a higher number of individuals (331) from 11 populations (Liangzihe, Helong, Lushuihe, Linjiang, Jiaohe, Hulin, Boli, Muling, Maoershan, Fangzheng and Wangqing). Group 1 comprised almost all of the *P. koraiensis* plant materials from Xiaoxinganling Mountains, whereas Group 2 comprised almost all of the individuals from Changbaishan Mountains, suggesting a relationship between genetic structure and geographical distribution of the populations.

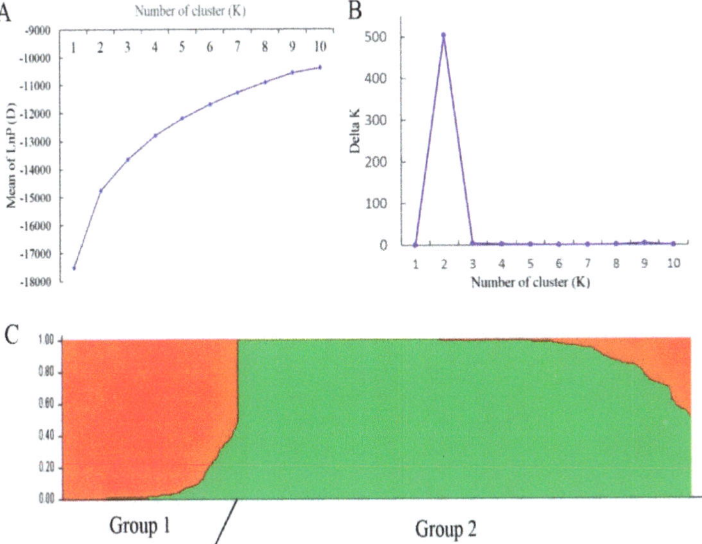

Figure 2. Population structure determined from 480 *Pinus koraiensis* individuals based on microsatellite data. (**A**) Estimation of population structure using mean lnP (D) with ten repetitions for K ranging from 1 to 10. (**B**) Estimation of population structure using delta K (ΔK) with the number of clusters (K) ranging from 1 to 10. (**C**) Estimation of population structure of 16 populations based on structure analysis.

To further analyze cluster patterns, principal component analysis (PCA) based on the pairwise genetic distance matrix of 15 EST-SSRs was performed; the results are shown in Figure 3. The 480 individuals from the 16 populations were roughly divided into two clusters according to the first two axes in the PCA plot. Principal axes 1 and 2 accounted for 22.99% and 12.46%, respectively, of the total genetic variation among the individuals, together accounting for 35.45% of the total genetic variation (Figure 3A). Five populations (Heihe, Liangshui, Zhanhe, Tieli and Hegang) were grouped into cluster 1, and the remaining populations (Liangzihe, Helong, Lushuihe, Linjiang, Jiaohe, Hulin, Boli, Muling, Maoershan, Fangzheng and Wangqing) were grouped into cluster 2. The same clustering was obtained in the STRUCTURE analysis using the same dataset, indicating marked genetic differentiation. Furthermore, the Neighbor-joining (NJ) dendrogram based on Nei's genetic distance clustered the 480 *P. koraiensis* individuals from the 16 populations into 2 clusters, consistent with the above results (Figures 3B and 4).

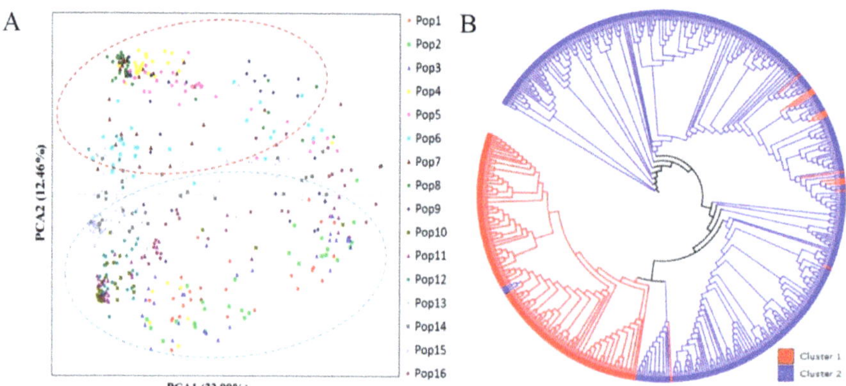

Figure 3. Genetic variation and relationships among the 16 sampled natural *Pinus koraiensis* populations in northeast China. (**A**) Principal coordinate analysis (PCA) based on pairwise genetic distance. (**B**) NJ dendrogram of 480 individuals based on Nei's (1983) genetic distance.

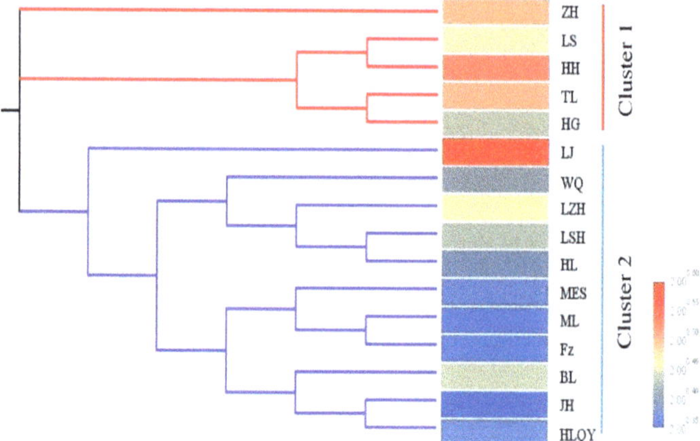

Figure 4. NJ dendrogram of 16 populations based on Nei's (1983) genetic distance and a heatmap of expected heterozygosity (He) of 16 populations.

2.5. Correlations between Genetic Distance and Geographic Distance

The genetic distance estimated based on molecular markers may be related to the distribution of the species under study and the geographic distance between individuals or populations. In this study, the geographic distance and genetic distance values ranged from 37.72 km to 825.45 km and from 0.02 to 0.83, respectively. To investigate the correlations between genetic distance and geographic distance, the Mantel test was carried out. The results showed that genetic distance was not significantly correlated with the geographic distance among the *P. koraiensis* populations (p = 0.26, R^2 = 0.01), indicating a lack of association between geographical distance and the genetic differentiation of *P. koraiensis* (Figure 5). Liangzihe and Hegang populations exhibited the lowest geographic distance and were not grouped in the same cluster. Therefore, there was no obvious isolation by genetic and geographical distance among the sampled populations.

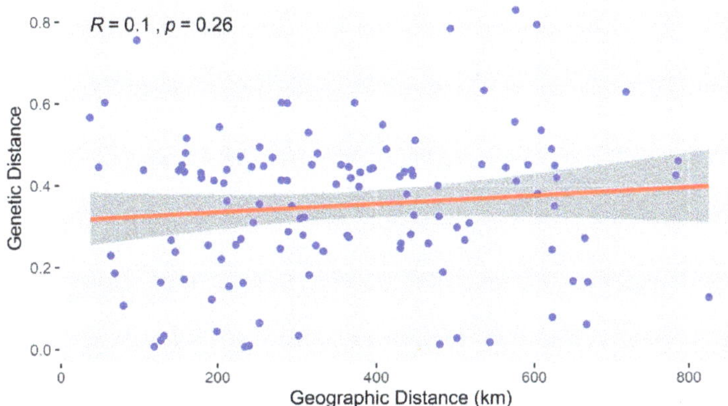

Figure 5. Correlation between genetic distance and geographic distance in *P. koraiensis* populations determined using the Mantel test.

3. Discussion

To understand the genetic differentiation of forest tree populations and contribute to the development of effective breeding strategies, comprehensive evaluations of natural germplasm resources of individual species are essential; such evaluations can accelerate breeding strategies and industrial development [36,37]. Naturally, *P. koraiensis* mainly grows in the cold temperate zone, especially in northeast China, and natural forests of this species have been shown to be sensitive to climate factors. Thus, to conserve genetic resources of this species, it is important to obtain data on its genetic diversity and population structure. In present study, we conducted a population genetic analysis using codominant molecular markers, representing the first such analysis in *P. koraiensis*. The results can help guide the genetic improvement and resource conservation of this important gymnosperm.

3.1. Genetic Diversity

Genetic diversity has been increasingly evaluated in species lacking a reference genome, including some conifers [38], endemic species [39] and endangered plants [40]. Studies of genetic diversity can provide insight into speciation and genetic variation within and among populations and can aid the development of conservation strategies. However, transcriptome data and molecular markers remain lacking for *P. koraiensis*; the available genetic data provide few markers suitable for the study of population genetics in this species. Evaluating the germplasm resources of this species represents the first step towards understanding the genetics of natural *P. koraiensis* populations. A high level of genetic diversity in natural *P. koraiensis* populations was detected in this study, with mean values of 10.33 and 0.521 for Na and He, respectively. High genetic diversity was observed in the Heihe population in the northern Xiaoxinganling Mountains, possibly due to less human disturbance in this region than in other areas. According to a previous study, a PIC value equal to or more than 0.5 indicates high genetic information for genetic markers. In the present study, the PIC values obtained for the multiallelic EST-SSR markers ranged from 0.142 to 0.833, with a mean value of 0.461, indicating a high level of genetic information among the 480 *P. koraiensis* individuals from the 16 natural populations. The genetic diversity of *P. koraiensis* obtained in the present study is higher than that reported for *Pinus bungeana* (Na = 3.70, He = 0.36) [41], *Pinus dabeshanensis* (He = 0.36) [42] and *Pinus yunnanensis* (Na = 4.10, He = 0.43) [43] but lower than that reported for *Pinus tabulaeformis* (Na = 6.52, He = 0.68) [44]. The genetic diversity of a species may vary with characteristics such as adaptability, pollination mechanism and population size [45–47]. The observed genetic diversity in the present study might be attributable to the genetic background, life history and population dynamics of *P. koraiensis*. Previous studies have found that *P. koraiensis* has a large population size, long life cycle, strong adaptability, long pollination distance and large genome size, and it has a complex genetic background, which allows it to generate high genetic diversity [48–52]. Natural selection under the changing environmental conditions is likely to lead to differences in genotype frequency among populations. In addition, previous studies have suggested that evaluations of genetic diversity are limited by low numbers of populations and molecular markers [53,54]. For instance, analyses of the genetic diversity of natural *P. koraiensis* populations have been conducted using a variety of molecular marker techniques. Kim et al. [53] analyzed allozyme loci variation and found a moderate level of genetic diversity among natural *P. koraiensis* populations in Korea. The genetic diversity of natural *P. koraiensis* populations detected by allozyme markers (He = 0.18) in Russia [54] was much lower than that identified using EST-SSRs markers in this study. These findings indicate that *P. koraiensis* maintains high genetic diversity worldwide. The level of genetic diversity detected in this study is similar to that detected based on nine EST-SSR markers in seven natural populations of *P. koraiensis* in northeast China (He = 0.610) [35]. Xiaoxinganling Mountain of China was considered as the distribution center for *P. koraiensis*, possessing abundant germplasm resources and ancient founding stocks and maintaining considerable numbers of individuals. In addition, the genetic diversity of *P. koraiensis* populations from Xiaoxinganling Mountains was higher than that of the Changbaishan Mountains populations, with high expected heterozygosity and abundant private alleles found for the former populations (Figure 4). All these results indicate that Xiaoxinganling Mountains may be the center of genetic diversity of *P. koraiensis* in China.

3.2. Population Genetic Differentiation

Detection of genetic differentiation is a key process in the genetic improvement of forest trees. Regarding the estimation of genetic differentiation, past studies have considered an Fst value higher than 0.15 but lower than 0.25 to indicate significant divergence [55–57]. In the present study, the genetic differentiation assessed by EST-SSRs among *P. koraiensis* populations ranged from 0.014 to 0.348, with a mean value of 0.177, indicating significant differentiation among populations in China. However, previous studies reported low genetic differentiation among populations as assessed by allozyme loci variation in Korea (Fst = 0.06) [53] and Russia (Fst = 0.015) [54] and by EST-SSRs in China (Fst = 0.02) [35]. In addition, Kim et al. [31] studied the genetic variation of *P. koraiensis* in Korea, Russia and China using allozymes and RAPDs and detected small differences among the three regions. Different degrees of genetic differentiation were observed in natural *P. koraiensis* populations in these countries, with low Fst values. The main reason for these differences is that only limited numbers of natural populations and molecular markers were analyzed. The genetic differentiation index (Fst) is correlated with gene flow (Nm). Generally, the greater the degree of differentiation, the weaker the gene flow, i.e., a lower gene migration rate among populations [57–60]. Natural *P. koraiensis* forests originated in Siberia and in northeast Asia have undergone regeneration, succession and migration over millions of years [61,62]. After the Quaternary glaciation, many species died out, but the *P. koraiensis* forests persisted into the present and underwent a range of changes and varying degrees of differentiation. In natural *P. koraiensis* populations, low levels of genetic differentiation have been observed in Korea [31], whereas high genetic differentiation has occurred in northeast China, which may have contributed to the rich *P. koraiensis* germplasm resources (representing approximately 60% of the world's total) and broad distribution area (more than 3000 hectares) in this country. The mean He (0.521) across all loci was greater than Ho (0.374), indicating a high heterozygosity among the sampled populations of *P. koraiensis*. This high heterozygosity is attributable to the fact that Pinus species exhibit cross-pollination and wind pollination. Furthermore, the AMOVA suggested that most of the genetic variation (more than 60%) in *P. koraiensis* exists within populations, with a small proportion occurring among populations; these findings are consistent with findings in other cross-pollinating species.

3.3. Population Structure and Gene Flow

Analyses of population structure can provide insight into population size, breeding system, extent of isolation and population migration or gene flow [63,64]. Furthermore, such analyses can help reveal the relationships between genetic variation and environmental stresses and enhance our understanding of evolution. Evaluating population structure is a key component of genome-wide association analysis (GWAS) and marker-assisted selection (MAS) [65]. *P. koraiensis* is mainly distributed in Xiaoxinganling and Changbaishan Mountains in northeast China, areas with a humid climate. Due to the environmental conditions, the germplasm resources of *P. koraiensis* from different locations display high phenotypic and genetic variation. The STRUCTURE analysis of population structure identified two groups (optimal K = 2) from the 16 natural populations, with 5 populations in one group and the remainder in another group. Similar results were obtained in the PCA and dendrogram (neighbor-joining tree) analysis, indicating genetic differentiation of *P. koraiensis* populations in China. Interestingly, individuals from Xiaoxinganling Mountains were clustered into one group, occupying a northern area, which makes them more like an ancestral group. Furthermore, the samples from Changbaishan Mountains and the adjacent ridge region were clustered into the other group; the populations corresponding to these samples are distributed in a southern area and exhibit different degrees of genetic differentiation and gene flow. However, some of the individuals from Xiaoxinganling Mountains were clustered into cluster 2, although the majority were clustered into cluster 1 (Figure 3A,B). The main potential reasons for this finding are as follows: (1) These two mountain regions are close to each other, and some hybridization events may occur; (2) for populations separated by a short spatial distance, the probability of gene flow is high, which will affect population genetic structure; and (3) pollen and seed dispersal occur over long distances in this species, which promote gene flow. The genetic structure of the natural *P. koraiensis* populations in China determined in this study is consistent with the current geographical distribution of these populations. Furthermore, the findings are consistent with previous studies showing that populations in similar geographical locations or environments tend to cluster into the same group [66,67]. In this study, although some admixture was detected by the STRUCTURE and PCA analyses, the population dendrogram also suggested two subgroups, with 5 populations in cluster 1 and 11 populations in cluster 2.

Gene flow among populations is closely related to geographical distance and effective population size and can generate new genetic combinations, potentially enhancing species resilience and persistence [68–70]. In plants, migration, or gene flow, is achieved via seeds, pollen and other propagules, and influences the genetic diversity and differentiation among independent evolutionary units [71,72]. We found that two genetically distinct populations (Zhanhe and Wangqing) exhibited segregation from other populations, which may be related to their geographic distance from other populations (approximately 565 km), limiting the level of gene flow between them. These independent units play an important role in maintaining the genetic diversity of this species. This interpretation is consistent with previous studies demonstrating that isolated populations of plants with long-distance pollination may have higher levels of genetic diversity than large contiguous populations [73–75]. Moreover, high levels of gene flow were found among Helong, Maoershan and Fangzheng populations. Thus, it can reduce the effects of artificial selection or genetic drift and promote the maintenance of genetic information. Similar results were obtained for *Camelina sativa* accessions [65]. Previous studies have also found that extensive gene flow can alter the gene frequencies in populations to affect genetic diversity and structure. In our study, although a strong correlation between gene flow and geographic distance between populations was observed, some degree of gene flow was also evident between geographically distant populations. In addition, geographic distance was not correlated with genetic distance in the natural *P. koraiensis* populations in this study, suggesting that geographic distribution may not be a determinant factor for the genetic structure of populations.

3.4. Conservation and Management Strategies

Evaluations of germplasm resources are needed to maintain abundant genetic variation and high levels of genetic diversity of some species of interest and establish sound conservation strategies. Our population genetic analysis revealed that the populations distributed in the Xiaoxinganling Mountains (Zhanhe, Heihe, Liangshui, Tieli and Hegang) exhibit high levels of genetic diversity and moderate levels of gene flow (Figure 4). These populations represent the core populations and have stronger environmental adaptability and evolutionary potential than the other populations, and they can be considered independent genetic units. Hence, measures such as in situ conservation should be implemented for conserving natural *P. koraiensis* resources. In addition, the marginal populations represent special germplasm resources. They are characterized by low genetic diversity but have high levels of genetic differentiation relative to the other natural populations. Habitat fragmentation can reduce gene flow among populations, leading to a loss of genetic diversity. In this study, the Helong population, which occurs in a marginal area, should be targeted for conservation measures, such as ex situ measures. In addition, the greatest level of population differentiation was observed between Helong and Liangshui populations, indicating that these populations can be considered independent units. Therefore, regulations and management strategies must be established to protect the natural habitat of this species and prohibit harvest. More importantly, a national-level core germplasm resource library of *P. koraiensis* should be established by the government, with the objectives of maintaining genetic variation, improving plant adaptability to environmental changes and developing new breeding materials. Under these measures, the existing natural *P. koraiensis* populations in China can be protected and be better used as a source of resources for genetic improvement in the future.

For forest management perspectives, efforts should be made to regulate timber harvesting in these population in order to reduce loss of genetic diversity. Particularly, the marginal populations, such as the Helong population, that are characterized by low genetic diversity should be given the highest management priority by enrichment planting of individuals from different populations to enhance the genetic diversity within this population. Forest management should also focus on suppression of wildfire in these forests as population fragmentation driven by wildfire can reduce gene flow among populations, leading to a loss of genetic diversity. In addition, these forests should be protected from pest and disease, such as the white pine blister rust diseases that affect the trees.

4. Materials and Methods

4.1. Plant Materials and Genomic DNA Extraction

In this study, 16 populations of *P. koraiensis* from Jilin Province (J) and Heilongjiang Province (H) were considered. A total of 480 samples were collected from these populations, which occur throughout the natural distribution areas in northeastern China (Table 5, Figure 6). The sampled populations were selected to represent the main distribution region and included the Xiaoxinanling Mountains group (G1, including populations P4, P5, P6, P7 and P8) and the Changbai Mountains group (G2, including P1, P2, P3, P9, P10, P11, P12, P13, P14, P15 and P16). The populations were distributed across the northernmost (Heihe, 49°24′47″, 126°36′47.16″), southernmost (Linjiang, 42°0′36″, 127°13′12″) and easternmost regions (Hulin, 45°46′12″, 132°58′48″) of natural *P. koraiensis* distribution in northeast China. The average altitude of the sampled individuals was 320 m. To obtain representative samples of populations, sampling was conducted with at least 200 m between sampled individuals. The number of sampled individuals per population ranged from 29 to 31, with an average number of 30. Fresh needle samples of *P. koraiensis* with no signs of pests or disease were collected from the middle portion of the crown and immediately frozen in liquid nitrogen and stored at −80 °C for subsequent genomic DNA extraction and PCR amplification. In addition, nucleic acids were extracted from needles using the improved cetyltrimethyl ammonium bromide (CTAB) method described by Li et al. [76]. DNA quality and concentration were evaluated using 1.0% agarose gel electrophoresis and the K5500 Plus microspectrophotometer (KAIAO Technology Development Co., Ltd., Beijing, China), respectively.

Table 5. Summary of *Pinus koraiensis* sampled populations in NE China.

Population	Population Codes	Group	Sample Size	Latitude (N)	Longitude (E)	Average Altitude (m)	Province
Zhanhe	P6	G1	30	48°1′23″	127°0′56.16″	358	Heilongjiang
Heihe	P4	G1	30	49°24′47″	126°36′47.16″	131	Heilongjiang
Liangshui	P5	G1	30	47°43′48″	128°55′12″	231	Heilongjiang
Tieli	P7	G1	29	47°58′48″	128°4′48″	213	Heilongjiang
Hegang	P8	G1	30	47°21′29″	129°33′50.03″	63	Heilongjiang
Liangzihe	P1	G2	30	47°1′54″	129°41′49.92″	500	Heilongjiang
Lushuihe	P3	G2	30	42°24′00″	127°28′59.88″	732	Jilin
Linjiang	P9	G2	30	42°0′36″	127°13′12″	342	Jilin
Helong	P2	G2	30	42°32′47″	129°1′3.72″	452	Jilin
Wangqing	P16	G2	31	43°19′26″	129°45′14.04″	232	Jilin
Muling	P13	G2	30	43°49′54″	129°45′19.08″	410	Jilin
Jiaohe	P10	G2	30	44°0′36″	127°25′12″	196	Jilin
Maoershan	P14	G2	30	45°16′22″	127°30′14.40″	536	Heilongjiang
Hulin	P11	G2	30	45°46′12″	132°58′48″	84	Heilongjiang
Boli	P12	G2	30	45°42′00″	130°25′12″	525	Heilongjiang
Fangzheng	P15	G2	30	45°49′48″	128°48′00″	111	Heilongjiang

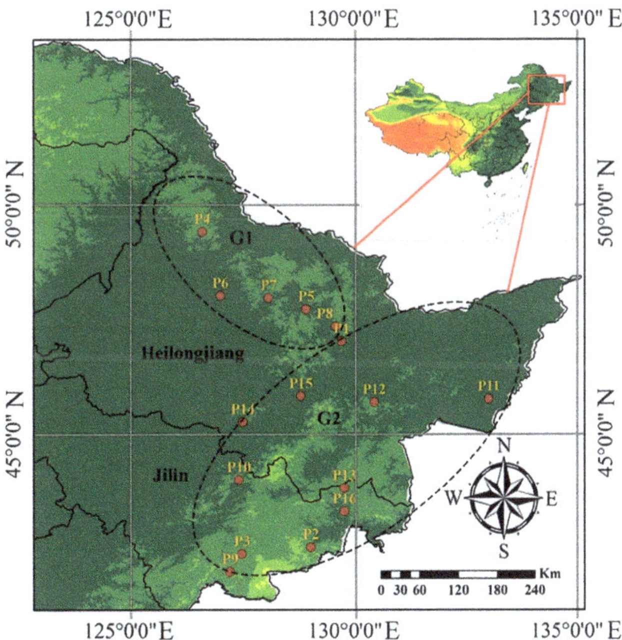

Figure 6. Distribution of *Pinus koraiensis* populations sampled in northeast China. G1 represents Xiaoxinganling Mountains geographical groups. G2 represents Changbaishan Mountains geographical groups.

4.2. PCR Amplification and SSR Analysis

Fifteen highly polymorphic and reproducible EST-SSR markers of *P. koraiensis* developed in our laboratory were selected in this study to detect polymorphisms in the 16 sampled *P. koraiensis* populations. The primers of *P. koraiensis* were developed as described by Li et at. [23]. Eight capillary electrophoresis templates were amplified with fifteen primers synthesized by Sangon Biotech (Shanghai, China), and universal M13 sequence (5′-TGTAAAACGACGGCCAGT-3′) labeled with four fluorescent dyes (TAMRA, FAM, HEX and ROX) was added at the 5′ end of the forward primers. DNA was diluted to a working concentration of 25 ng/μL. To detect SSR loci, polymerase chain reaction (PCR) was performed in a total volume of 20 μL containing 10 μL 2× Super PCR Mix (Beijing Genomics Institute Tech Solutions (Bejing Liuhe) Co., Ltd., Beijing, China), 2 μL template DNA, 0.8 μL forward primer (1 μM), 3.2 μL reverse primer (1 μM), 1 μL M13 primer with fluorescent label and 3 μL ddH$_2$O. The PCR amplification conditions were as follows: 94 °C for 5 min followed by 30 cycles at 94 °C for 30 s, 57 °C for 30 s, and 72 °C for 30 s; followed by 8 cycles at 94 °C for 30 s, 55 °C for 30 s, and 72 °C for 30 s; followed by final extension at 72 °C for 10 min. The PCR products were subjected to 1.0% agarose gel electrophoresis and then analyzed by high performance capillary electrophoresis (HPCE) using an ABI 3730XL DNA Sequencer (Applied Biosystems, Foster City, CA, USA) to detect fragment size. The original sequence data were analyzed using GeneMapper (version 4.1) software.

4.3. Data Analysis

GeneMapper was used to obtain the microsatellite allele data, and the Microsatellite toolkit v 3.1.14 was used to convert the data into the necessary format for analysis. The genetic diversity analysis was conducted using GENALEX software version 6.50 [77] with the following parameters: number of alleles (Na), effective number of alleles (Ne), observed (Ho) and expected (He) heterozygosity, number of rare alleles (NRA), Shannon diversity index (I), Hardy–Weinberg equilibrium (HWE), F-statistics (Fis, Fit and Fst) and Nei's genetic distance. The TBtools software [78] was used to plot the heatmap of expected heterozygosity (He). In addition, we calculated the polymorphism information content (PIC) values of each SSR primer using the PICcalc program [79]. Gene flow (Nm) was calculated as Nm = (1 − Fst)/4 × Fst and used to measure the degree of gene exchange among or within the 16 populations. ALREQUIN software (version 3.5) [80] was used to analyze the level and sources of molecular genetic variation via AMOVA based on the evolutionary distances among and within the sampled populations and the observed genetic clusters. The total genetic variation was divided into three components: among groups, among populations within groups and within populations.

To evaluate the population genetic structure of *P. koraiensis*, a Bayesian clustering algorithm was performed in STRUCTURE software (version 2.3) [81] with the following settings: K-values from 1 to 10, with ten runs per K value and a burn-in period and number of Markov chain Monte Carlo (MCMC) reps after burn-in of 100,000 iterations and 100,000, respectively. The optimal K value for the number of populations was determined based on the delta-K values calculated by the Evanno method [82], using an algorithm of the online tool of STRUCTURE HARVESTER [83], where a clear peak was observed in the plot of delta K. In addition, principal component analysis (PCA) was performed to evaluate the genetic relationships among different populations using GENALEX software version 6.50. Based on the Nei's genetic distance (1983), a Neighbor-joining (NJ) phylogenetic tree of the populations was constructed using PowerMarker software (version 3.25) [84] and annotated and visualized using the online tool interactive Tree Of Life (iTOL) [85]. Geographic distance among populations was calculated as described in Li et al.'s study [76]. Finally, to detect the gene flow among the 16 populations, a relative migration network was constructed using the 'diveRsity' [86] package of R software (version 3.5.0) [87].

5. Conclusions

This study investigated the genetic diversity and population structure of natural *P. koraiensis* populations in northeast China, and proposed some conservation strategies for this valuable conifer species. This study is the first comprehensive report of the genetic diversity of natural *P. koraiensis* populations in China. We found that the existing *P. koraiensis* populations in China maintain high levels of genetic diversity, which provide a foundation for germplasm innovation and genetic improvement of *P. koraiensis*. The population genetic analysis in this study identified two independent genetic units (Liangshui and Helong populations) that exhibit high degrees of genetic differentiation. The populations distributed in the Xiaoxinganling Mountains are highly genetically diverse and may represent the central population of natural *P. koraiensis* in China. Furthermore, the genetic structure of *P. koraiensis* populations identified in this study is consistent with the geographical distribution of these populations in China. These results have significance for the protection of natural *P. koraiensis* germplasm resources in China as well as for developing improved genotypes through breeding. Our findings provide genetic information useful for future genome-wide association studies (GWAS) and marker-assisted selection (MAS) and genomic selection (GS) studies. It is, therefore, recommended to further conduct research on genetic improvement for timber and cone production using marker-assisted selection and/or genomic selection as well as genotype by environment interaction studies should be carried out to identify suitable site-specific genotypes.

Author Contributions: Conceptualization, X.Z. and X.L.; methodology, X.L., Y.L. and M.Z.; validation, X.L., Y.X. and M.T.; resources, X.L.; writing—original draft preparation, X.L.; writing—review and editing, X.L., M.T., and X.Z.; supervision, X.Z.; project administration, X.Z.; funding acquisition, X.Z. All authors have read and agreed to the published version of the manuscript.

Funding: This research study was supported by the Fundamental Research Funds for the Central Universities (Northeast Forestry University) (No. 2572017DA02).

Institutional Review Board Statement: Not applicable.

Informed Consent Statement: Not applicable.

Data Availability Statement: Data for this study can be made available with reasonable request to the authors.

Conflicts of Interest: The authors declare that the research was conducted in the absence of any commercial or financial relationships that could be construed as a potential conflict of interest.

References

1. Choi, D.-S.; Jin, H.-O.; Lee, C.-H.; Kim, Y.-C.; Kayama, M. Effect of soil acidification on the growth of Korean pine (*Pinus koraiensis*) seedlings in a granite-derived forest soil. *Environ. Sci. Int. J. Environ. Physiol. Toxicol.* **2005**, *12*, 33–47.
2. Jia, Y.; Zhu, J.; Wu, Y.; Fan, W.-B.; Zhao, G.-F.; Li, Z.-H. Effects of Geological and Environmental Events on the Diversity and Genetic Divergence of Four Closely Related Pines: *Pinus koraiensis*, *P. armandii*, *P. griffithii*, and *P. pumila*. *Front. Plant Sci.* **2018**, *9*, 1264. [CrossRef]
3. Wang, F.; Chen, S.; Liang, D.; Qu, G.-Z.; Chen, S.; Zhao, X. Transcriptomic analyses of *Pinus koraiensis* under different cold stresses. *BMC Genom.* **2020**, *21*, 1–14. [CrossRef]
4. Ma, J.L. The evolution of Korea Pine forest. *J. Northeast. For. Univ.* **1997**, *25*, 67–71.
5. Wang, H.; Shao, X.-M.; Jiang, Y.; Fang, X.-Q.; Wu, S.-H. The impacts of climate change on the radial growth of *Pinus koraiensis* along elevations of Changbai Mountain in northeastern China. *For. Ecol. Manag.* **2013**, *289*, 333–340. [CrossRef]
6. Zhu, J.; Wang, K.; Sun, Y.; Yan, Q. Response of *Pinus koraiensis* seedling growth to different light conditions based on the assessment of photosynthesis in current and one-year-old needles. *J. For. Res.* **2014**, *25*, 53–62. [CrossRef]
7. Fan, Y.; Moser, W.K.; Cheng, Y. Growth and Needle Properties of Young *Pinus koraiensis* Sieb. et Zucc. Trees across an Elevational Gradient. *Forests* **2019**, *10*, 54. [CrossRef]
8. Li, X.; Liu, X.T.; Wei, J.T.; Li, Y.; Tigabu, M.; Zhao, X.Y. Genetic improvement of *Pinus koraiensis* in China: Current situation and ffuture prospects. *Forests* **2020**, *11*, 148. [CrossRef]
9. Zhang, J.; Zhou, Y.; Zhou, G.; Xiao, C. Composition and Structure of *Pinus koraiensis* Mixed Forest Respond to Spatial Climatic Changes. *PLoS ONE* **2014**, *9*, e97192. [CrossRef]

10. Wang, X.C.; Zhang, M.H.; Ying, J.; Li, Z.S.; Zhang, Y.D. Temperature signals in tree-ring width and divergent growth of Ko-rean pine response to recent climate warming in northeast Asia. *Trees* **2016**, *31*, 1–13.

11. Seo, Y.-W.; Balekoglu, S.; Choi, J.-K. Growth pattern analysis by stem analysis of Korean white pine (*Pinus koraiensis*) in the central northern region of Korea. *For. Sci. Technol.* **2014**, *10*, 220–226. [CrossRef]

12. Zhang, S.T.; Zhang, L.G.; Wang, L.; Zhao, Y.H. Total phenols, flavonoids, and procyanidins levels and total antioxidant activity of different Korean pine (*Pinus koraiensis*) varieties. *J. For. Res.* **2018**, *30*, 1743–1754. [CrossRef]

13. Lai, F.; Xie, H.; Wang, J. Functions of biology and disease control with Pycnogonid from pine bark extract. *Chin. Tradit. Herb. Drugs* **2005**, *36*, 127–131.

14. Liu, M.; Mao, Z.J.; Li, Y.; Sun, T.; Li, X.H.; Huang, W.; Liu, R.P.; Li, Y.H. Response of radial growth of *Pinus koraiensis* in broad-leaved Korean pine forests with different latitudes to climatical factors. *Ying Yong Sheng Tai Xue Bao J. Appl. Ecol.* **2016**, *27*, 1341–1352.

15. Kang, K.; Harju, A.; Lindgren, D.; Nikkanen, T.; Almqvist, C.; Suh, G. Variation in effective number of clones in seed orchards. *New For.* **2001**, *21*, 17–33. [CrossRef]

16. Han, S.U.; Kang, K.S.; Kim, C.S.; Kim, T.S.; Song, J.H. Effect of top-pruning in a clonal seed orchard of *Pinus koraiensis*. *Ann. For. Res.* **2008**, *51*, 155–156.

17. Park, J.M.; Kwon, S.H.; Lee, H.J.; Na, S.J.; El-Kassaby, Y.A.; Kang, K.S. Integrating fecundity variation and genetic related-ness in estimating the gene diversity of seed crops: *Pinus koraiensis* seed orchard as an example. *Can. J. Forest. Res.* **2016**, *47*, 366–370. [CrossRef]

18. Guan, G.Y. Effects of different scion types on grafting survival rate and growth potential of *Pinus koraiensis*. *Prot. For. Sci. Technol.* **2019**, *000*, 44–45.

19. Xia, D.; Yang, S.W.; Yang, C.P.; Lv, Q.Y.; Liu, G.F.; Zhang, P.G. Study of *Pinus koraiensis* provenance test (1): Provenance re-gionalization. *J. Northeast. Univ.* **1991**, *19*, 122–128.

20. Wang, Y.X.; Dong, Y.H.; Yang, H.; Wang, H.Z.; Xu, J.M.; Shan, M.G.; Zhang, G.L. Open pollination progeny test of clonal seed orchard of *Pinus koraiensis*. *For. Sci. Technol.* **2005**, *34*, 20–21.

21. Liang, D.; Ding, C.; Zhao, G.; Leng, W.; Zhang, M.; Zhao, X.; Qu, G. Variation and selection analysis of *Pinus koraiensis* clones in northeast China. *J. For. Res.* **2017**, *29*, 611–622. [CrossRef]

22. Chen, M.-M.; Feng, F.; Sui, X.; Li, M.-H.; Zhao, D.; Han, S. Construction of a framework map for *Pinus koraiensis* Sieb. et Zucc. using SRAP, SSR and ISSR markers. *Trees* **2010**, *24*, 685–693. [CrossRef]

23. Li, X.; Liu, X.; Wei, J.; Li, Y.; Tigabu, M.; Zhao, X. Development and Transferability of EST-SSR Markers for *Pinus koraiensis* from Cold-Stressed Transcriptome through Illumina Sequencing. *Genes* **2020**, *11*, 500. [CrossRef]

24. Kwon, S.; Kim, S.; Kim, J.; Kang, W.; Park, K.-H.; Kim, C.-B.; Girona, M.M. Predicting Post-Fire Tree Mortality in a Temperate Pine Forest, Korea. *Sustainability* **2021**, *13*, 569. [CrossRef]

25. Li, J.Q.; Wang, L.H.; Zhan, Q.W.; Liu, Y.L.; Yang, X.C. Transcriptome characterization and functional marker development in Sorghum sudanense. *PLoS ONE* **2016**, *11*, e0154947. [CrossRef] [PubMed]

26. Zhu, H.; Song, P.; Koo, D.-H.; Guo, L.; Li, Y.; Sun, S.; Weng, Y.; Yang, L. Genome wide characterization of simple sequence repeats in watermelon genome and their application in comparative mapping and genetic diversity analysis. *BMC Genom.* **2016**, *17*, 1–17. [CrossRef]

27. Zhao, X.; Li, C.; Wan, S.; Zhang, T.; Yan, C.; Shan, S. Transcriptomic analysis and discovery of genes in the response of Arachis hypogaea to drought stress. *Mol. Biol. Rep.* **2018**, *45*, 119–131. [CrossRef] [PubMed]

28. Liu, Y.; Geng, Y.; Song, M.; Zhang, P.; Hou, J.; Wang, W. Genetic Structure and Diversity of Glycyrrhiza Populations Based on Transcriptome SSR Markers. *Plant Mol. Biol. Rep.* **2019**, *37*, 401–412. [CrossRef]

29. Gouy, M.; Rousselle, Y.; Chane, A.T.; Anglade, A.; Royaert, S.; Nibouche, S.; Costet, L. Genome wide association mapping of agro-morphological and disease resistance traits in sugarcane. *Euphytica* **2014**, *202*, 269–284. [CrossRef]

30. Huang, Z.; Zhao, N.; Qin, M.; Xu, A. Mapping of quantitative trait loci related to cold resistance in *Brassica napus* L. *J. Plant Physiol.* **2018**, *231*, 147–154. [CrossRef]

31. Kim, Z.S.; Hwang, J.W.; Lee, S.W.; Yang, C.; Gorovoy, P.G. Genetic Variation of Korean Pine (*Pinus koraiensis* Sieb. et Zucc.) at Allozyme and RAPD Markers in Korea, China and Russia. *Silvae Genet.* **2005**, *54*, 235–246. [CrossRef]

32. Acheampong, A.; Leveque, N.; Tchapla, A.; Heron, S. Application of SRAP in the genetic diversity of *Pinus koraiensis* of different provenances. *Afr. J. Biotechnol.* **2010**, *8*, 1000–1008.

33. Chen, J.Y. Analysis of Genetic Diversity of *Pinus koraiensis* Plantation in CaoHekou Forest Farm by ISSR Marker. *Bull. Bot. Res.* **2009**, *29*, 633–636.

34. Feng, F.-J.; Han, S.-J.; Wang, H.-M. Genetic diversity and genetic differentiation of natural *Pinus koraiensis* population. *J. For. Res.* **2006**, *17*, 21–24. [CrossRef]

35. Tong, Y.W.; Lewis, B.J.; Zhou, W.M.; Mao, C.R.; Wang, Y.; Zhou, L.; Yu, D.P.; Dai, L.M.; Qi, L. Genetic Diversity and Population Structure of Natural *Pinus koraiensis* Populations. *Forests* **2020**, *11*, 39. [CrossRef]

36. Halbauer, E.-M.; Bohinec, V.; Wittenberger, M.; Hansel-Hohl, K.; Gaubitzer, S.; Sehr, E.M. Genetic diversity of flax accessions originating in the Alpine region: A case study for an ex situ germplasm evaluation based on molecular marker. *Euphytica* **2017**, *213*, 120. [CrossRef]

37. Krishnan, A.G.; John, R.; Cyriac, A.; Sible, G. Estimation of genetic diversity in nutmeg (*Myristica fragrans* Houtt.) selections using RAPD markers. *Int. J. Plant Sci.* **2017**, *12*, 102–107. [CrossRef]

38. Thomas, L. Genetic variation in Turkish Red Pine (*Pinus brutia* Ten.) seed stands as determined by RAPD markers. *Silvae Genet.* **2017**, *53*, 169–175.

39. Chung, M.Y.; Moon, M.O.; López-Pujol, J.; Chung, J.M.; Chung, M.G. Genetic diversity in the two endangered endemic species Kirengeshoma koreana (Hydrangeaceae) and Parasenecio pseudotaimingasa (Asteraceae) from Korea: Insights into population history and implications for conservation. *Biochem. Syst. Ecol.* **2013**, *51*, 60–69. [CrossRef]

40. Sanaa, A.; Boulila, A.; Boussaid, M.; Ben Fadhel, N. Genetic diversity, population structure and linkage disequilibrium analysis in the endangered Tunisian *Pancratium maritimum* Linnaeus (Amaryllidaceae) populations. *Afr. J. Ecol.* **2016**, *54*, 379–382. [CrossRef]

41. Duan, D.; Jia, Y.; Yang, J.; Li, Z.H. Comparative transcriptome analysis of male and female conelets and development of mi-crosatellite markers in *Pinus bungeana*, an Endemic Conifer in China. *Genes* **2017**, *8*, 393. [CrossRef]

42. Zhang, Z.; Wang, H.; Chen, W.; Pang, X.; Li, Y. Genetic diversity and structure of native and non-native populations of the endangered plant Pinus dabeshanensis. *Genet. Mol. Res.* **2016**, *15*, 15027937. [CrossRef]

43. Xu, Y.L.; Cai, N.H.; Woeste, K.; Kang, X.Y.; He, C.Z.; Li, G.Q.; Chen, S.; Duan, A.A. Genetic diversity and population structure of *Pinus yunnanensis* by simple sequence repeat markers. *Forest. Sci.* **2016**, *62*, 38–47. [CrossRef]

44. Wang, M.B.; Hao, Z.Z. Rangewide Genetic Diversity in Natural Populations of Chinese Pine (*Pinus tabulaeformis*). *Biochem. Genet.* **2010**, *48*, 590–602. [CrossRef]

45. Ferrer, M.M.; Eguiarte, L.E.; Montaña, C. Genetic structure and outcrossing rates in *Flourensia cernua* (Asteraceae) growing at different densities in the South-western Chihuahuan Desert. *Ann. Bot.* **2004**, *94*, 419–426. [CrossRef]

46. Hellmann, J.J.; Pineda-Krch, M. Constraints and reinforcement on adaptation under climate change: Selection of genetically correlated traits. *Biol. Conserv.* **2007**, *137*, 599–609. [CrossRef]

47. Wang, Z.; Kang, M.; Liu, H.; Gao, J.; Zhang, Z.; Li, Y.; Wu, R.; Pang, X. High-Level Genetic Diversity and Complex Population Structure of Siberian Apricot (*Prunus sibirica* L.) in China as Revealed by Nuclear SSR Markers. *PLoS ONE* **2014**, *9*, e87381. [CrossRef]

48. Zu, Y.; Yu, J.; Wang, A. Study on pollination characteristics of natural population of *Pinus koraiensis*. *Acta Ecol. Sin.* **2000**, *20*, 430–433.

49. Zhang, Z.; Zhang, H.G.; Mo, C.; Zhang, L. Transcriptome sequencing analysis and development of EST-SSR markers for *Pinus koraiensis*. *Sci. Silvae Sin.* **2015**, *51*, 114–120.

50. Omelko, A.; Ukhvatkina, O.; Zhmerenetsky, A. Disturbance history and natural regeneration of an old-growth Korean pine-broadleaved forest in the Sikhote-Alin mountain range, Southeastern Russia. *For. Ecol. Manag.* **2016**, *360*, 221–234. [CrossRef]

51. Parsons, B.J.; Newbury, H.J.; Jackson, M.T.; Ford-Lloyd, B.V. Contrasting genetic diversity relationships are revealed in rice (*Oryza sativa* L.) using different marker types. *Mol. Breed.* **1997**, *3*, 115–125. [CrossRef]

52. Chen, W.; Hou, L.; Zhang, Z.; Pang, X.; Li, Y. Genetic Diversity, Population Structure, and Linkage Disequilibrium of a Core Collection of Ziziphus jujuba Assessed with Genome-wide SNPs Developed by Genotyping-by-sequencing and SSR Markers. *Front. Plant Sci.* **2017**, *8*, 575. [CrossRef] [PubMed]

53. Kim, Z.S.; Lee, S.W.; Lim, J.H.; Hwang, J.W.; Kwon, W. Genetic diversity and structure of natural populations of *Pinus koraiensis* (Sieb. et Zucc.) in Korea. *For. Genet.* **1994**, *1*, 41–49.

54. Potenko, V.V.; Velikov, A.V. Allozyme Variation and mating system of coastal populations of *Pinus koraiensis* sieb. et Zucc. In Russia. *Silvae Genet.* **2001**, *50*, 117–122.

55. Wright, S. The interpretation of population structure by F-Statistics with special regard to systems of mating. *Evolution* **1965**, *19*, 395–420. [CrossRef]

56. Holsinger, K.E.; Weir, B.S. Genetics in geographically structured populations: Defining, estimating and interpreting FST. *Nat. Rev. Genet.* **2009**, *10*, 639–650. [CrossRef] [PubMed]

57. Gerlach, G.; Jueterbock, A.; Kraemer, P.; Deppermann, J.; Harmand, P. Calculations of population differentiation based on GST and D: Forget GST but not all of statistics! *Mol. Ecol.* **2010**, *19*, 3845–3852. [CrossRef]

58. Waples, R.S. Separating the wheat from the chaff: Patterns of genetic differentiation in high gene flow species. *J. Hered.* **1998**, *89*, 438–450. [CrossRef]

59. Raaijeveld-Smit, F.J.L.K.; Eebee, T.J.C.B.; Riffiths, R.A.G.; Oore, R.D.M.; Schley, L. Low gene flow but high genetic diversity in the threatened Mallorcan midwife toad Alytes muletensis. *Mol. Ecol.* **2005**, *14*, 3307–3315. [CrossRef] [PubMed]

60. Schmidt, A.M.; Jacklyn, P.; Korb, J. Isolated in an ocean of grass: Low levels of gene flow between termite subpopulations. *Mol. Ecol.* **2013**, *22*, 2096–2105. [CrossRef]

61. Zhao, F.; Yang, J.; He, H.S.; Dai, L. Effects of Natural and Human-Assisted Regeneration on Landscape Dynamics in a Korean Pine Forest in Northeast China. *PLoS ONE* **2013**, *8*, e82414. [CrossRef] [PubMed]

62. Bao, L.; Kudureti, A.; Bai, W.; Chen, R.; Wang, T.; Wang, H.; Ge, J. Contributions of multiple refugia during the last glacial period to current mainland populations of Korean pine (*Pinus koraiensis*). *Sci. Rep.* **2015**, *5*, 18608. [CrossRef]

63. Schoettle, A.W.; Goodrich, B.A.; Hipkins, V.D.; Richards, C.; Kray, J. Geographic patterns of genetic variation and population structure in *Pinus aristata*, Rocky Mountain bristlecone pine. *Can. J. For. Res.* **2011**, *42*, 23–37. [CrossRef]

64. Shamseldeen, E.; Sallam, A.; Belamkar, V.; Emara, H.A.; Nower, A.A.; Salem, K.F.; Poland, J.; Baenziger, P.S. Genetic diversity and population structure of F3:6 nebraska winter wheat genotypes using genotyping-by-sequencing. *Front. Genet.* **2018**, *9*, 76.

65. Luo, Z.N.; Brock, J.; Dyer, J.M.; Kutchan, T.; Schachtman, D.; Augustin, M.; Ge, Y.F.; Fahlgren, N.; Abdel-Haleem, H. Genetic diversity and population structure of a camelina sativa spring panel. *Front. Plant Sci.* **2019**, *10*, 184. [CrossRef] [PubMed]
66. Wang, H.; Vieira, F.G.; Crawford, J.E.; Chu, C.; Nielsen, R. Asian wild rice is a hybrid swarm with extensive gene flow and feralization from domesticated rice. *Genome Res.* **2017**, *27*, 1029–1038. [CrossRef]
67. Peng, L.P.; Cai, C.F.; Zhong, Y.; Xu, X.X.; Xian, H.L.; Cheng, F.Y.; Mao, J.F. Genetic analyses reveal independent domestication origins of the emerging oil crop *Paeonia ostii*, a tree peony with a long-term cultivation history. *Sci. Rep.* **2017**, *7*, 5340. [CrossRef]
68. Chung, M.Y.; Chung, M.G. Large effective population sizes and high levels of gene flow between subpopulations of *Lilium cernuum* (Liliaceae). *Biochem. Syst. Ecol.* **2014**, *54*, 354–361. [CrossRef]
69. Bartkowska, M.P.; Wong, A.Y.C.; Sagar, S.P.; Zeng, L.; Eckert, C.G. Lack of spatial structure for phenotypic and genetic variation despite high self-fertilization in *Aquilegia canadensis* (Ranunculaceae). *Heredity* **2018**, *121*, 605–615. [CrossRef]
70. Herman, A.; Brandvain, Y.; Weagley, J.; Jeffery, W.R.; Keene, A.C.; Kono, T.J.Y.; Bilandžija, H.; Borowsky, R.; Espinasa, L.; O'Quin, K.; et al. The role of gene flow in rapid and repeated evolution of cave-related traits in Mexican tetra, *Astyanax mexcanus*. *Mol. Ecol.* **2018**, *27*, 4397–4416. [CrossRef]
71. Yerka, M.K.; De Leon, N.; Stoltenberg, D.E. Pollen-Mediated Gene Flow in Common Lambsquarters (*Chenopodium album*). *Weed Sci.* **2012**, *60*, 600–606. [CrossRef]
72. Hernando, R.C.; Ken, O.; Mauricio, Q.; Fuchs, E.J.; Antonio, G.R. Contrasting patterns of population history and seed-mediated gene flow in two endemic costa rican oak species. *J. Hered.* **2018**, *109*, 530–542.
73. Aegisdóttir, H.H.; Patrick, K.; Jürg, S. Isolated populations of a rare alpine plant show high genetic diversity and considerable population differentiation. *Ann. Bot.* **2009**, *104*, 1313–1322. [CrossRef]
74. Kaljund, K.; Jaaska, V. No loss of genetic diversity in small and isolated populations of Medicago sativa subsp. falcata. *Biochem. Syst. Ecol.* **2010**, *38*, 510–520. [CrossRef]
75. Dostálek, T.; Münzbergová, Z.; Plačková, I. High genetic diversity in isolated populations of Thesium ebracteatum at the edge of its distribution range. *Conserv. Genet.* **2013**, *15*, 75–86. [CrossRef]
76. Li, X.; Li, M.; Hou, L.; Zhang, Z.; Pang, X.; Li, Y. De Novo Transcriptome Assembly and Population Genetic Analyses for an Endangered Chinese Endemic *Acer miaotaiense* (Aceraceae). *Genes* **2018**, *9*, 378. [CrossRef]
77. Peakall, R.; Smouse, P.E. GenAlEx 6.5: Genetic analysis in Excel. Population genetic software for teaching and research—An update. *Bioinformatics* **2012**, *28*, 2537–2539. [CrossRef] [PubMed]
78. Chen, C.J.; Xia, R.; Chen, H.; He, Y.H. TBtools, a Toolkit for Biologists integrating various HTS-data handling tools with a user-friendly interface. *Mol. Plant* **2020**, *13*, 1194–1202. [CrossRef]
79. Sándor, N.; Péter, P.; István, C.; Mousapour, G.A.; Géza, H.; János, T. PICcalc: An online program to calculate polymorphic information content for molecular genetic studies. *Biochem. Genet.* **2012**, *50*, 670–672.
80. Hamrick, J.L.; Godt, M.J.W. Effects of life history traits on genetic diversity in plant species. *Philos. Trans. R. Soc. B Biol. Sci.* **1996**, *351*, 1291–1298. [CrossRef]
81. Falush, D.; Stephens, M.; Pritchard, J.K. Inference of Population Structure Using Multilocus Genotype Data: Linked Loci and Correlated Allele Frequencies. *Genetics* **2003**, *164*, 1567–1587. [CrossRef]
82. Evanno, G.; Regnaut, S.; Goudet, J. Detecting the number of clusters of individuals using the software structure: A simulation study. *Mol. Ecol.* **2005**, *14*, 2611–2620. [CrossRef]
83. Earl, D.A.; Vonholdt, B.M. Structure harvester: A website and program for visualizing STRUCTURE output and implementing the Evanno method. *Conserv. Genet. Resour.* **2011**, *4*, 359–361. [CrossRef]
84. Liu, K.; Muse, S.V. PowerMarker: An integrated analysis environment for genetic marker analysis. *Bioinformatics* **2005**, *21*, 2128–2129. [CrossRef] [PubMed]
85. Letunic, I.; Bork, P. Interactive Tree Of Life (iTOL): An online tool for phylogenetic tree display and annotation. *Bioinformatics* **2007**, *23*, 127–128. [CrossRef]
86. Keenan, K.; Mcginnity, P.; Cross, T.F.; Crozier, W.W.; Prodöhl, P.A. DiveRsity: An R package for the estimation and exploration of population genetics parameters and their associated errors. *Methods Ecol. Evol.* **2013**, *4*, 782–788. [CrossRef]
87. Ihaka, R.; Gentleman, R. R: A Language for Data Analysis and Graphics. *J. Comput. Graph. Stat.* **1996**, *5*, 299–314.

Review

Underutilized Fruit Crops of Indian Arid and Semi-Arid Regions: Importance, Conservation and Utilization Strategies

Vijay Singh Meena [1,†], Jagan Singh Gora [2,†], Akath Singh [3], Chet Ram [2], Nirmal Kumar Meena [4], Pratibha [5], Youssef Rouphael [6], Boris Basile [6,*] and Pradeep Kumar [3,*]

[1] ICAR—National Bureau of Plant Genetic Resources, Regional Station—Jodhpur, ICAR-CAZRI Campus, Jodhpur 342003, Rajasthan, India; vijay.meena1@icar.gov.in
[2] ICAR—Central Institute for Arid Horticulture, Bikaner 334006, Rajasthan, India; jagan.gora@icar.gov.in (J.S.G.); chet.ram2@icar.gov.in (C.R.)
[3] ICAR—Central Arid Zone Research Institute, Jodhpur 342003, Rajasthan, India; Akath.Singh@icar.gov.in
[4] Department of Fruit Science, College of Horticulture and Forestry, Agriculture University, Kota, Jhalawar 326023, Rajasthan, India; nirmalchf@gmail.com
[5] Rajasthan College of Agriculture, MPUAT, Udaipur 313001, Rajasthan, India; prolaniyaskr23@gmail.com
[6] Department of Agricultural Sciences, University of Naples Federico II, 80055 Portici, Italy; youssef.rouphael@unina.it
* Correspondence: boris.basile@unina.it (B.B.); pradeep.kumar4@icar.gov.in (P.K.)
† These authors contributed equally to this work.

Citation: Meena, V.S.; Gora, J.S.; Singh, A.; Ram, C.; Meena, N.K.; P.; Rouphael, Y.; Basile, B.; Kumar, P. Underutilized Fruit Crops of Indian Arid and Semi-Arid Regions: Importance, Conservation and Utilization Strategies. *Horticulturae* **2022**, *8*, 171. https://doi.org/10.3390/horticulturae8020171

Academic Editor: Esmaeil Fallahi

Received: 31 December 2021
Accepted: 15 February 2022
Published: 18 February 2022

Publisher's Note: MDPI stays neutral with regard to jurisdictional claims in published maps and institutional affiliations.

Abstract: Nowadays, there is a large demand for nutrient-dense fruits to promote nutritional and metabolic human health. The production of commercial fruit crops is becoming progressively input-dependent to cope with the losses caused by biotic and abiotic stresses. A wide variety of underutilized crops, which are neither commercially cultivated nor traded on a large scale, are mainly grown, commercialized and consumed locally. These underutilized fruits have many advantages in terms of ease to grow, hardiness and resilience to climate changes compared to the major commercially grown crops. In addition, they are exceptionally rich in important phytochemicals and have medicinal value. Hence, their consumption may help to meet the nutritional needs of rural populations, such as those living in fragile arid and semi-arid regions around the world. In addition, local people are well aware of the nutritional and medicinal properties of these crops. Therefore, emphasis must be given to the rigorous study of the conservation and the nutritional characterization of these crops so that the future food basket may be widened for enhancing its functional and nutritional values. In this review, we described the ethnobotany, medicinal and nutritional values, biodiversity conservation and utilization strategies of 19 climate-resilient important, underutilized fruit crops of arid and semi-arid regions (Indian jujube, Indian gooseberry, lasora, bael, kair, karonda, tamarind, wood apple, custard apple, jamun, jharber, mahua, pilu, khejri, mulberry, chironji, manila tamarind, timroo, khirni).

Keywords: climate resilient; arid zone fruits; adaptation; nutritional quality

1. Introduction

The world population (7.87 billion) is currently growing at a rate of 1.03% per year and is expected to reach around 9.6 billion in 2050. India has 1.38 billion people accounting for 17.5% of the world population, with a meager 2.4% of the world surface area [1]. Nowadays, the greatest challenge is to provide this burgeoning population with stable, safe and nutritious quality food. In the current Global Hunger Index (GHI), India stands at position 101 of 116 countries; this presents a gloomy situation in combating malnutrition, eventually affecting the socio-economic progress [2]. The World Health Organization (WHO) has also indicated that hungriness is the most serious problem worldwide, particularly for African countries and India. Therefore, 195 nations have decided to adopt sustainable development goals (SDG) for addressing the serious malnutrition problems with a holistic approach by the year 2030 [3]. Consumer awareness about the health benefits of fruits offers great thrust for their regular consumption as part of a balanced diet. Worldwide demand of nutrient-dense fruit has increased immensely in recent years not only for enhancing people's nutritional status but also for their positive effects on immune and metabolic health. This is particularly interesting considering the COVID-19 pandemic scenario.

In India, major fruit crops, such as mango, banana, citrus, guava and apple, account for more than 72% of the total area under fruit crops, while indigenous (native) fruit crops contribute only 6.56% of the area (0.437 mha) with quite high productivity (11.47 tons/ha) [4]. Climate change is inducing a rise in air temperatures, UV radiation levels and in the frequency of extreme events, such as drought or flood, which, especially in arid or semi-arid areas, can result in an intensification of the negative impact of salinity, mineral deficiency/toxicity and of diseases and insect-pest attacks on crops [5–8]. Consequently, climate change represents a great threat to obtaining the sustainable production of major commercial fruits [5]. Under such environmental conditions, the fulfillment of the consumers' choice and nutritional food security at an affordable and sustainable level is a major concern for the researchers as well as the growers.

Under the given circumstances, specific growing areas may be utilized for exploiting the potential of underutilized crops producing edible fruits that meet the food and nutritional demand of local population. It is necessary to explore some biotic and abiotic resistant/tolerant native underutilized fruit crops that could be resilient to certain climatic variations and adapt to a wide range of agro-climatic conditions. The indigenous fruit crops are not only proven to be superior in terms of wider adaptability to environmental conditions but are also known for their nutritional value [9]. However, a limited amount of research has been carried out for the development of production protocols and utilization of these underutilized fruit species. Moreover, the limited number of identified varieties, the low availability of quality planting materials and the inadequate availability of suitable cultural and post-harvest management practices are still major limitations challenging the systematic cultivation of these underutilized crops.

The vegetation of arid areas includes a large number of edible fruit-bearing and food-producing species. In the Indian arid zone, around 30 plant species are known for their different edible uses, and around 19 of them bear edible fruits and possess horticultural importance [10]. Many of the underutilized fruit crops can be used as fresh fruit but also for culinary and medicinal purposes providing important nutrients, and some of them also have ornamental values. Local people are aware of their medicinal and nutritional properties. Indeed, most indigenous underutilized fruit crops, such as ber, kair, aonla, lasora and phalsa, are richer in minerals, antioxidants and phytonutrients compared to many commercial fruit crops. Moreover, these underutilized fruits are not very popular and are sold at very low prices in the local markets because of the lack of (a) people's awareness of their nutritive values, (b) consumption habits, (c) limited research and (d) developmental policies by the government agencies for their potential exploitation.

Considering the importance of these tree crop species in traditional medicine, their nutritional richness and wide adaptability, the Government of India, under their centrally sponsored scheme, i.e., Mission on Integrated Development of Horticulture (MIDH, then 'National Horticulture Mission', NHM) during 2005–2006 gave a special impetus to establish orchards of underutilized fruit species. This paper reviews the importance of 19 underutilized fruit crops endemic to Indian arid and semi-arid conditions, and their adaptation mechanism to stress conditions, genetic diversity, ethnobotany, medicinal and nutritional values and possible ways for their conservation and potential exploitation for improving nutritional and socio-economic security of the regions.

2. Characteristics and Potential Uses of Indigenous Underutilized Fruit Crops

Abiotic stresses caused by environmental factors are the most common yield-limiting factors globally, and they cause up to 70% of the yield losses in major fruit crops [11,12].

In more detail, the individual potential yield losses induced by the different climatic adversities were reported to be the following: high temperature: 40–50%; salinity: 20%; drought: 17%; low temperature: 15% [13]. Arid and semi-arid regions are considered the hotspot for abiotic stresses, such as extreme temperatures, intense solar radiation, salinity, drought and nutrient deficiency, where the commercial fruit crops either fail to grow or struggle to express their potential performance. Under such climatic conditions, the integration of arid-zone underutilized fruit crops can be a better strategy to sustain the crop productivity under stress due to their typical morphological, physiological, anatomical and biochemical xerophytic characteristics that allow them to perform optimally under harsh climates. Therefore, adaptive traits such as those that increase the overall resilience and resistance to suboptimal environmental conditions do not necessarily result in a yield penalty. It is generally assumed that adaptive traits ensure yield stability in specific conditions, being fitness typically measured in terms of fertility, fruits and seeds. For instance, these traits include phenology shifts (flowering/ripening in a specific period of the year) and/or morphological characteristics (root/shoot ratio, leaf macro and/or micro-morphological traits, etc.) that allow specific genotypes to escape environmental stresses (not necessarily involving an active and metabolically costly response to stress). This can result in the capacity of these genotypes to have fruits reaching ripening compared to those that did not have any adaptive trait.

In order to cope with abiotic stresses, the arid-zone underutilized fruit crops, such as ber (*Zizyphus* spp.), aonla (*Emblica officinalis*), bael (*Aegle marmelos*), jamun (*Syzigium* spp.) and wood apple (*Feronia limonia*), have modified and/or developed their organs to assure vital morpho-physiological functions (i.e., strong deep root system, a high root-to-shoot ratio for reaching into deeper moist soil layers and uptake more water and nutrients) [14]. Similarly, crops such asber, bael, lasora (*Cordia mixa*) and pilu (*Salvadora persica*) have round, thick and barked stems for easier water storage and reduced cuticle transpiration. Some crops such as kair (*Capparis decidua*), lasora, aonla and pilu have synchronized flowering and fast fruit development during the season characterized by larger moisture availability [15,16]. Crops such as ber, phalsa and bael exhibit leaf shedding/dormancy for reducing water loss in summer and for protecting the plants from frost in winter [17–19]. Similarly, other underutilized crops possess numerous morphological characters, such as spines instead of leaves (ber), scanty foliage (kair), spiny cladodes (prickly pear), mucilaginous sap for reduced transpiration loss (kair, lasora, pilu, bael, etc.), small-sized and thick leaves, fur/hairiness and waxy coating on the leaf surface and sunken and deep stomata, for water-saving through the reduction in transpiration rate and heat shocks (ber, phalsa, lasora, fig), and selective or reduced absorption of cation (Na^+) and anions (Cl^-, SO_4^{2-}) [19]. These characteristics are also associated with the accumulation of osmolytes, compatible organic and inorganic solutes (proline, phenolics, flavonoids, soluble sugars, glycine, betaine, etc.), and biosynthesis of enzymatic and non-enzymatic antioxidants, heat shock proteins and drought-responsive genes to maintain cell turgor, allowing better survival under the adverse conditions of arid and semi-arid environments [20,21]. In addition, the genetic basis of the adaptive traits deserves to be studied because this information could be used in future breeding programs that may also involve novel tools, such as genome editing.

These underutilized fruit crops may represent the next generation of futuristic crops, which could enhance the farmer's income through sustainable production systems even under a climate-change scenario. The characteristics and the potential uses of the 19 underutilized fruit crops of arid and semi-arid regions are separately described below, and a summary of their main traits is provided in Table 1.

2.1. Indian Jujube (*Ziziphus mauritiana L.*)

The Indian jujube (ber) belongs to the family *Rhamnaceae*, and it is known as the king of arid-zone fruits or as poor man's apple. The ber tree is fast growing and has a spreading canopy and a short bole; branches are slender, downy, brown bold spines in pairs [22]. The ber tree is extremely drought-hardy due to the deep taproot system and xerophytic characteristics, such as (a) dormancy (leaf shedding) during the peak period of hot summer preventing transpiration, (b) waxy and hairy leaves, (c) thick bark [14]. It grows well even in marginal or poor soils where most other commercial fruit trees either fail to grow or have very poor performance. The jujube seeds contain saponins, jujubogenin and obelin lactone [23,24]. Jujube wood is utilized as fuel or charcoal making and its leaves are used as fodder for sheep and goats [25].

The fruit has a spongy, sweet, tasty pulp and is an excellent source of vitamins C, A, B, carotenoids, protein, Ca, P, K, Rb, Br, La and sugars (fructose, glucose and galactose) [26]. The smoke of its burning leaves is also utilized to cure cutaneous, cough and cold. Ber fruit is mostly consumed as fresh within 4–5 days after harvest due to the short shelf life. Thus, it is necessary to develop a value-added product at a farmer-field or industry level, and there is the need to work on the diversification and popularization of jujube products. It is the only fruit crop that can give good returns even under rainfed conditions due to its wide adaptability under a large variety of soils, water availability conditions and climates (with the exception of heavy frosts) in arid and semi-arid regions. In addition to nutritional and economic health, some jujube cultivars, such as Dragon, Mushroom, So and Teapot, are known for their landscape values, such as unique fruit shape, fruit color and tree shape, and are planted in gardens and backyards due to their dwarf habit and compact canopy [27,28].

2.2. Indian Gooseberry (Emblica officinalis G.)

The Indian gooseberry is an indigenous and important minor fruit. It belongs to the family *Euphorbiaceae*, and it is grown in diverse soil and climatic conditions of India [29]. The medicinal and therapeutic properties of aonla are considered as 'amritphal' or a wonder fruit for health [30]. The aonla fruit is 3 and 160-times richer in protein and vitamin C compared to apples, respectively [31]. It is the richest source of Vitamin C (500–1800 mg/100 g) among the fruits after Barbedos cherry, and the content in leucoanthocyanins polyphenols, pectin, iron, calcium and phosphorus makes its fruit largely used in Ayurvedic medicines for making *Triphala* and *Chyavanprash* [30,32].

As a result of the intensive research carried out since the beginning of the 21st century and the development of 30 varieties, this fruit species is grown commercially in some areas, and it was proven to be a fruit crop potentially suitable also for arid ecosystems. It is hardy, productive and highly remunerative even when managed without much care under drought and saline areas of arid and semi-arid regions. The aonla fruit is highly perishable, acidic and astringent; consumers do not prefer them for fresh consumption [33], whereas aonla fruit is generally used to prepare a number of delicious, processed food products, such as preserve, candy, jelly, toffee, pickle, leather, squash, juice, RTS beverage, cider, shreds, dried powder and ayurvedic tonics, such as Chayvanprash, Triphala, Amrit Kalash and Amol Ki Rasayan [34,35].

2.3. Lasora (Cordia myxa L.)

Cordia, locally known as Gonda, Lasora, Lehsua, Indian cherry, Assyrian Plum or Bird's Nest Tree, belongs to the *Boraginaceae* family and is grown across India except for the high hills and the temperate climates [36]. *Cordia* is a fast-growing tree with a beautiful inverted dome/umbrella crown, utilized as an avenue tree and ornamental furniture; ovate, alternate and stalked leaves used as fodder during hot summer when green grasses are not available and also used as rearing lac insect [37,38]. Trees bear white color hermaphrodite flowers in March and drupaceous green unripe fruit ready for harvesting from April to June. It is mostly used as green fresh vegetables and pickles, especially in the lean period when the availability of conventional vegetables is limited [39]. The fruit is considered as a naturally rich source of antioxidants, i.e., carotenoids, ascorbic acid, phenols, and minerals, crude fiber, protein, ascorbic acid, ash and vitamins, which represent essential nutrients for human health and for curing certain human ailments (improve digestion, birdlime, anti-tumor, anti-helmentic, diuretic, demulcent and expectorant; improve hair growth) [40–42].

2.4. Bael [Aegle marmelos L. (Correa)]

Bael is the only species of the genus *Aegle*, which belongs to the family *Rutaceae* [43]; it is one of the oldest indigenous fruits known by various names in different parts of India, such as billi, Bengal quince, stone apple, golden apple and Japanese bitter orange [44]. Bael has a wide distribution in various ranges of edaphic-climatic conditions due to its ability to withstand heat, drought and low-temperature poor-nutrient soil [45]. It is deciduous, medium-sized, slender, gum bearing with a cauliflorous fruiting habit, deep taproot system, bold thorny branches and trifoliate leaves [46]. Its trifoliate leaves resemble a trident, so people offer them to Lord Shiva Lingam to get rid of worry and suffering [47]. Bael can be used as avenue and ornamental trees (golden color ripen fruit); shells of the dried fruit after removing pulp are used as fashioned cups, small containers, ornamental pills, snuff boxes, etc. [48].

The bael fruit is a rich source of riboflavin used to cure beriberi, and unripe fruit is suggested to treat diarrhea and dysentery, whereas the marmelosin in fruit has therapeutic properties being a good remedy for stomach ailments [49]. However, all plant parts of bael contain various compounds with medicinal values, e.g., coumarins, alkaloids, sterols and essential oils, that have analgesic, antipyretic, anti-inflammatory, anti-antifungal, microfilaria, hypoglycemic, anti dyslipidemic, antiproliferative, wound healing, insecticidal and anti-fertility abilities [50]. Bael fruit is consumed only in processed products, such as powder, preserve, nectar, toffee [51]. These products have had high market demand during the COVID-19 pandemic period due to its ayurvedic medicinal values. Their current price in the market is high and for this reason, bael is becoming a remunerative crop for farmers of arid and semi-arid areas.

2.5. Kair [Capparis decidua (Forsk.)]

Capparis decidua Forsk belongs to the *Capparidaceae* family, and it is locally known as Kair, Ker, KarilTeent, Della, and Neptiin. It is an indigenous, multipurpose small woody perennial much-branched, leafless bushy shrub widely grown without much care on farm boundaries, orans, gochars and wastelands tracts of arid and semi-arid regions [52]. Its xerophytic characteristics, such as deep root system, scanty foliage, mucilaginous sap and tough conical spine, make it an ideal plant for stabilizing sand dunes and controlling soil erosion by wind during the hot desiccating summer in the Thar desert of western Rajasthan [53]. However, it easily survives in desert conditions characterized by temperatures ranging from −8 to +48 °C or more, drought, saline and poor nutrients soil ecological conditions [54]. In general, kair is naturally propagated through seeds, root suckers, hardwood cuttings and tissue culture, but the plant survival rate is very low [55]. Kair plants produce pink, red and white flowers in the axil of the spine three times a year, but the main flowering flush occurs in March–April, and fruit matures just before the monsoon [56].

The kair fruit is used as a vegetable, pickles and condiments. Dried fruit is an important ingredient of a traditional vegetable of Rajasthan known as 'Panchkutta'. Its fruit is rich in proteins, carbohydrates, fiber and minerals (Ca, P and Fe). It is used in medicine for sedation, anticonvulsant asthma, inflammation and cough, since it contains isocodonocarpine, α- and β-amyrin, taraxasterol, erythrodiolalkaloids in plant organs [57].

2.6. Karonda (Carissa carandas L.)

Karonda is a plant of Indian origin, belonging to the family *Apocynaceae* and locally known as Christ's thorn. Karonda is a hardy, evergreen, spiny, low-growing bushy multiple branched shrub grown for bio-fencing/live-fencing in gardens, orchards or in very small-scale plantations in Rajasthan, Gujarat and Uttar Pradesh [58,59]. Its drought-hardy nature is due to xerophytic features, and the plant offers 5–8 kg fruit yields without much care and management in arid and semi-arid regions. Karonda produces flowers in January–February and June–July and fruit ripen in 60–90 days after fruit set. Depending on the genotypes, white, green, purple and pinkish-red colored fruits are common, due to which it is also used as an ornamental plant in gardens [38]. Immature fruit is usually used for producing pickles and chutney but occasionally is used as a vegetable, while fully ripe fruit is consumed fresh or processed to produce candies and colored extracts used as natural food colorant [60].

Karonda fruit is considered the richest source of iron (39 mg per 100 g), contains a fair amount of vitamin C and is used to cure of anemia and scurvy [61]. In addition, they are a good source of calcium, magnesium and phosphorus and have high antioxidant activity [62]. The mature fruit is suitable for making pickles and jellies due to the high content of pectin. They can also be exploited for making jams, squashes, syrups and chutneys, which have high market demand.

2.7. Tamarind (Tamarindus indica L.)

Tamarindus inidica is a dicotyledonous, monotypic, long-lived, semi-evergreen fruit plant belonging to the family *Leguminosae* [63]. It has a wide range of adaptability, and it is an ideal tree for avenue plantation as a roadside, backyard and agroforestry systems. It bears terminal and lateral drooping bisexual flowers in May–June and forms fruit as pendulous pods ten months after fruit set [64]. Tamarind fruit pulp and seeds contain tartaric acid, reducing sugar, tannin, pectin, cellulose, fiber, potassium, calcium phosphorous and other minerals, such as sodium, iron and zinc [65]. The fruit pulp is the chief source for souring sauces, curries, chutneys, beverages, food colorants and it is considered a great delicacy [66]. All its parts are valuable for food, fodder, timber, fuel, textile, nutritional and pharmaceutical industries, such as fluoride remover [67,68]. Tamarind trees are planted as roadside avenue trees in the Banaras Hindu University, Varanasi, the largest university campus of India.

2.8. Wood Apple [Feronia limonia (L.) Swingle]

Wood apple is an indigenous fruit tree that is also known as kainth, elephant apple and monkey fruit [69,70]. Systematic block plantation in the form of orchards of wood apple is uncommon, whereas it is mostly found in isolation as a stray plant in the plains of Southern Maharashtra, Uttar Pradesh, West Bengal, Madhya Pradesh and Chhattisgarh states of India. The wood apple is a small-to-moderate size, glabrous, deciduous tree with thorny branches, rough and spiny bark and it is able to grow on saline, poor and neglected lands normally unsuitable for fruit cultivation [71]. It is the only species of the *Citrus* family that can tolerate both drought and salinity stress. Its flowering starts from February to May, and fruit matures October to December depending on the moisture availability [14,72].

The fruit of wood apple is a berry with rough, hard-shelled, large, globose, woody pericarp and sweetish aromatic edible pulp [73]. Several organs of the *F. limonia* tree have excellent therapeutic and functional properties: leaves (diuretic, anti-microbial and stomach disorders), roots and bark (insecticidal and snakebite), spines (liver and menorrhagia), gum (diarrhea and diabetes) and fruit pulp (skin cancer, diarrhea, sore throat, Jaundice and gastropathy) [74–76]. The fruit is processed as powder, preserve, squash, sherbet, beverage, jam, cream, leather, wine, toffee, candy, RTS, pickle and capsules [77,78].

2.9. Custard Apple (Annona squamosa L.)

Custard apple is one of the drought-hardy fruit plants belonging to the family *Annonaceae*, which is commercially cultivated in a limited area of the Indian Deccan plateau region. The light, gravel and small pebbles soil is also suitable for its cultivation. Custard apple plants are small, semi-deciduous shrubs with simple leaves, cauliflorous flowering, bisexual and protogynous flowers, superior ovary, fruit etaerio of berries [79]. Its flower is borne mostly in new flushes after the shedding of old leaves commencing from March to August with a peak in April–May [80]. The fruit is climacteric, it may be symmetrically heart-shaped, lopsided or irregular, and the interspaces between the protuberances become yellow at full maturity [81].

The demand of custard apple fruit is increasing in domestic and international markets thanks to their sensory, therapeutic and nutritional properties, as well as their pleasant flavor. Custard apple fruit contains vitamins A, B, C, E, and K1, essential minerals, antioxidants and polyunsaturated fatty acids. They are antimalarial, antifeedant, immuno-suppressive, cytotoxic, diterpenes and are used to treat HIV [82,83]. Moreover, a range of cosmetic products using custard apple is available in the market, such as perfumery, soaps, pimple creams, essential oils, hair lotions, ayur slim capsules, cold balms, anti-stress massage oil, pain massage oils, and foot care creams [84].

2.10. Jamun (Syzygium cumunii Skeels)

Jamun belongs to the *Myrtaceae* family is an Indigenous evergreen hardy fruit tree that naturally grows in neglected and marshy areas. Deep loamy, well-drained soils and dry weather during the flowering and fruiting period are ideal conditions for its cultivation. Jamun flower panicles emerge at the leaf axil during March–April. Fruit is borne in clusters (10–40 fruit) and are round or oblong, single-seeded berries with a single sigmoid type development pattern, and are non-climacteric [85].

Jamun is rich in biochemical compounds, e.g., anthocyanins, myrecetin, ellagic acid, isoquercetin, glucoside, kaemferol, and it is used for its anti-inflammatory, neuropsycho, anti-microbial, anti-HIV, nitric oxide or free radical scavenging, anti-fertility and anti-ulcerogenic activities [86]. Glycosides in the seed, jambolin or antimellin, are considered to have anti-diabetic properties by halting the conversion of starch into sugar [87,88]. Ripe jamun fruit is used to prepare many products, such as squashes, juices, jam, jelly, pickles and wines. In Goa and the Philippines, the fermented fruit of jambolans is used to produce Brandy and a distilled liquor called 'jambava' [89].

2.11. Jharber (Ziziphus nummularia Burm. f.)

Jharber is a 1–2 m tall, perennial, deciduous and thorny shrub with drooping branches that in nature occupies almost all the habitats of extremely arid environments (crop and grazing lands, sandy-saline, rocky, degraded pastures) [10]. Anatomical features, such as the presence of papilla, crypt stomata, epidermis with a thick outer wall and thick cuticle and deep taproot system provide the best tolerance to drought, salinity and high-temperature stresses, making this species adaptable to extreme arid regions [90,91]. It flowers in July–August, and fruit ripens in November–December; the fruit is small-sized drupes with a globose-ovoid shape, dark red color and little edible pulp of sub acidic taste [22].

The jharber dried fruit contains triterpenoids, alkaloids and saponins and are used in medicine for their anticancer, stomachic, sedative, blood purifier, anti-obesity, antipyretic, anodyne, refrigerant, pectorial, anti-anemia, vomiting and styptic properties [92]. Leaves of jharber, locally called 'Pala', are often used as fodder for camels, cattle, goats and sheep, being rich in crude fiber, crude protein, calcium and phosphorus [93]. In addition, some plant organs have local medicinal uses. For instance, (a) the leaves are used to obtain poultices to heal wounds or they are used to cure asthma, fever, gum bleeding and liver problems; (b) the bark is used to treat diarrhea; (c) the roots are used as a decoction to cure fever, whereas its powder is adopted to treat ulcers and wounds; (d) the fruit is laxative and antiemetic; (e) the seeds are sedative [94].

2.12. Mahua [Madhuca longifolia (Koenig)]

Mahua is an indigenous deciduous tree belonging to the family *Sapotaceae* and is characterized by medium-to-large-sized canopy, grey-black cracked bark, milky and short trunk and many-branched [95]. The flowers are grouped in dense clusters with long pedicels and have a coriaceous (leathery, stiff and tough) calyx and a tubular, cream-colored, scented caduceus corolla. The fruit is a pinkish-yellow berry with 1–4 recalcitrant seeds [96]. The trees are heterozygous and cross-pollinated. It is a multipurpose tree, which fulfills the three basic requirements of tribal people (food, fodder and fuel) [97]. Mahua flowers are edible and highly nutritive, being a good source of sugars, vitamins, proteins, minerals and fats, and they are used as a sweetener to prepare numerous traditional dishes, such as barfi, kheer, halwa and meethi puri, in the tribal belts of Madhya Pradesh, Rajasthan,

Gujarat, Orissa, Jharkhand, Chhattisgarh and Andhra Pradesh [98,99]. Mahua dry flowers are also fermented to produce wine, brandy, ethanol, acetone and lactic acids [100]. Mahua is also used in medicine for its hepatoprotective, antiburn, anti-skin disease and wound healing, emollient, bone healing, swelling gum, anti-ulcer, anti-snake bite, milk production stimulation in lactating women, anti-bronchitis, anti-diabetic, diuretic, immune system stimulating, digestive, antioxidant, energetic and glucose booster activities [101,102].

2.13. Pilu (Salvadora persica L.)

Pilu is also known as kharijal, meetajal, mustard or salt bush, toothbrush tree, and belongs to the family *Salvadoraeceae*. It is a perennial, evergreen, large, much-branched shrub or tree widely found in Gujarat, Rajasthan, Haryana and Punjab and is suitable for the forestation of ravines, saline and alkaline lands as shelterbelts/windbreaks due to its hardy xerophytic nature [103,104]. Pilu possesses a number of potentially therapeutic compounds, namely salvadoricine, salvadourea, β-sotisterol, trimethyl amine, thioglucoside, di-benzyl thiourea, rutin, potash, chlorine, sulfur, etc. [105].

Pilu's fibrous branches are a natural toothbrush (Miswak) and thus are used for oral hygiene [106]. They are also used in a number of important medicines, e.g., antiseptics, abrasives, detergents, astringents, fluorides, enzyme inhibitors, dental diseases, anti-tumors, anti-leprosy, anti-ulcers, anti-gonorrhea, and antiscorbutic products [107–109]. Moreover, the fruit is a source of sweeteners and are used for producing fermented drinks; the tender shoots are eaten as a salad, and the seeds are rich in C12 and C14 acids used in the soap and detergent industry [110].

2.14. Khejri [Prosopis cineraria (Druce.) L.]

Khejri or Jand/shami belongs to the *Leguminosae* family and is considered as the wonder tree, nature's gift, the king of desert and the golden tree. It is a desert dwelling tree that is the lifeline tree of the Indian *Thar* desert because each and every part of the tree is used to improve the socio-economic life of the local people [111,112]. Khejri is an evergreen, slow-growing tree with exfoliated bark, rounded canopy, small and mucilaginous leaves and a strong deep taproot system that can reach extraordinary depths (up to 53 m or more). It was reported to be drought and salinity tolerant (10.0 to 25.0 EC dSm^{-1}) [113,114]. It is native to Arabia and the Indian *Thar* desert [115], and it is extensively distributed in the *Thar* desert of the Indian states of Rajasthan Haryana, Gujarat, Punjab and Delhi [116].

It is a multipurpose tree as it provides a vegetable pod, flour, cattle fodder, fuel, timber, gum, resin and medicine [117]. It is also used as fencing/windbreak, avenue tree, on farm boundaries in water deficit areas, topiary, bonsai and screening trees in home gardening, and forest restoration in arid landscapes [38,118]. It is a high litter accumulating tree and improves soil fertility through fixing atmospheric nitrogen, and these effects result in the increase in the soil content of organic matter, soluble calcium and available phosphorus and in a reduction in soil pH [119,120]. Moreover, the Khejri tree is considered a productivity booster in inter-cropping and companion cropping systems thus it is highly suitable for agroforestry systems in arid and semi-arid regions. Khejri green leaves have a very high nutritional value containing crude protein (11.9–18.0%), crude fiber (13–22%), nitrogen-free extract (43.5%), ash (6–8%), ether extract (2.9%), calcium (2.1%) and phosphorus (0.4%) [121]. The unripe pods ('Sangri'), an important ingredient of the *Panchkutta* vegetable, is nutritionally rich in crude protein (18%), fat (2.0%), carbohydrate (56%), crude fiber (26%), phosphorus (0.4%), calcium (0.4%) and iron (0.2%), and is consumed as a green or dry vegetable, pickles and flour [122]. All the organs of Khejri have therapeutic properties: flowers (Patuletin glycoside patulitrin, rutinsitosterolluteolin and prosogerin A and B) are used for their anti-diabetic activity and for treating Lewis lung carcinoma [123]; leaves (spicigerine, campesterol and Tricosan-1-α) for mouth ulcers [124]; pod and seed (Prosogerin, linoleic acid and prosophylline) for preventing protein malnutrition and calcium iron deficiency in blood, asthma, piles and leprosy, etc. [112,125,126].

2.15. Mulberry (Morus alba L.)

Mulberry belongs to the *Moraceae* family, is native to South-West Asia and has wide geographical distribution from temperate to tropical climates [127]. The *Morus* genus comprises 16 deciduous species out of 24 species, and *Morus alba, M. nigra,* and *M. rubra* are mainly grown in commercial orchards [128]. It is a perennial, woody, fast-growing, deciduous tree with alternate, simple and often lobed leaves, catkin inflorescence and composite sorosis fruit [129]. Most of its commercial cultivation has aimed to produce silk and shelterbelts rather than fruit. However, the latter are rich in vitamins, minerals, dietary fiber, sugar, amino acids, carotenoids, flavonoids and phytosterols. They are used as a functional food in the forms of *masala*, herbal tea, marmalades, juices, yogurt, biscuits, smoothies, capsules or as natural dyes, cosmetics oil and dietary food products, such as pekmez, kome and pestil [130,131]. The various suggested pharmacological uses are for obesity, cardiac diseases, diabetes, hypercholesterolemia, tumors, oxidative stress, brain damage and for their anti-fungal, anti-aging, anxiolytic and hepato-protective activities [132–135].

2.16. Chironji (Buchana nialanzan)

Chironji belongs to the family *Anacardiaceae*, which originates from the Indian subcontinent [136]. It has no specific requirements in terms of soil and climate, and it is naturally found in the arid and semi-arid forests of Jharkhand, Chhattisgarh, Madhya Pradesh, Rajasthan, Gujarat and Uttar Pradesh [137]. Chironji is a medium-sized, subdeciduous/evergreen plant with a straight trunk and coriaceous leaves [138]. It is a highly heterozygous, cross-pollinated plant with a strong tendency to alternate bearing [138]. The fruit can be eaten both raw and roasted. Its kernel contains fats (59.0%), proteins (19.0–21.6%), carbohydrates (12.1%), fiber (3.8%), phosphorus (528.0 mg), calcium (279.0 mg), iron (8.5 mg) and vitamins [139]. It has the potential capacity to cure various diseases, such as snakebite, dysentery, diarrhea, asthma, burning sensation of body, fever, ulcers, cold and Alzheimer's, and it has anti-diabetic and antihyperlipidemic activity [81,140–142].

2.17. Manila tamarind [Pithecellobium dulce (Roxb.) Benth.]

Manila tamarind is commonly known as Madras thorn Monkey pod and Jungle jalebi and belongs to the *Fabaceae* family. It is a multipurpose, fast-growing, medium-sized thorny tree used as live fencing, animal fodder, hardwood timber, windbreak and a potential source of lac culture. Its fruit has a sweet acidic taste and high content of dietary fiber, proteins, Ca, Fe, P, unsaturated fatty acids and antioxidants [143–145]. Manila fruit is used to treat toothaches, mouth ulcers, sore gums, dysentery, chronic diarrhea, stress, aging symptoms and dark skin spots [146–148].

2.18. Timroo (Diospyros melanoxylon Roxb.)

Timroo or tendu belongs to the family *Ebenaceae* and is native to India and Sri Lanka [36]. It is found in endemic conditions within limited areas of Gujarat, Madhya Pradesh, Rajasthan, Jharkhand, Bihar, Chhattisgarh and Tamil Nadu [149]. It is a long-lived, deciduous, dioecious, seedless parthenocarpic berry fruit. Its leaves are commercially used for *bidis* making (indigenous, traditional cigarette), agricultural implements and furniture. Most importantly, it is used as an indicator for high sulfur dioxide concentration [150,151]. Timroo fresh fruit has high total phenolic content, flavonoids, scavenging activity, antioxidants and β-carotene content as equal or more to guava, plum, star fruit, mango, kiwi and apple fruit [152–154]. Its bark extracts are used to treat dyspepsia, diarrhea, and smallpox (burnt bark) by ethnomedicine practitioners [155–157].

2.19. Khirni (Manilkara hexendra L.)

Khirni/rayan belongs to the *Sapotaceae* family, and it is a native to India, evergreen, medium-sized, slow-growing fruit plant with a spreading canopy [36]. It is a wild plant found in the arid and semi-arid to tropical climate as an avenue tree and can be used as bonsai due to the evergreen, dense foliage and dwarf habit [158]. It bears flowers in February–March, whereas fruit ripen in May–June, and it is commercially used as rootstock for sapota to exploit its tolerance to salinity and drought [159]. Its bark, seeds and fruit are rich sources of tannins, oil and vitamin A, respectively [160]. Khirni fruit and bark are used for numerous medicinal purposes, such as curing fever, flatulence, stomach disorder, leprosy, ulcers, opacity of the cornea, dyspepsia, urethrorrhea and bronchitis [161,162].

Table 1. Summary of the important characteristics of 19 underutilized fruit crops.

Common Names	Species	Drought Tolerant	Marginal/ Poor Soils	Vitamins	Mineral Elements	Antioxidants	Medicinal Properties	Shelf-Life	Consumed Form
Aonla (Indian Gooseberry)	*Emblica officinalis* G.	Yes [163]	Yes [163]	C [31]	Ca, Fe, P [30,32]	Leucoanthocyanins, gallic acid, ascorbic acid [30,32]	Yes [30]	Perishable [33]	Raw, processed [34,35]
Bael	*Aegle marmelos* L. (Correa)	Yes [45]	Yes [45]	B_1, B_2, A, C [164]	Fe, Ca, K, P [165,166]	Marmelosin, psoralen [50,166]	Yes [49,50]	Very low perishability [167]	Processed [51,168]
Ber (Indian Jujube)	*Ziziphus mauritiana* L.	Yes [23,42]	Yes [53]	C, A, B [11,72,85]	Ca, P, K, Rb, Br, La [45,48,74]	Carotenoids [102]	Yes [8]	Perishable [68,89]	Raw, dry, processed [1,26,103]
Chironji	*Buchanania lanzan*	Yes [137]	Yes [137,169]	B_1, B_2, C [170]	P, Ca, Fe [139]	Polyphenolics [170,171]	Yes [81,141,142]	Highly perishable (fruit) very low perishability (kernel) [172]	Raw, processed [139]
Custard apple	*Annona squamosa* L.	Yes [173]	Yes [173]	A B_1, B_2, B_3, C, E [82,83]	K, Mg, Ca, Zn, Fe [174]	Carotenoid, flavonoids [83,84,174]	Yes [82,83]	Perishable [175]	Raw, processed [84]
Jamun	*Syzygium cumunii* Skeels	Yes [176]	Yes [176]	C, B, E [177,178]	K, Na, Mg, Ca, Fe [179]	Ascorbic acid, phenolics [86,178]	Yes [87,88]	Highly perishable [176]	Raw, processed [89]
Jharber	*Ziziphus nummularia* Burm. f.	Yes [10,91]	Yes [10]	C, B group, A [22]	K, P, Ca, Fe, Na [180]	Phenolics and ascorbic acid [92]	Yes [92]	Very low perishability [22]	Raw, dry [22]
Kair	*Capparis decidua* (Forsk.)	Yes [53,54]	Yes [52]	A, C, E [57,181]	AI, P, Na, Mg, Fe, Ca [57,182]	Rutin, tocopherols, carotinoids [182–184]	Yes [57]	Very low perishability [56]	Processed [56]

Table 1. *Cont.*

Common Names	Species	Drought Tolerant	Marginal/ Poor Soils	Vitamins	Mineral Elements	Antioxidants	Medicinal Properties	Shelf-Life	Consumed Form
Karonda	*Carissa carandas* L.	Yes [185,186]	Yes [187]	C, A [61,62]	Fe, Ca, Mg, P [61,62]	Phenolics, flavonoids, anthocyanins [61,62]	Yes [188]	Moderately perishable [186,187]	Raw, processed [60]
Khejri	*Prosopis cineraria* (Druce.) L.	Yes [113,114]	Yes [113,114]	K_1, A, C [189]	Ca, P, Fe, Zn [121,122]	Phenolics, carotenoids, saponin [123]	Yes [123,126]	Moderately perishable [126]	Dry, processed [126]
Khirni	*Manilkara hexandra* L.	Yes [159]	Yes [158]	A, C, E [160]	Ca, Fe, Zn, Cu, Se [189]	Quercetin, myricetin, rutin [160]	Yes [161,162]	Highly perishable [190]	Raw, processed [191]
Lasora	*Cordia myxa* L.	Yes [192]	Yes [193]	C [10]	Ca, P, Zn, Fe [41,194,195]	Polyphenols, flavonoids [41,194,196]	Yes [197]	Perishable [198]	Processed [39,199]
Mahua	*Madhuca longifolia* Koenig	Yes [200,201]	Yes [200]	C, A [201]	Ca, P [202]	Ascorbic acid [98]	Yes [101,102]	Very low perishability [201]	Processed [99,100]
Manila tamarind	*Pithecellobium dulce* (Roxb.) Benth.	Yes [148]	Yes [148]	C, B6, B1 [203]	K, P, Ca, Fe, Zn [204]	Anthocyanins, polyphenolics [145,205]	Yes [146,147]	Moderately perishable [206]	Raw, processed [207]
Mulberry	*Morus alba* L.	Yes [127]	Yes [127]	C, E, K [208]	Fe, Cu, Mg, K, Se, Na [208]	Zeaxanthin, resveratrol [208]	Yes [132,135]	Highly perishable [208]	Raw, processed [130,131]
Pilu	*Salvadora persica* L.	Yes [103,104]	Yes [103,104]	E, C, A [209]	K, Cl, Na, S, Fe [210]	Polyphenols, flavonoids. carotenoids [209]	Yes [105,108]	Highly perishable [110]	Processed [110]

Table 1. *Cont.*

Common Names	Species	Drought Tolerant	Marginal/ Poor Soils	Vitamins	Mineral Elements	Antioxidants	Medicinal Properties	Shelf-Life	Consumed Form
Tamarind	*Tamarindus indica* L.	Yes [211]	Yes [211]	C, K, B$_6$ [65]	K, Ca, P, Na, Fe, Zn [65]	Polyphenols, flavonoids, carotenoids [67,68]	Yes [67,68]	Very low perishability [66,212]	Raw, processed [66]
Timroo	*Diospyros melanoxylon* Roxb.	Yes [213]	Yes [213]	C, A [214]	K, Ca, P [214]	Polyphenolics, beta-carotene [152–154]	Yes [155,157]	Perishable [214]	Raw, processed [215]
Wood apple	*Feronia limonia* L.	Yes [71]	Yes [71]	A, B$_2$, C [216]	Ca, P, Fe[216]	Phenolics [216,217]	Yes [74–76]	Very low perishability [218]	Processed [77,78]

3. Diversity and Conservation of Genetic Resources of Indigenous Underutilized Fruit Crops

India is regarded as a mega-diverse country, as it houses 11.18% of the world's recorded plant species and over 2.4% world's surface area [219]. Out of 34 global biodiversity hotspots, India shares four biodiversity hotspots (Western Ghats and Sri Lanka, Himalayas, Indo-Burma, Sundal) and one-third of the higher plant species present in these areas are endemic. The conservation and management of this biodiversity is considered to be mandatory for the use of humankind [220].

A significant part of the flora and fauna biodiversity is threatened by many factors related to climate change [25]. The indigenous underutilized fruit resources have remained on the back foot due to the introduction of ethnic fruit species and the advancement in the cultivation of existing major fruit species. Furthermore, the ever-increasing population undoubtedly is exerting huge pressure on the degradation of biodiversity (ecosystems, genes and species), causing severe genetic losses. Most of the endangered fruit species are conserved, domesticated, improved only by traditional societies/farmers and tribal people. However, in the last decades, a systemic effort laid on the exploration, conservation and their utilization by the different institutions [221].

The main Indian statutory body responsible for plant genetic resource collection and conservation, the ICAR-National Bureau of Plant Genetic Resources (ICAR-NBPGR), New Delhi, India, has conserved about 440,000 accessions of 1900 species in its gene bank and about 3520 accessions of recalcitrant species in a cryo gene bank since its inception (1976). The ICAR-NBPGR has in-situ and ex-situ conservation at its 10 regional stations and 59 designated National Active Germplasm sites (NAGS) across the country. The present status of underutilized fruit crops conservation in India is 1717 at ICAR-NBPGR and its collaborative centers (Table 2), whereas the major research institutions working on underutilized fruit crops in the country are maintaining 1127 accessions in their field gene banks (Table 3). Furthermore, 357 accessions belonging to eleven underutilized fruit species are being cryopreserved in gene banks [149].

Table 2. Status of collection and conservation of indigenous fruit species undertaken by different institutions of India.

Crop	No. of Accessions
Ber	487
Aonla	159
Bael	57
Karonda	50
Timroo	24
Manila Tamarind	24
Mahua	153
Khirni	74
Phalsa	36
Pilu	207
Jamun	198
Tamarind	248
	1717

Table 3. Underutilized fruit species' germplasm being conserved in field gene banks at different institutions in India.

Name of Crop	CIAH, Bikaner	RS, HAU, Bawal	NBPGR, Jodhpur	PAU, Abohar	CAZRI, Jodhpur
Ziziphus mauritiana	318	47	26	34	40
Ziziphus rotundifolia	22	-	-	-	-
Emblica officinalis	50	6	-	-	-
Punica granatum	154	-	-	-	-
Carissa carrandus		4	-	-	13
Cordia myxa	65	30	17	-	-
Aegle marmelos	17	10	5	-	-
Grewia subineaqualis	06	04	04	-	-
Capparis decidua	06	22	22	-	20
Syzigium cuminii	50	-	-	-	-
Tamarindus indica	25	-	-	-	-
Madhuca latifolia	50	-	-	-	-
Buchanania lanzan	30	-	-	-	-
Manilkara hexandra	30	-	-	-	-

RS: Regional station; HAU: Hisar Agriculture University, PAU: Punjab Agriculture University.

4. Strategies for the Improvement and Promotion of Underutilized Fruit Crops

The underutilized fruit species play a crucial role in mitigating nutritional insecurity and poverty in those rural or tribal areas of the country where the availability of fruits is either low or not accessible to them. In recent years, concerted efforts have been laid by some of the public sector research organizations in the improvement of underutilized fruit crops. As a result, some varieties have been developed through selection and hybridization for higher adaptability with good yield and quality traits under arid and semi-arid environments (Table 4). The locally adapted species serve as a reservoir of stress-related genes that are a potential source for the improvement of stress-tolerant varieties in future breeding programs. Several production technologies have been standardized by different institutions, e.g., propagation methods (rootstock selection, grafting/budding), plant spacing, canopy management, nutrient and water management, crop regulation, plant protection and post-harvest management, and value addition (Table 4).

Table 4. Recent advances made for the cultivation and utilization of underutilized fruit crops.

Species	Improved Varieties	Production Technologies	Major Value-Added Products	Ornamental and Other Values	Institute Involved
Aonla	NA-7, NA-10, NA-6, Anand-1, Anand-2, Laksmi-52, Goma Aishwaria [222,223]	Standarization of propagation (patch budding), high-density planting, canopy management, value addition, integrated nutrient management [223]	Candy, Chyawanpras, Shreds, Candy, Preserve, Squash, RTS, Pickle, Jelly, Leather, Toffee, aonla powder [34,35,223]	Avenue plantation [30]	ICAR-CIAH, Bikaner; ANDUAT, Ayodhya
Bael	Goma Yashi, NB-5, NB-7, NB-9, Pant Aparna, Pant Sujata, Pant Shivani, Thar Divya, Thar Neelkanth, Thar Shristhi, Thar Prikriti, Thar Shivangi [224]	Standarization of propagation (patch budding), detoppng for promotion of scion wood in mother plant, planting geometry and high-density planting, plant architectural engineering, water and nutrient management, fruit drop and cracking management, bael-based cropping system [224]	Squash, preserve, candy [51,223,225]	Avenue plantation, windbreak plantation [48]	ICAR-CIAH, Bikaner; ANDUAT, Ayodhya; GBPUAT, Pantnagar
Ber	Gola, Umran, Goma Kirti, Katha, Seb, Thar Sevika, Thar Bhuvraj, CAZRI Ber 2018 [22,226]	Standarization of propagation (T-budding), insitu budding, top-working, pruning and training system, high-density planting system, water and nutrient management, fruit fly and stone weevil management [22,227]	Osmodehydrated ber, Canned ber, Jam, Pickle [22]	Ornamental purpose, windbreak, bio-fencing, furniture [27,28]	ICAR-CIAH, Bikaner; NBPGR, New Delhi; HAU, Hisar; ICAR-CAZRI, Jodhpur
Chironji	Thar Priya [228,229]	Standardization of propagation (soft wood grafting), insitu grafting, training and pruning, nutrient management, processing, value addition [228]	Dried seed [139]	Avenue plantation [137]	ICAR-CIAH, Bikaner (Raj); NIFTEM, Sonipat (Haryana)

Table 4. *Cont.*

Species	Improved Varieties	Production Technologies	Major Value-Added Products	Ornamental and Other Values	Institute Involved
Custard apple	Arka Neelanchal Vikram, Arka Sahan, Balanagar [230]	Standardization of propagation (whip grafting), high-density planting system, canopy management, flower regulation, processing, value-added products [230–232]	Puree, jam, RTS, Juice, Frozen pulp [233,234]	Kitchen gardening [230]	ICAR-IIHR, Bangalore
Jamun	Goma Prinynka [235]	Standardization of propagation (patch budding and soft wood grafting), high-density planting system, canopy management, processing, value addition [235–237]	Jamun Juice, RTS, Squash, Nectar, Jam, Vinegar, Wine, Jelly, Cider, Syrup [236,237]	Avenue and windbreak plantation [236,237]	ICAR-CIAH, Bikaner (Raj); ICAR-CAZRI, Jodhpur; ICAR-IIHR, Bangalore
Jharber	-	Dehydrated products [238]	Churan, Bar, Toffee [238]	Bio-fencing, wind break, forest and soil restoration [238]	ICAR-CIAH, Bikaner
Kair	CZJK-3 and CZJK5 [239]	Standarization of propagation (root cutting and tissue culture), post-harvest management, value addition [55,56,239]	Vegetable, Pickle panchkutta [240]	Bio-fencing, wind break, forest and soil restoration, ornamental value [52,53]	ICAR-CIAH, Bikaner; ICAR-CAZRI, Jodhpur
Karonda	Thar Kamal, Konkan Bold, Maru Gaurav, Pant Manohar, Pant Sudarshan, Pant Suvarna [185,241]	Standardization of propagation (cutting/air layering), high-planting system, training and pruning, nutrient and water management, value addition [185,241]	Murabba, Jam, Jelly, Pickle, Chutney [185,241,242]	Bio-fencing in kitchen garden, orchards, windbreak, ornamental [38,58,59]	ICAR-CIAH, Bikaner (Raj); ICAR-CAZRI, Jodhpur; GBPUAT, Pantnagar (UK)
Khejri	Thar Sobha, Thar Amruta [243]	Standardization of propagation (patch budding), water and nutrient management, canopy management, khejri-based model (HBCPSMA) [243,244]	Vegetables, Pickle, Biscuits [122]	Bio-fencing, wind break, forest and soil restoration, bonsai [38,118]	ICAR-CIAH, Bikaner

Table 4. *Cont.*

Species	Improved Varieties	Production Technologies	Major Value-Added Products	Ornamental and Other Values	Institute Involved
Khirrni	Thar Rituraj [245]	Standardization of propagation (cleft grafting), plant spacing, canopy management, cropping system, value addition [225,245]	Dehydrated fruit, fruit bar, RTS, Jam [225]	Avenue plantation [158]	ICAR-CIAH, Bikaner(Raj)
Lasora	Thar Bold, Maru Samaridhi, Karan Lasora [246–248]	Standarization of propagation (patch budding), canopy management, defoliation, integrated pest management, value addition [246–248]	Pickle, Beverage, Chutney [39]	Avenue and border plantation as windbreak [37,38]	ICAR-CIAH, Bikaner; ICAR-CAZRI, Jodhpur; SKNAU, Jobner
Mahua	Thar Madhu [249]	Standardization of propagation (soft wood grafting), canopy management, nutrient and water management, post-harvest management, value addition (alcoholic beverage) [249]	Alcohol, Bakery, Vinegar, Syrup, Wine [98–100]	Avenue and ornamental plantation [249]	ICAR-CIAH, Bikaner (Raj); NIFTEM, Sonipat (Haryana)
Manila tamarind	PKM 1 [250]	Standardization of propagation (cutting, micropropagation), value addition [251,252]	Dried seed, Oil extraction [207]	Bio-fencing, windbreak and shelterbelts [252]	ICAR-CIAH, Bikaner(Raj); TNAU, Tamilnadu
Mulberry	Thar Lohit, Thar Harit, Delhi Local [253]	Standardization of propagation (cutting), mixed farming system, nutrient management, training and pruning system by Tamil Nadu Agriculture University [253,254]	Squash, RTS [130,131]	Avenue plantation, furniture [128]	ICAR-CIAH, Bikaner
Pilu	-	Value-added products [106,110]	Squash, RTS, Miswak [106,110]	Avenue, wind break, forest and soil restoration [103,104]	ICAR-CAZRI, Jodhpur

Table 4. *Cont.*

Species	Improved Varieties	Production Technologies	Major Value-Added Products	Ornamental and Other Values	Institute Involved
Tamarind	Periyakulam 1 (PKM1) [255]	Standardization of propagation (root/stem cutting, wedge grafting), water and nutrient management, insect-pest management, processing, value-added products [255–257]	Jam, Jelly, Syrup, Pulp Powder, Candy, Seed kernel powder [66]	Avenue plantation, wind break, ornamental as bonsai purpose [255]	TNAU, Tamilnadu, ICAR-CIAH, Bikaner
Timroo	-	-	-	Avenue tree [149]	ICAR-CIAH, Bikaner(Raj)
Wood apple	Thar Guarav [258]	Standardization of propagation (soft wood grafting), high-density planting system, processing, and value addition [258,259]	Chutney, Pickle, Frozen puree, sauce [77,78]	Boundry plantation of farms [258]	ICAR-CIAH, Bikaner

In addition to the medicinal and nutritional value of fruits, these underutilized fruit crops have multipurpose utilities, such as ornamental, avenue tree, rootstocks, bio-fencing, windbreak/shelter tree, furniture, screening in backyard gardening, forest restoration and as social and economic plants [28,38]. The promotion of these underutilized fruit crops can be accelerated by providing training and demonstrations of the developed technologies to the end-users. Emphasis needs to be directed towards the developmental activities, such as the establishment of planting nurseries for ensuring the supply of quality planting materials and processing units for scale-up of their values added products at commercial levels through entrepreneurship or self-help groups or farmer producer organizations (FPO), etc. The adoption of these crops can be promoted by planting them on community lands, in the premises of religious places, gardens, parks, etc., where, besides providing recreation or ornamental values, these would provide nutrient-rich fruits to the people (Table 1). The government, through its various schemes, such as MIDH, tribal sub-plan (TSP), scheduled caste sub-plan (SCSP), are providing quality planting materials for the establishment of commercial orchards of different fruit crops, including underutilized fruits. These crops are less prone to insect pests and diseases. Pesticide application is almost negligible; thus they can be fitted well into organic farming. The underutilized fruit crops are expected to get special attention under the recently launched centrally sponsored scheme *'Prakritik kheti'* (Natural farming) that aimed at promoting the cultivation of crops and fruit species in their natural habitats. Furthermore, in order to promote the importance of indigenous species, the government has been giving major emphasis through a slogan 'vocal for local'.

In addition, it is required that the development of policies for incentives to on-farm conservation of biodiversity, and the recognition through felicitation and monetary supports to the people and societies involved in the conservation and utilization of such important indigenous underutilized fruit species. Furthermore, the inclusion of course curricula about indigenous fruit species at the school level will create awareness among the children. For creating awareness, extension specialists can organize special awareness camps/campaigns and exhibitions at micro and macrolevels, conveying themes of unexploited-underutilized fruit crops. Furthermore, the use of mass media, such as radios, televisions, newspapers and other printed and electronic media platforms, can play an effective role in creating awareness about the significance of underutilized fruit crops among the growers and other stakeholders, as well as consumers.

5. Conclusions

Underutilized fruit crops play an important role for their therapeutic properties due to their significant medicinal and nutritional value and can be considered as future horticultural assets to help nations assuring nutrition and food security, besides providing recreational, social and environmental significance. Being hardy and adapted locally, these species may serve the purpose of enhancing sustainable farm income under the harsh arid and semi-arid environments, such as waste lands (jharber, kair, pilu), marginal or saline soil and water conditions (ber, aonla, bael, karonda, etc.), rocky terrains (custard apple, timroo, jamun, etc.) and also in the backyard or kitchen gardening (karonda, custard apple, khejri, mulberry, etc.) and avenue plantation (tamarind, lasora, khirni, jamun, mulberry, khejri, etc.). Some fruit crops possess high significance in food processing industries (aonla, bael, karonda, tamarind, etc.). Their horticultural development is moving quite fast, and in the future, there will be greater technological adoption, extension and policy planning, both in traditional horticultural enterprises as well as in the commercial sectors. Special efforts are needed in terms of research and development of suitable specific packaging practices and of superior varieties. Considering the potential of the underutilized fruit crops, the emphasis shall be made to substantiate the efforts being made towards eradicating global malnutrition by their direct introduction in other arid and semi-arid regions of the world so as to achieve the goal of sustainable development (1 to 3) of the United Nations.

Author Contributions: Conceptualization, V.S.M., P.K., J.S.G. and A.S.; methodology, V.S.M., C.R., N.K.M., J.S.G., A.S. and P.; writing—original draft preparation, V.S.M., J.S.G., N.K.M., A.S., P. and P.K.; visualization and illustration of Tables, J.S.G., P.K., C.R., A.S., Y.R., B.B. and P.; writing—review and editing, P.K., A.S., Y.R. and B.B.; draft finalization, A.S., P.K., Y.R. and B.B. All authors have read and agreed to the published version of the manuscript.

Funding: There was no external funding involved in the study.

Institutional Review Board Statement: Not applicable.

Informed Consent Statement: Not applicable.

Data Availability Statement: Not applicable.

Acknowledgments: The authors thankfully acknowledge A.K. Singh, Principal Scientist, ICAR-Central Institute for Arid Horticulture, CHES, Godhra, Gujarat, for providing some of the pictures of fruit crops, and the Director, ICAR-NBPGR, New Delhi, ICAR-CAZRI, Jodhpur, for extending logistic support for conducting the study.

Conflicts of Interest: The authors declare no conflict of interest.

References

1. U.N. World Population Prospects. UN.org. 2021. Available online: https://population.un.org/wpp/ (accessed on 11 December 2020).
2. Grebmer, V.K.J.; Bernstein, R.; Alders, O.; Dar, R.; Kock, F.; Rampa, M.; Wiemers, K.; Acheampong, A.; Hanano, B.; Higgins, R.; et al. *Global Hunger Index—One Decade to Zero Hunger: Linking Health and Sustainable Food Systems*; Welthungerhilfe: Bonn, Germany, 2020.
3. Anonymous. *International Food Policy Research Institute, Global Nutrition Report 2016: From Promise to Impact: Ending Malnutrition by 2030*; International Food Policy Research Institute: Washington, DC, USA, 2016.
4. NHB. *Indian Horticulture Database-2019*; National Horticulture Board, Ministry of Agriculture and Farmer Welfare, Government of India: New Delhi, India, 2019.
5. Gora, J.S.; Verma, A.K.; Singh, J.; Chaudhary, D.R. Climate Change and Production of Horticultural Crops. In *Agriculture Impact of Climate Change*; CRC Press: Boca Raton, FL, USA, 2019; pp. 45–61.
6. Chatzistathis, T.; Fanourakis, D.; Aliniaeifard, S.; Kotsiras, A.; Delis, C.; Tsaniklidis, G. Leaf Age-Dependent Effects of Boron Toxicity in Two *Cucumis melo* Varieties. *Agronomy* **2021**, *11*, 759. [CrossRef]
7. Mumivand, H.; Shayganfar, A.; Tsaniklidis, G.; Bistgani, Z.E.; Fanourakis, D.; Nicola, S. Pheno-Morphological and Essential Oil Composition Responses to UVA Radiation and Protectants: A Case Study in Three Thymus Species. *Horticulturae* **2022**, *8*, 31. [CrossRef]
8. Sanwal, S.K.; Mann, A.; Kumar, A.; Kesh, H.; Kaur, G.; Rai, A.K.; Kumar, R.; Sharma, P.C.; Kumar, A.; Bahadur, A.; et al. Salt Tolerant Eggplant Rootstocks Modulate Sodium Partitioning in Tomato Scion and Improve Performance under Saline Conditions. *Agriculture* **2022**, *12*, 183. [CrossRef]
9. Berwal, M.K.; Haldhar, S.M.; Ram, C.; Shil, S.; Kumar, R.; Gora, J.S.; Singh, D. *Calligonum polygonoides* L. as Novel Source of Bioactive Compounds in Hot Arid Regions: Evaluation of Phytochemical Composition and Antioxidant Activity. *Plants* **2021**, *10*, 1156. [CrossRef] [PubMed]
10. Rathore, M. Nutrient content of important fruit trees from arid zone of Rajasthan. *J. Hortic. For.* **2009**, *1*, 103–108.
11. Mantri, N.; Patade, V.; Penna, S.; Ford, R.; Pang, E. Abiotic stress responses in plants: Present and future. In *Abiotic Stress Responses in Plants*; Springer: New York, NY, USA, 2012; pp. 1–19.
12. Zorb, C.; Geilfus, C.-M.; Dietz, K.J. Salinity and crop yield. *Plant Biol.* **2019**, *21*, 31–38. [CrossRef] [PubMed]
13. Ashraf, M.; Harris, P.J. *Abiotic Stresses: Plant Resistance through Breeding and Molecular Approaches*; Haworth Press: New York, NY, USA, 2005.
14. Chundawat, B.S. *Arid Fruit Culture*; Oxford and IBH Publishing Co. Pvt. Ltd.: New Delhi, India, 1990.
15. Pareek, S.; Mukherjee, S.; Paliwal, R. Floral biology of ber—A Review. *Agric. Rev.* **2007**, *28*, 277–282.
16. Adhikary, T.; Kundu, S.; Chattopadhayay, P.; Ghosh, B. Flowering Intensity and Sex Ratio of Ber (*Ziziphus mauritiana* Lamk.). *Int. J. Bio-Res. Stress Manag.* **2019**, *10*, 107–112. [CrossRef]
17. Singh, R.S.; Nath, V.; Tewari, J.C. Lasora—A promising fruit tree for arid ecosystem. *Deco Mirror* **1996**, *3*, 5–11.
18. Thind, S.K.; Mahal, J.S. *Package of Practices for Cultivation of Fruits*; Additional Director of Communication for Punjab Agricultural University: Ludhiana, India, 2021.
19. Saroj, P.L. Global Scenario for Arid and Semi-Arid Fruit Production for Combating Hunger. In Proceedings of the National Conference on Arid Horticulture for Enhancing Productivity and Economic Empowerment Held at ICAR-CIAH, Bikaner, India, 27–29 October 2018; pp. 1–14.
20. Anjum, S.A.; Xie, X.; Wang, L.C.; Saleem, M.F.; Man, C.; Lei, W. Morphological, physiological and biochemical responses of plants to drought stress. *Afr. J. Agric. Res.* **2011**, *6*, 2026–2032.

21. Bhargava, R.; Berwal, M.K. *Adaptation for Drought Tolerance in Arid Horticultural Crops*; CIAH/Tech./Pub. No. 66; ICAR-Central Institute for Arid Horticulture: Bikaner, India, 2018; p. 30.

22. Meghwal, P.R.; Khan, M.A.; Tewari, J.C. *Ber: Growing ber (Ziziphus mauritiana Lam) for Sustainable Income and Employment in Arid and Semi-Arid Regions*; Director, ICAR-Central Arid Zone Research Institute: Jodhpur, India, 2007; p. 26.

23. Maciuk, A.; Lavaud, C.; Thépenier, P.; Jacquier, M.J.; Ghédira, K.; Zèches-Hanrot, M. Four new dammaranesaponins from *Zizyphus lotus*. *J. Nat. Prod.* **2004**, *67*, 1639–1643. [CrossRef] [PubMed]

24. Abdoul-Azize, S. Potential Benefits of Jujube (*Zizyphus lotus* L.) Bioactive Compounds for Nutrition and Health. *J. Nutr. Metabol.* **2016**, 2867470. [CrossRef] [PubMed]

25. Feyssa, D.H.; Njoka, J.T.; Asfaw, Z.; Nyangito, M.M. Wild edible fruits of importance for human nutrition in semi-arid parts of East Shewa Zone, Ethiopia: Associated indigenous knowledge and implications to food security. *Pak. J. Nutr.* **2011**, *10*, 40–50. [CrossRef]

26. Anjum, M.A.; Rauf, M.A.; Bashir, H.; Ahmad, A. The evaluation of biodiversity in some indigenous Indian jujube (*Ziziphus mauritiana*) germplasm through physico-chemical analysis. *Acta Sci. Pol. Hortorum Cultus* **2018**, *17*, 39–52. [CrossRef]

27. Liu, M.; Wang, M. *Germplasm Resources of Chinese Jujube*; China Forestry Publ. House: Beijing, China, 2009. (In Chinese)

28. Yao, S.; Heyduck, R. Ornamental Jujube Cultivar Evaluation in the Southwestern United States. *HortTechnol.* **2018**, *28*, 557–561. [CrossRef]

29. Tiwari, J.P.; Mishra, D.S.; Mishra, K.K.; Mishra, N.K. Aonla. In *Medicinal and Aromatic Crops*; Jitendra, S., Ed.; Avishkar Publishers and Distributors: Jaipur, India, 2007; pp. 112–123.

30. Ganachari, A.; Thangavel, K.; Ali, S.M.; Nidoni, U.; Ananthacharya, A. Physical properties of Aonla fruit relevant to the design of processing equipments. *Int. J. Eng. Sci. Technol.* **2010**, *2*, 7562–7566.

31. Barthakur, N.N.; Arnold, N.P. Chemical analysis of Emblica (*Emblica officinalis* Gaertn) and its potential as food source. *Hortic. Sci.* **1991**, *47*, 99–105. [CrossRef]

32. Rajkumar, N.V.; Theres, M.; Kuttan, R. *Emblica officinalis* fruits afford protection against experimental gastric ulcers in rates. *Pharm. Biol.* **2001**, *39*, 375–380. [CrossRef]

33. Priya, M.D.; Khatkar, B.S. Effect of processing methods on keeping quality of aonla (*Emblica officinalis* Gaertn.) preserve. *Int. Food Res. J.* **2013**, *20*, 617–622.

34. Choptra, R.N. ; *Chopra's Indigenous Drugs of India*, 2nd ed.; U.N. Dhur and Sons Private Ltd.: Calcutta, India, 1958.

35. Gopalan, C. *Nutritive Value of Indian Foods*; National Institute of Nutrition, Indian Council of medical research: Hyderabad, India, 2002; p. 53.

36. Stewart, J.L.; Brandis, D. *The Forest Flora of North-West and Central India*; Singh, B., Singh, M.P., Eds.; New Connaught Place: Dehradun, India, 1992; p. 602.

37. Ahuja, S.C.; Ahuja, S.; Ahuja, U. Nutraceautical wild Fruits of India-Lasora (Cordia)-History, Origin and Folklore Scholars. *Acad. J. Biosci.* **2020**, *8*, 187–209.

38. Reddy, V.R.; Kumar, R.; Saroj, P.L. *Indigenous Ornamental Flora for Arid Landscapes*. 2019, pp. 105–111. Available online: https://www.researchgate.net/publication/336937366_Indigenous_ornamental_flora_for_arid_landscapes (accessed on 10 February 2022).

39. Bhatnagar, P.; Singh, J.; Meena, C.B. Lasoda that blooms on tree trunk—A report. *HortFlora Res. Spectr.* **2016**, *5*, 175–176.

40. Wassel, G.; El-Menshawi, B.; Saeed, A.; Maharan, G.; El-Merzabain, M. Screening of selected plants for pyrrolizidine alkaloids and antitumor activity. *Pharmazie* **1987**, *42*, 709. [PubMed]

41. Jamkhande, P.G.; Barde, S.R.B.; Patwekar, S.L.; Tidke, P.S. Plant profile, phytochemistry and pharmacology of *Cordia dichotoma* (Indian cherry). *Asian Pac. J. Trop. Biomed.* **2013**, *3*, 1009–1012. [CrossRef]

42. Meghwal, P.R.; Singh, A. *Lasoda or Gonda*; Breeding of Underutilized Fruit Crops; Jaya Publishing House: New Delhi, India, 2015; pp. 247–253.

43. Gopal, M. *India through the Ages*; Guatam, K.S., Ed.; Ministry of Information and Broadcasting, Government of India: New Delhi, India, 1990; p. 79.

44. Mani, A.; Singh, A.K.; Jain, N.; Misra, S. Flowering, Fruiting and Physiochemical Characteristics of Bael (*Aegle marmelos* Correa) Grown in Northern Districts of West Bengal. *Curr. J. Appl. Sci. Technol.* **2002**, *23*, 1–8. [CrossRef]

45. Singh, A.K.; Singh, S.; Saroj, P.L.; Krishna, H.; Singh, R.S.; Singh, R.K. Research status of bael (*Aegle marmelos*) in India: A review. *Indian J. Agric. Sci.* **2019**, *89*, 1563–1571.

46. Singh, A.K.; Singh, S.; Saroj, P.L. *Bael (Production Technology)*; Technical Bulletin; ICAR-CIAH: Rajasthan, India, 2018; Volume 67, pp. 1–53.

47. Slathia, P.S.; Paul, N.; Gupta, S.K.; Sharma, B.C.; Kumar, R.; Kher, S.K. Traditional uses of under-utilized tree species in sub-tropical rainfed areas of Kathua, Jammu & Kashmir. *Indian J. Tradit. Knowl.* **2016**, *16*, 164–169.

48. Manandhar, N.P. *Plants and People of Nepal*; Timber Press: Portland, OR, USA, 2002; ISBN 0-88192-527-6.

49. Hasan, M.A.; Singh, S.R.; Majhi, D.; Devi, H.L.; Singh, Y. Significance of minor fruits in health care. In Proceedings of the International Convention of Botanicals in Integrated Health Care, Organized by Agri-horticultural Society of India, Kolkata, India, 26–28 December 2010; pp. 162–166.

50. Neeraj, V.B.; Johar, V. Bael (*Aegle marmelos*) Extraordinary Species of India: A Review. *Int. J. Curr. Microbiol. Appl. Sci.* **2017**, *6*, 1870–1887.

51. Dashora, L.K. Genetic resources of sub-tropical underutilized fruits. Proceedings of National Seminar on Tropical and Subtropical Fruits, Navsari Agricultural University, Navsari, India, 9–11 January 2013; pp. 54–61.

52. Meghwal, P.R.; Tiwari, J.C. Kair (*Capparis decidua* (Forsk.) Edgew—A multipurpose woody species for arid regions. *For. Trees Livelihoods* **2002**, *12*, 313–319. [CrossRef]

53. Muthana, K.D. Capparis decidua. In *Trees for Dry/and*; Hocking, D., Ed.; Oxford and IBH publishing Co. Pvt. Ltd.: New Delhi, India, 1993; pp. 141–143.

54. Vyas, G.; Sharma, R.; Kumar, V.; Sharma, T.B.; Khandelwal, V. Diversity analysis of *Capparis decidua* (Forssk.) Edgew. Using biochemical and molecular parameters. *Genet. Res. Crop. Evol.* **2005**, *6*, 905–911. [CrossRef]

55. Pareek, O.P. Horticultural development in arid regions. *Indian Hortic.* **1978**, *23*, 25–30.

56. Singh, D.; Singh, R.K. Kair (*Capparis decidua*): A potential ethnobotanical weather predictor and livelihood security shrub of the arid zone of Rajasthan and Gujarat. *Indian J. Tradit. Knowl.* **2011**, *10*, 146–155.

57. Goyal, M.; Nagori, B.P.; Sasmal, D. Sedative and anticonvulsant effects of an alcoholic extract of *Capparis decidua*. *J. Nat. Med.* **2009**, *63*, 375–379. [CrossRef] [PubMed]

58. Sawant, B.R.; Desai, U.T.; Ranpise, S.A.; More, T.A.; Sawant, S.V. Genotypic and Phenotypic variability in Karonda (*Carissa caranda* L.). *J. Maharashtra Agric. Univ.* **2002**, *27*, 266–268.

59. Kumar, D.; Pandey, V.; Nath, V. Karonda (*Carissa conjesta*)-An Underutilized Fruit Crop. In *Underutilized and Underexploited Horticultural Crops*; Peter, K.V., Ed.; New India Publishing Agency: New Delhi, India, 2007; Volume 1, pp. 313–325.

60. Singh, A.K.; Singh, P. Power of significance of difference among fruit and seed size parameters of karonda (*Carrisa carandus* Linn.). *Ann. Agric. Res.* **1998**, *19*, 6671.

61. Gopalan, C.; Ramasastri, B.V.; Balasubramanian, S.C. *Nutritive Value of Indian Foods*; Narasinga Rao, B.S., Deosthale, Y.G., Pant, K.C., Eds.; National Institute of Nutrition: Hyderabad, India, 1989.

62. Panda, D.; Panda, S.; Pramanik, K.; Mondal, S. Karonda (*Carissa* spp.): An Underutilized Minor Fruit Crop with Therapeutic and Medicinal Use. *Int. J. Econ. Plants* **2014**, *1*, 36–41.

63. Storrs, A.E.G. *Know your Trees. Some Common Tree Found in Zambia*; Regional Soil Conservation Unit (RSCU): Nairobi, Kenya, 1995.

64. Coronel, R.E. *Tamarindus indica*, L. In *Plant Resources of South East Asia, Wagenongen: Pudoc No.2 Edible Fruits and Nuts*; Verheij, E.W.M., Coronel, R.E., Eds.; PROSEA Foundation: Bogor, Indonesia, 1991; pp. 298–301.

65. Soong, Y.Y.; Barlow, P.J. Antioxidant activity and phenolic content of selected fruit seeds. *Food Chem.* **2004**, *88*, 411–417. [CrossRef]

66. El-Siddig, K.; Gunasena, H.P.M.; Prasad, B.A.; Pushpakumara, D.K.N.G.; Ramana, K.V.R.; Vijayan, P.; Williams, J.T. *Tamarind (Tamarindus indica L.)*; Southampton Centre for Under-utilised Crops: Southampton, UK, 2006; p. 188.

67. Pugalenthi, M.; Vadivel, V.; Gurumoorthi, P.; Janardhanan, K. Compositive nutritional evaluation of little known legumes, *Tamarindus indica, Erythirna indica* and *Sesbania bispinosa*. *Trop. Subtrop. Agroecosyst.* **2004**, *4*, 107–123.

68. Ronke, R.A.; Saidat, O.G.; Abdulwahab, G. Coagulation-Flocculation Treatment of Wastewater Using Tamarind Seed Powder. *Int. J. ChemTech Res.* **2016**, *9*, 771–780.

69. Suresh, C.; Reddy, D.; Harinath, Y.; Naik, B.R.; Seshaiah, K.; Reddy, A.V.R. Development of wood apple shell (*Feronia acidissima*) powder biosorbent and its application for the removal of Cd (II) from aqueous solution. *Sci. World J.* **2014**, 154809. [CrossRef]

70. Hiwale, S. *Wood Apple (Feronia limonia Linn.) Sustainable Horticulture in Semiarid Dry Lands*; Springer: Berlin/Heidelberg, Germany, 2015; pp. 225–235.

71. Troup, R.S. *The Silviculture of Indian Trees, Vol. III*; Lauraceae to Coniferae. Pub. under the authority of His Majesty's secretary of state for India in Council; Clarendon Press: Oxford, UK, 1921; pp. 101–103.

72. Bhore, D.P. *Lecture on Dryland Horticulture Advances in Arid Zone Fruits*; Mahatma Phule Agricultural University: Rahuri, India, 1988.

73. Arora, R.K.; Anjula, P. *Wild Edible Plants of India-Diversity, Conservation and Use*; National Bureau of Plant Genetic Resources: New Delhi, India, 1996; pp. 229–259.

74. Lim, T. *Limonia acidissima Edible Medicinal and Non-Medicinal Plants*; Springer: Dordrecht, The Netherlands; Belin/Heidelberg, Germany; London, UK; New York, NY, USA, 2012; Volume II: Fruits, pp. 884–889.

75. Pandey, S.; Satpathy, G.; Gupta, R.K. Evaluation of nutritional, phytochemical, antioxidant and antibacterial activity of exotic fruit "*Limonia acidissima*". *J. Pharmacogn. Phytochem.* **2014**, *3*, 81–88.

76. Singh, A.; Singh, S.; Yadav, V.; Sharma, B. Genetic variability in wood apple (*Feronia limonia*) from Gujarat. *Indian J. Agric. Sci.* **2016**, *86*, 1504–1508.

77. Vidhya, R.; Narain, A. Development of preserved products using under exploited fruit, wood apple (*Limonia acidissima*). *Am. J. Food Technol.* **2011**, *6*, 279–288. [CrossRef]

78. Singhania, N.; Kajla, P.; Bishnoi, S.; Barmanray, A.; Ronak. Development and storage studies of wood apple (*Limonia acidissima*) chutney. *Int. J. Chem. Stud.* **2020**, *8*, 2473–2476. [CrossRef]

79. Nalwadi, V.C.; Sulikeri, G.S.S.; Singh, C.D. Floral biology study of custard apple under Dharwar condition. *Progress. Hortic.* **1975**, *7*, 15–24.

80. Sahoo, S.C.; Panda, J.M.; Mohanty, D. Studies on floral biology of custard apple under Bhubaneshwer condition. *Orissa J. Hortic.* **2000**, *28*, 43–45.

81. Patil, R.N.; Rothe, S.P. *Buchanania lanzan*: An enormous medicinal value. *IJARIIE* **2017**, *3*, 2395–4396.

82. Liu, K.; Li, H.; Yuan, C.; Huang, Y.; Chen, Y.; Liu, J. Identification of phenological growth stages of sugar apple (*Annona squamosa* L.) using the extended BBCH-scale. *Sci. Hortic.* **2015**, *181*, 76–80. [CrossRef]

83. Senthil, R.; Silambarasan, R. Annona: A new biodiesel for diesel engine: A comparative experimental investigation. *J. Energy Inst.* **2015**, *88*, 459–469. [CrossRef]

84. Zahid, M.; Mujahid, M.; Singh, P.K.; Farooqui, S.; Singh, K.; Shahla, P.; Arif, M. *Annona squamosa* LINN. (Custard apple): An aromatic medicinal plant fruit with immense nutraceutical and therapeutic potentials. *Int. J. Pharmacogen. Sci. Res.* **2018**, *9*, 1745–1759.

85. Bailey, L.H.; Bailey, E.J. *Hortus Thirds—A Concise Dictionary of Plants Cultivated in the United States and Canada*; MacMillan Publishing Co.: New York, NY, USA, 1978; pp. 457–458.

86. Sagrawat, H.; Mann, A.; Kharya, M. Pharmacological Potential of Eugenia Jambolana: A Review. *Pharmacogn. Mag.* **2006**, *2*, 96–104.

87. Ratsimamanga, A.; Loiseau, A.; Ratsimamanga, S. Action of a Hypoglycemic Agent Found in the Young Bark of *Eugenia jambolania* [sic] (Myrtaceae) on Induced Hyperglycemia of the Rabbit and Continuation of Its Purification. *Comptes Rendus Hebd. Seances L'academie Des. Sci. D Sci. Nat.* **1973**, *277*, 2219–2222.

88. Giri, J.; Sathidevi, T.; Dushyanth, N. Effect of jamun seed extract on alloxan induced diabetes in rats. *J. Diabetic Assoc. India* **1985**, *25*, 115–119.

89. Dastur, J. *Useful Plants of India and Pakistan*, 2nd ed.; D.B. Taraporevala Sons & Co.: Mumbai, India, 1943.

90. Saxena, S.K. Morphology and Ecology. In *Bordi (Ziziphus nummularia) a Shrub of the Indian arid Zone Its Role in Silvipasture*; Mann, H.S., Saxena, S.K., Eds.; Central Arid Zone Research Institute: Jodhpur, India, 1981; pp. 3–9.

91. Padaria, J.C.; Yadav, R.; Tarafdar, A.; Lone, S.A.; Kumar, K.; Sivalingam, P.N. Molecular cloning and characterisation of drought stress responsive abscisic acid stress ripening (Asr1) gene from wild jujube, *Ziziphus nummularia* wight and Arn. *Mol. Biol. Rep.* **2016**, *43*, 849–859. [CrossRef]

92. Kumar, S.; Garg, V.K.; Kumar, N.; Sharma, P.K.; Chaudhary, S.; Upadhyay, A. Pharmacognostical studies on the leaves of *Ziziphus nummularia* (Burm. F.). *Eur. J. Exper. Biol.* **2011**, *1*, 77–83.

93. Bohra, H.C.; Ghosh, P.K. Nutritive value of pala for ruminants. In *Bordi, a Shrub of Indian Arid Zone—Its Role in Silvipasture*; Mann, H.S., Saxena, S.K., Eds.; Central Arid Zone Research Institute: Jodhpur, India, 1981; pp. 42–47.

94. Singh, S.; Meghwal, P.R. Socio-economic and horticultural potential of *Ziziphus* species in arid regions of Rajasthan India. *Genet. Resour. Crop Evol.* **2020**, *67*, 1301–1313. [CrossRef]

95. Behl, P.N.; Sriwasrawa, G.S. *Herbs Useful in Dermatological Therapy*; CBS Publishers and Distributors: New Delhi, India, 2002; Volume 2, pp. 94–95.

96. Wakte, K.V.; Kad, T.D.; Zanan, R.L.; Nadaf, A.B. 2011. Mechanism of 2-acetyl-1-pyrroline biosynthesis in *Bassia latifolia* Roxb. flowers. *Physiol. Mol. Biol. Plants* **2011**, *17*, 231–237. [CrossRef] [PubMed]

97. Patel, M.; Pradhan, R.C.; Naik, S. Physical properties of fresh mahua. *Int. UnionGeod. Geophys.* **2011**, *25*, 303–306.

98. Patel, M. Biochemical Investigations of Fresh Mahua (*Madhuca indica*) Flowers for Nutraceuticals. Ph.D. Thesis, Indian Institute of Technology, New Delhi, India, 2008.

99. Behera, S.; Ray, R.C.; Swain, M.R.; Mohanty, R.C.; Biswal, A.K. Traditional and Current Knowledge on the Utilization of Mahua (L.) Flowers *Madhuca latifolia* by the Santhal Tribe in Similipal Biosphere Reserve, Odisha, India. *Ann. Trop. Res.* **2016**, *38*, 94–104. [CrossRef]

100. Pinakin, D.J.; Kumar, V.; Kumar, A.; Gat, Y.; Suri, S. Mahua a boon for pharmacy and food industry. *Curr. Res. Nutr. Food Sci.* **2018**, *6*, 371–381. [CrossRef]

101. Devi, N.; Sangeetha, R. *Madhuca longifolia* (Sapotaceae) A review of its phytochemical and pharmacological profile. *Int. J. Pharmacogen. Biosci.* **2016**, *7*, 106–114.

102. Shrirao, A.V.; Kochar, N.I.; Chandewar, A.V. *Madhuca longifolia* (Sapotaceae): A review of its traditional use and phytopharmacological profile. *Res. Chron. Health Sci.* **2017**, *3*, 45–50.

103. DeCraene, L.R.; Wanntorp, L. Floral development and anatomy of Salvadoraceae. *Ann. Bot.* **2009**, *104*, 913–923.

104. Gautam, G.K.; Vidyasagar, G.; Dwivedi, S.C. *Study on Medicinal Plants from Indian Origin—A Text Book of Indian Medicinal Plants*; Lambert Academic Publication: Chisinau, Republic of Moldova, 2012; pp. 5–7.

105. Malik, S.; Ahmed, S.S.; Haider, S.I.; Muzaffer, A. Salvadoricine—A new indole alkaloid from the leaves of *Salvadora persica*. *Tetrahedron Lett.* **1986**, *28*, 163–164. [CrossRef]

106. WHO, World Health Organization. *Prevention of Oral Diseases*; WHO offset publication No. 103; World Health Organization: Geneva, Switzerland, 1987; p. 61.

107. Mathur, S.; Shekhawat, G.S.; Batra, A. Micropropagation of *Salvadorapersica* via cotyledonary nodes. *Indian J. Biotechnol.* **2002**, *1*, 197–200.

108. Bukar, A.; Danfillo, I.S.; Adeleke, O.A.; Ogunbodede, E.O. Traditional oral health practices among Kanuri women of Borno State, Nigeria. *Odontostomatol. Trop.* **2004**, *27*, 25–31.

109. Khatak, S.; Khatak, A.A.; Siddqui, N.; Vasudeva, A.; Aggarwal, P.; Aggarwal, M. *Salvadora persica*. *Pharmacogn. Rev.* **2010**, *4*, 209–214. [CrossRef]

110. Rao, C.K.; Geetha, B.L.; Suresh, G. *Red List of Threatened Vascular Plant Species in India: Compiled from the 1997 IUCN Red List of Threatened Plants*; ENVIS, Botanical Survey of India, Ministry of Environment & Forests: Kolkata, India, 2003.

111. USNAS (United States National Academy of Sciences). *Firewood Crops: Shrub and Tree Species for Energy Production*; National Academy Press: Washington, DC, USA, 1980; pp. 150–151.

112. Liu, Y.; Singh, D.; Nair, M.G. Pods of Khejri (*Prosopis cineraria*) consumed as a vegetable showed functional food properties. *J. Funct. Foods* **2012**, *4*, 116–121. [CrossRef]

113. Arshad, M.; Ashraf, M.; Arif, N. Morphological variability of *Prosopis cineraria* (L.) Druce, from the Cholistan Desert, Pakistan. *Genet. Res. Crop. Evol.* **2006**, *53*, 1589–1596. [CrossRef]

114. Ramoliya, P.J.; Patel, H.M.; Joshi, J.B.; Pandey, A.N. Effect of salinization of soil on growth and nutrient accumulation in seedlings of Prosopis cineraria. *J. Plant Nutr.* **2006**, *29*, 283–303. [CrossRef]

115. Khan, M.A.R. *The Indigenous Trees of the United Arab Emirates*; Dubai Municipality: Dubai, United Arab Emirates, 1999; p. 150.

116. Pasiecznik, N.M.; Harris, P.J.C.; Smith, S.J. *Identifying Tropical Prosopis Species: A Field Guide*; Association HDR; HDRA Publishing: Coventry, UK, 2004; p. 26.

117. Khasgiwal, P.C.; Mithal, B.M. Studies on *Prosopis spicigera* gum. Part II: Emulsifying properties and HLB value. *Indian J. Pharm.* **1970**, *32*, 82–85.

118. Singh, G.; Singh, B.; Tomar, U.K.; Sharma, S. *A Manual for Dryland Afforestation and Management*; Published for Arid Forest Research Institute; Scientific Publishers (India): Jodhpur, India, 2016; ISBN 978-81-7233-978-4. Available online: www.afri.icfre.org (accessed on 21 December 2021).

119. Mann, H.S.; Shankarnarayan, K.A. The role of Prosopis cineraria in an agropastoral system in western Rajasthan. In *Browse in Africa*; Le Houerou, H.N., Ed.; International Livestock Center for Africa: Addis Ababa, Ethiopia, 1980; pp. 437–442.

120. Yadav, R.S.; Yadav, B.L.; Chhipa, B.R. Litter dynamics and soil properties under different tree species in a semi-arid region of Rajasthan, India. *Agrofor. Syst.* **2008**, *73*, 1–12. [CrossRef]

121. Pareek, O.P.; Nath, V. Variability in Horticultural traits of *Khejri* in *Thar Desert*. *IPGRI Newsl. Asia Pac. Ocean.* **1997**, *24*, 23–25.

122. Samadia, D.K. Thar Shobha: New khejri variety. *Indian Hortic.* **2016**, *61*, 12–13.

123. Sharma, N.; Garg, V.; Paul, A. Antihyperglycemic, antihyperlipidemic and antioxidative potential of *Prosopis cineraria* bark. *Indian J. Clin. Biochem.* **2010**, *25*, 193–200. [CrossRef]

124. Malik, A.; Kalidhar, S.B. Phytochemical examination of *Prosopis cineraria* L. Druce leaves. *Indian J. Pharm. Sci.* **2007**, *69*, 576–578.

125. Persia, F.A.; Rinaldini, E.; Hapon, M.B.; Gamarra-Luques, C. Overview of Genus Prosopis Toxicity Reports and its Beneficial Biomedical Properties. *J. Clin. Toxicol.* **2016**, *6*, 326–333.

126. Afifi, H.S.A.; Al-Rub, I.A. Prosopis cineraria as an unconventional legumes, nutrition and health benefits. In *Legume Seed Nutraceutical Research*; Jimenez-Lopez, J.C., Clemente, A., Eds.; Intech Open: London, UK, 2018; Volume 1, pp. 69–86.

127. Khan, M.A.; Rahman, A.A.; Islam, S.; Khandokhar, P.; Parvin, S.; Islam, M.B.; Hossain, M.; Rashid, M.; Sadik, G.; Nasrin, S.; et al. A comparative study on the antioxidant activity of methanolic extracts from different parts of *Morus alba* L. (Moraceae). *BMC Res. Notes* **2013**, *6*, 24. [CrossRef] [PubMed]

128. Yigit, D.; Akar, F.; Baydas, A.; Buyukyildiz, M. Elemental composition of various mulberry species. *Asian J. Chem.* **2010**, *22*, 3554–3560.

129. Vijayan, K.; Chakraborti, S.P.; Ghosh, P.D. Screening of mulberry (*Morus* spp.) for salinity tolerance through in vitro seed germination. *Indian J. Biotechnol.* **2004**, *3*, 47–51.

130. Ercisli, S.; Orhan, E. Chemical composition of white (*Morus alba*), red (*Morus rubra*) and black (*Morus nigra*) mulberry fruits. *Food Chem.* **2007**, *103*, 1380–1384. [CrossRef]

131. Singhal, B.K.; Khan, M.A.; Dhar, A.; Baqual, F.M.; Bindroo, B.B. Approaches to industrial exploitation of mulberry (*Mulberry sp.*) fruits. *J. Fruit Ornam. Plant Res.* **2009**, *18*, 83–99.

132. Dat, N.T.; Binh, P.T.; Quynh, T.P.; Van Minh, C.; Huong, H.Y.; Lee, J.J. Cytotoxic prenylated flavonoids from *Morus alba*. *Fitoterapia.* **2010**, *81*, 1224–1227. [CrossRef] [PubMed]

133. Gupta, G.; Imran, K.; Firoz, A. Anxiolytic activity of moralbosteroid, a steroidal glycoside isolated from *Morus alba*. *Phytopharmacology* **2013**, *4*, 347–353.

134. Zheng, S.; Liao, S.; Zou, Y.; Qu, Z.; Shen, W.; Shi, Y. Mulberry leaf polyphenols delay aging and regulate fat metabolism via the germline signaling pathway in *Caenorhabditis elegans*. *AGE* **2014**, *36*, 9719. [CrossRef] [PubMed]

135. Kiran, T.; Yuan, Y.Z.; Andrei, M.; Fang, Z.; Jian, G.Z.; Zhao, J.W. 1-Deoxynojirimycin, its potential for management of non-communicable metabolic disorders. *Trends Food Sci. Technol.* **2019**, *89*, 88–99.

136. Hemavathy, J.; Prabhakar, J.V. Lipid Composition of Chironji (*Buchanania lanzan*) Kernel. *J. Food Compos. Anal.* **1988**, *1*, 366–370. [CrossRef]

137. Pandey, G.P. Effects of Gaseous Hydrogen Fluoride on the leaves of Terminalia *tomentosa* and *Buchanania lanzan* Trees. *J. Environ. Pollut.* **1985**, *37*, 323–334. [CrossRef]

138. Singh, S.; Singh, A.K.; Appara, V.V. Genetic diversity in chironji (*Buchanania lanzan*) under semi-arid ecosystem of Gujarat. *Indian J. Agric. Sci.* **2006**, *76*, 695–698.

139. Khare, C.P. *Indian Medicinal Plants: An Illustrated Dictionary*; Springer: Berlin/Heidelberg, Germany; New York, NY, USA, 2007; p. 104.

140. Reddy, D.S. Assessment of nootropic and amnestic activity of centrally acting agents. *Indian J. Pharmaol.* **1997**, *29*, 208–221.

141. Gandhi, H.R. Diabetes and coronary artery disease Importance of risk factors. *Cardiol. Today* **2001**, *1*, 31–134.

142. Chitra, V.; Dharani, P.P.; Pavan, K.K.; Narayana, R.A. Wound healing activity of alcoholic extract of *Buchanania lanzan* in albino rats. *Int. J. ChemTech Res.* **2009**, *1*, 1026–1031.

143. Brewbaker, J.L. *Pithecellobium dulce*—Sweet and thorny. In *NFT Highlights*; Nitrogen Fixing Tree Association (NFTA): Oahu, HI, USA, 1992; No.92-01.

144. Franco, O.L.; Melo, F.R. Osmoprotectants—A plant strategy in response to osmotic stress. *Russ. J. Plant Physiol.* **2000**, *47*, 137–144.

145. Shukla, S.; Jain, S. Incorporation of manila tamarind (*Pithecellobium dulce*) pulverize as a source of antioxidant in Muffin cake. *J. Food Sci. Res.* **2017**, *8*, 386–390. [CrossRef]

146. Megala, J.; Geetha, A. Free radical-scavenging and H+, K+-ATPase inhibition activities of *Pithecellobium dulce*. *Food Chem.* **2010**, *121*, 1120–1128. [CrossRef]

147. Arul, S.S.; Muthukumaran, P. Analgesic and anti-inflammatory activities of leaf extract of Pithecellobium dulce Benth. *Int. J. PharmTech Res.* **2011**, *3*, 337–341.

148. Hiwale, S. Non-Traditional Crops: Manila Tamarind (*Tamarindus indica* L.). In *Sustainable Horticulture in Semiarid Dry Lands*; Springer: New Delhi, India, 2015.

149. Malik, S.K.; Choudhury, R.; Dharial, O.P.; Bhandari, D.C. *Genetic Resources of Tropical Underutilized Fruits of India*; NBPGR: New Delhi, India, 2010; Volume 17.

150. Singh, M.P.; Singh, J.K.; Mohanka, R.; Sah, R.B. *Forest Environment and Biodiversity*; Daya Publishing House: Delhi, India, 2007; p. 214.

151. Behera, M.C.; Nath, M.R. Financial valuation of non-timber forest products flow from tropical dry deciduous forests in Boudh district, Orissa. *Int. J. Farm Sci.* **2014**, *2*, 83–94.

152. Guo, C.; Yang, J.; Wei, J.; LiY, X.J.; Jiang, Y. Antioxidant activities of peel, pulp and seed fractions of common fruits as determined by FRAP assay. *Nutr. Res.* **2003**, *23*, 1719–1726. [CrossRef]

153. Ribeiro, S.M.R.; Barbosa, L.C.A.; Queiroz, J.H.; Knodler, M.; Schieber, A. Phenolic compounds and antioxidant capacity of Brazilian mango (*Mangifera indica* L.) varieties. *Food Chem.* **2008**, *110*, 620–626. [CrossRef]

154. Bhat, R.; Karim, A.A. Antioxidant capacity and phenolic content of selected tropical fruits from Malaysia, extracted with different solvents. *Food Chem.* **2009**, *115*, 785–788.

155. Mallavadhani, U.V.; Panda, A.K.; Rao, Y.R. Pharmacology and Chemotaxonomy of Diospyros. *Phytochemistry* **1970**, *49*, 901–951. [CrossRef]

156. Orwa, C.; Mutua, A.; Kindt, R.; Jamnadass, R.; Simons, A. *Agroforest Tree Database: A Tree Species Reference and Selection Guide Version 4.0*; World Agroforestry Centre ICRAF: Nairobi, Kenya, 2009.

157. Hmar, B.Z.; Mishra, S.; Pradhan, R.C. Physico-Chemical, Mechanical and Antioxidant Properties of Kendu (*Diospyros Melanoxylon* Roxb.). *Curr. Res. Nutr. Food Sci.* **2017**, *5*, 214–222. [CrossRef]

158. Anonymous. Collection of specimen bonsai plants at government sunder nursery, CPWD, Nizamuddin, New Delhi. In *Handbook of Horticulture*; Central Public Works Department Ministry of Housing and Urban Affairs, Government of India: New Delhi, India, 2020. Available online: https://cpwd.gov.in/Publication/CPWDHortHhandbook.pdf (accessed on 8 December 2021).

159. Bose, T.K. *Fruits of India—Tropical and Subtropical*; Naya Prokash: Calcutta, India, 1985.

160. Xian-Zi, T.S. *Manilkara* Adanson, Fam. Pl. 2: 166. 1763, nom. cons. *Flora China* **1996**, *5*, 206.

161. Raju, V.S.; Reddy, K.N. Ethno-medicine for dysentery and diarrhoea from Khammam district of Andhra Pradesh. *Indian J. Tradit. Knowl.* **2005**, *4*, 443–447.

162. Chanda, S.; Parekh, J. Assessment of antimicrobial potential of *Manilkara hexandra* leaf. *Pharmacogn. J.* **2010**, *2*, 448–455.

163. Chadha, K.L. *Handbook of Horticulture*; Indian Council of Agricultural Research: New Delhi, India, 2013; pp. 140–142.

164. Bhardwaj, R.L.; Nandal, U. Nutritional and therapeutic potential of bael (*Aegle marmelos*) fruit juice: A review. *Nutr. Food Sci.* **2015**, *45*, 895–919. [CrossRef]

165. Sharma, K.; Chauhan, E.S. Nutritional and phytochemical evaluation of fruit pulp powder of *Aegle marmelos*. *J. Chem. Pharm. Res.* **2017**, *10*, 809–814.

166. Sarkar, T.; Salauddin, M.; Chakraborty, R. In-depth pharmacological and nutritional properties of bael (*Aegle marmelos*): A critical review . *J. Agric. Food Res.* **2020**, *2*, 100081. [CrossRef]

167. Singh, A.K.; Chaurasiya, A.K. Post Harvest Management and value addition in Bael (*Aegle marmelos* Corr.). *Int. J. Interdiscip. Multidiscip. Stud.* **2014**, *1*, 65–77.

168. Singh, A.; Sharma, H.K.; Kaushal, P.; Upadhyay, A. Bael (*Aegle marmelos Correa*) products processing: A review. *Afr. J. Food Sci.* **2014**, *8*, 204–215.

169. Malik, S.K.; Chaudhury, R.; Panwar, N.S.; Dhariwal, O.P.; Choudhary, R.; Susheel, K. Genetic resources of chironji (*Buchanania lanzan* S): A socio-economic important tree species of central Indian population. *Genet. Resour. Crop Evol.* **2012**, *59*, 615–623. [CrossRef]

170. Phogat, N.; Bisht, V.; Purwar, S. Chironji (*Buchanania lanzan*) Wonder Tree: Nutritional and Therapeutic Values. *Int. J. Curr. Microbiol. Appl. Sci.* **2020**, *9*, 3033–3042.

171. Warokar, A.S.; Ghante, M.H.; Duragkar, N.J.; Bhusari, K.P. Anti-inflammatory and antioxidant activities of methanolic extract of Buchananialanzan Kernel. *Indian J. Pharm. Educ. Res.* **2010**, *44*, 363–368.

172. Singh, J.; Naik, R.K.; Patel, S.; Mishra, N.K. Design and Development of Chironji (*Buchanania lanzan*) Decorticator. *Int. J. Eng. Res. Technol.* **2016**, *5*, 46–51.

173. Rajput, C.B.S. Custard apple. In *Fruits of India Tropical and Subtropical*; Bose, T.K., Ed.; Naya Prakash pub.: Calcutta, India, 1985; pp. 479–486.
174. Kumar, M.; Changan, S.; Tomar, M.; Prajapati, U.; Saurabh, V.; Hasan, M.; Sasi, M.; Maheshwari, C.; Singh, S.; Dhumal, S.; et al. Custard Apple (*Annona squamosa* L.) Leaves: Nutritional Composition, Phytochemical Profile, and Health-Promoting Biological Activities. *Biomolecules* **2021**, *11*, 614. [CrossRef] [PubMed]
175. Tanuja; Rana, D.K.; Rathi, D.S. Shelf Life Enhancement of Custard Apple (*Annona squamosa* L.) under Sub-Tropical Conditions of Garhwal Hills. *Int. J. Curr. Microbiol. Appl. Sci.* **2021**, *10*, 3175–3180.
176. Mishra, D.S.; Singh, A.K.R.; Singh, S.; Singh, A.K.; Swamy, G.S.K. Jamun. In *Tropical and Sub Tropical Fruit Crops: Crop Improvement and Varietal Wealth, Part II*; Ghosh, S.N., Ed.; Jaya Pub. House: Delhi, India, 2014; pp. 375–387.
177. Shahnawaz, M.; Sheikh, S.A.; Nizamani, S.M. Determination of Nutritive Values of Jamun Fruit (*Eugenia jambolana*) Products. *Pak. J. Nutr.* **2009**, *8*, 1275–1280. [CrossRef]
178. Carr, A.C.; Frei, B.A. Toward a new recommended dietary allowance for vitamin C based on antioxidant and health effects in humans. *Am. J. Clin. Nutr.* **1999**, *69*, 1086–1107. [CrossRef]
179. Nath, V.; Kumar, D.; Pandey, V. *Fruits for the Future, Vol 1: Well Versed Arid and Semi Arid Fruits*; Satish Serial Publishing House: Delhi, India, 2008; p. 264.
180. Anonymous. *Composition of Ziziphus nummularia Leaves on Dry Weight Basis*; The Wealth of India: New Delhi, India, 1976; p. 120.
181. Chaturvedi, Y.; Nagar, R. Level of beta carotene and effects of processing on selected fruits and vegetables of the arid zone of India. *Plant Foods Hum. Nutr.* **2001**, *56*, 127–132. [CrossRef]
182. Kumar, S.; Sharma, R.; Kumar, V.; Govind, K.; Vyas, G.K.; Rathore, A. Combining molecular-marker and chemical analysis of *C. decidua* in the Thar Desert of Western Rajasthan (India). *Int. J. Trop. Biol.* **2013**, *61*, 311–320.
183. Abra, H.H.; Ali, M. Phytochemistry and bioactivities of a harsh terrain plant: *Capparis decidua* (Forsk) Edgew. *Nat. Prod. Ind. J.* **2011**, *7*, 222–229.
184. Ghangro, I.H.; Ghangro, A.B.; Channa, M.J. Nutritional assessment of non conventional vegetable *C. decidua* flower. *Rawal Med. J.* **2015**, *40*, 214–216.
185. Tripathi, P.C.; Karumakaran, G.; Sankar, V.; Senthilkumar, R. *Karonda: A Potential Fresh Fruit of Future*; Central Horticultural Experiment Station Indian Institute of Horticultural Research: Chettalli, India, 2014.
186. Chadha, K.L. Karonda. In *Handbook of Horticulture*; Directorate of Information and Publications of Agriculture, Indian Council of Agricultural Research: New Delhi, India, 2003; p. 200.
187. Singh, S.K.; Singh, A.K.; Meghwal, P.R.; Singh, A.; Swamy, G.S.K. Karonda. In *Tropical and Sub Tropical Fruit Crops: Crop Improvement and Varietal Wealth Part.1*; Ghosh, S.N., Ed.; Jaya Publishing House: New Delhi, India, 2014; Volume 13, pp. 393–402.
188. Dahot, M.U. Free Radicals, Antioxidants, and Human Disease: Curiosity, Cause, or Consequence. *Lancet* **1993**, *52*, 253–265.
189. Bhumi, B.; Shukla, Y.M.; Patel, V.H. Nutritional Profile of an Underutilized Indian Fruit: Rayan [*Manilkara hexandra* (Roxb.) Dubard. *Indian J. Agric. Biochem.* **2016**, *29*, 51–53.
190. Morais, P.L.D.; Oliveira, L.C.; Alves, R.E.; Alves, J.D.; Paiva, A. Amadurecimiento de sapoti (*Manilkara zapota* L.) submetidoao 1-metilciclopropeno. *Rev. Bras. Frutic.* **2006**, *28*, 369–373. [CrossRef]
191. Lata, K.; Singh, S.K.; Kumar, R.; Jhajuria, S. Potential of Dry Khirni (*Manilkara hexandra* Roxb.) Fruits as Nutritional Substitute. *J. Krishi Vigyan.* **2019**, *8*, 231. [CrossRef]
192. Meghwal, P.R.; Singh, A.; Kumar, P.; Morwal, B.R. Diversity, distribution and horticultural potential of *Cordia myxa* Roxb.: A promising underutilized fruit species of arid and semi-arid regions of India. *Genet. Resour. Crop Evol.* **2014**, *61*, 1633–1643. [CrossRef]
193. Meghwal, P.R.; Singh, A.; Singh, D. Research status of lasora (*Cordia myxa* L.) in India—A review. *Curr. Hortic.* **2021**, *9*, 15–19. [CrossRef]
194. Meghwal, P.R.; Singh, D.; Swami, S.; Singh, A. Assessment of Nutritional and Free Radical Scavenging Activity of Selected Genotypes of *Cordia Myxa* L.: A Potential Underutilized Fruit Crop of Indian Arid Zone. 2021. Available online: https://doi.org/10.21203/rs.3.rs-811273/v1 (accessed on 10 December 2021).
195. Aberoumand, A. Preliminary Evaluation of Some Phytochemical and Nutrients Constituents of Iranian *Cordia myxa* Fruits. *Int. J. Agric. Food Sci.* **2011**, *1*, 30–33.
196. Musa, A.; Mostafa, E.M.; Al-Sanea, M.M.; Ahmed, S.R.; Mostafa-Hedeab, G.; Abdelgawad, M.A. Insights studies for certain natural fda approved polyphenolics and repurposing for COVID-19. *Int. J. Res. Pharm. Sci.* **2020**, *11*, 1390–1395. [CrossRef]
197. Sason, R.; Anita, S. The Phytochemical and Pharmacological Properties of *Cordia dichotoma*: A Review. *Ayushdhara* **2015**, *2*, 155–161.
198. Haq, M.A.; Alam, M.J.; Hasnain, A.A. Novel Edible Coating to increase Shelf life of Chilgoza (*Pinus geradiana*). *Food Sci. Technol.* **2013**, *50*, 306–311.
199. Vidyasagar, G.; Jadhav, A.G.; Narkhede, S.P.; Narkhede, S.B. Isolation and Comparative Evaluation of *Cordia dichotoma* Forst. Mucilage as a Binding Agent with Standard Binder. *J. Chem. Pharm. Res.* **2010**, *2*, 722–726.
200. Bisht, V.; Solanki, N.; Dalal, N. Mahua an important Indian species: A review. *J. Pharmacogn. Phytochem.* **2018**, *7*, 3414–3418.
201. Malik, S.K.; Choudhary, R.; Panwar, N.S.; Dhariwal, O.P.; Deswal, R.P.S.; Pathak, N.; Chaudhary, R. Prospect traditional importance of Indian butter tree and genetic resources management. *Indian Hortic.* **2021**, *66*, 34–37.

202. Kureel, R.S.; Kishor, R.; DevDutt, A.P. *Mahua—A Potential Tree Born Oilseed*; National Oilseed and Development Board; Ministry of Agriculture Govt. of India: Gurgaon, India, 2009; pp. 1–27.
203. Pio-Leon, J.F.; Diaz-Camacho, S.; Montes-Avila, J. Nutritional and nutraceutical characteristics of white and red *Pithecellobium dulce* (Roxb.) Benth fruits. *Fruits* **2013**, *68*, 397–408. [CrossRef]
204. Ambasta, S.P. (Ed.) *The Useful Plants of India*; Council of Scientific & Industrial Research: New Delhi, India, 1986; p. 918.
205. Ung, L.; Pattamatta, U.; Carnt, N.; Wilkinson-Berka, J.L.; Liew, G.; White, A.J. Oxidative stress and reactive oxygen species: A review of their role in ocular disease. *Clin. Sci.* **2017**, *131*, 2865–2883. [CrossRef]
206. Rao, G.N.; Nagender, A.; Satyanarayana, A.; Rao, D.G. Preparation, chemical composition and storage studies of quamachil (*Pithecellobium dulce* L.) aril powder. *J. Food Sci. Technol.* **2011**, *48*, 90–95. [CrossRef] [PubMed]
207. Parrotta, J.A. *Pithecellobium dulce (Roxb.) Benth. Guamuchil, Madras thorn. Leguminosae (Mimosoideae) Legume Family*; USDA Forest Service, Southern Forest Experiment Station, Institute of Tropical Forestry: New Orleans, LA, USA, 1991; Volume 5, p. 40.
208. Jan, B.; Parveen, R.; Zahiruddin, S. Nutritional constituents of mulberry and their potential applications in food and pharmaceuticals: A review. *Saudi J. Biol. Sci.* **2021**, *28*, 3909–3921. [CrossRef]
209. Kumari, A.; Parida, A.K.; Rangani, J.; Panda, A. Antioxidant Activities, Metabolic Profiling, Proximate Analysis, Mineral Nutrient Composition of Salvadora persica Fruit Unravel a Potential Functional Food and a Natural Source of Pharmaceuticals. *Front. Pharmacol.* **2017**, *8*, 61. [CrossRef]
210. Ene-Obong, H.N.; Okudu, H.O.; Asumugha, U.V. Nutrient and phytochemical composition of two varieties of Monkey kola (*Cola parchycarpa* and *Cola lepidota*): An underutilised fruit. *Food Chem.* **2016**, *193*, 154–159. [CrossRef] [PubMed]
211. Satya, S.S.; Narina; Catanzaro, C.J. Tamarind (*Tamarindus indica* L.), an Underutilized Fruit Crop with Potential Nutritional Value for Cultivation in the United States of America: A Review. *Asian Food Sci. J.* **2018**, *5*, 1–15.
212. Obulesu, M.; Sila, B. Color changes of tamarind (*Tamarindus indica* L.) pulp during fruit development, ripening, and storage. *Int. J. Food Proper.* **2011**, *14*, 538–549. [CrossRef]
213. Chavan, S.B.; Uthappa, A.R.; Sridhar, K.B.; Keerthika, A.; Handa, A.K.; Newaj, R. Trees for life: Creating sustainable livelihood in Bundelkhand region of central India. *Curr. Sci.* **2016**, *111*, 994–1002. [CrossRef]
214. Jamil, Z.; Mohite, A.M.; Sharma, N. Phyto-chemical and nutritional profiling of tendu fruit (*Diospyros melanoxylon* Roxb.) And evaluation of its shelf stability. *Plant Arch.* **2021**, *21*, 656–664. [CrossRef]
215. Kala, C.P. Harvesting and supply chain analysis of ethnobotanical species in the Pachmarhi biosphere reserve of India. *Am. J. Environ. Protec.* **2013**, *1*, 20–27. [CrossRef]
216. Bag, S.K.; Srivastav, P.P.; Mishra, H.N. Optimization of process parameters for foaming of Bael (*Aegle marmelos* L.) fruit pulp. *Food Bioprocess Technol.* **2011**, *4*, 1450–1458. [CrossRef]
217. Nithya, N.; Saraswathi, U. In vitro antioxidant and antibacterial efficacy of *Feronia elephantum* Correa fruit. *Indian J. Nat. Prod. Resour.* **2010**, *1*, 301–305.
218. Moazzem, M.S.; Sikder, M.B.H.; Zzaman, W. Shelf-Life Extension of Wood Apple Beverages Maintaining Consumption-Safe Parameters and Sensory Qualities. *Beverages* **2019**, *5*, 25. [CrossRef]
219. Gautam, P.L.; Singh, A.K.; Srivastava, M.; Singh, P.K. Protection of Plant Varieties and Farmers' Rights: A Review. *Indian J. Genet. Plant Breed.* **2012**, *25*, 9–30.
220. Dhillon, B.S.; Tyagi, R.A.; Lal, A.; Saxena, S. (Eds.) *Plant Genetic Resource Management*; Narosa Pulishing House: New Delhi, India, 2004.
221. Marie, H. Why conserve crop diversity. *Indian J. Plant Genet. Resour.* **2016**, *29*, 258–260.
222. Singh, G. *Checklist of Commercial Varieties of Fruits*; Department of Agriculture and Cooperation Ministry of Agriculture Krishi Bhawan: New Delhi, India, 2012. Available online: https://agritech.tnau.ac.in/horticulture/pdf/tech_bulletin/national/Checklist_of_CommercialFruits-18-01-13.pdf (accessed on 10 December 2021).
223. Singh, A.K.; Singh, S.; Saroj, P.L. Scientific cultivation of aonla. *Kerala Karshakan* **2020**, *10*, 40–47.
224. Singh, A.K.; Singh, S.; Saroj, P.L.; Singh, G.P. Improvement and production technology of bael (*Aegle marmelos*) in India—A review. *Curr. Hortic.* **2021**, *9*, 3–14. [CrossRef]
225. Singh, A.K.; Singh, S.; Saroj, P.L.; Mishra, D.S.; Yadav, V.; Kumar, R. Cultivation of underutilized fruit crops in hot semi-arid regions: Developments and challenges—A review. *Curr. Hortic.* **2020**, *8*, 12–23. [CrossRef]
226. Meghwal, P.R.; Vashistha, B.B.; Singh, A.; Prasad, R.N. CAZRI Ber 2018: New early variety of ber. *Indian J. Hortic.* **2019**, *64*, 11–13.
227. Meghwal, P.R.; Kumar, P. Effect of supplementary irrigation and mulching on vegetative growth, yield and quality of ber. *Indian J. Hortic.* **2014**, *71*, 571–573.
228. Singh, S.; Singh, A.K.; Rao, V.V.A.; Bhargawa, R. Thar Priya: A new chironji variety. *Indian Hortic.* **2015**, *61*, 1–6.
229. Krishna, H.; Saroj, P.L.; Maheshwari, S.K.; Singh, R.S.; Meena, R.K.; Chandra, R.; Parashar, A. Underutilized fruits of arid & semi-arid regions for nutritional and livelihood security. *Int. J. Minor Fruits Med. Aromat. Plants* **2019**, *5*, 1–14.
230. Rymbai, H.; Patel, N.L.; Patel, C.R.; Reddy, A.G.K.; Hiwale, S.S.; Chovatia, R.S.; Varu, D.K. Custard Apple. In *Crop Improvement and Varietal Wealth, Tropical and Sub Tropical Fruit crops*; Jaya Publishing House: New Delhi, India, 2019; Volume 7, pp. 237–268.
231. Encina, C.L.; Padilla, I.M.G.; Carzola, J.M.; Caro, E. Tissue culture in cherimoya. *Acta Hortic.* **1999**, *497*, 289–294. [CrossRef]
232. George, A.P.; Nissen, R.J. Propagation of Annona Species: A Review. *Sci. Hortic.* **1987**, *33*, 75–85. [CrossRef]
233. Shrivastava, R.; Dubey, S.; Dwivedi, A.P.; Pandey, C.S.; Banafar, R.N.S. Effect of recipe treatment and storage period on biochemical composition of custard apple (*Annona squamosa* L.) nectar. *Progress. Hortic.* **2013**, *45*, 110–114.

234. Khodifad, B.C.; Kumar, N.; Vyas, D.K.; Seth, N.; Prem, M. Pre and Post-harvest Practices, Processing and Value Addition of Custard Apple. *Intl. J. Food. Ferment. Technol.* **2016**, *6*, 219–231. [CrossRef]
235. Singh, S.; Singh, A.K.; Mishra, D.S.; Appa Rao, V.V. Effect of shoot pruning on yield and fruit quality of jamun cv. Goma Priyanka. *Indian J. Arid Hortic.* **2017**, *12*, 100–102.
236. Singh, S.; Singh, A.K.; Bagle, B.G. Propagating jamun successfully. *Indian Hortic.* **2007**, *52*, 31–33.
237. Singh, A.K.; Bajpai, A.; Singh, V.K.; Ravishankar, H.; Tandon, D.K. *The Jamun (Syzygium cuminii Skeels)*; Central Institute for Subtropical Horticulture (ICAR): Rehmankhera, India, 2009; pp. 1–28.
238. Meghwal, P.R.; Singh, A. Underutilized Fruits Research in Arid Regions: A Review. *Ann. Arid Zone.* **2016**, *56*, 23–36.
239. Mahla, H.R.; Singh, J.P. Assessment of in-situ Variability in Kair (*Capparis decidua*) Germplasm for Utilization in Genetic Improvement through ex-situ Conservation. *Ann. Arid Zone.* **2013**, *52*, 109–112.
240. Singh, D.; Bhardwaj, R.; Chaudhary, M.K.; Meena, M.L.; Wangchu, L. Panchkutta: A Unique Indigenous Food of Thar Desert for Biodiversity Conservation and Nutritional Security. In Proceedings of the Indigenous and traditional food systems in Asia and the Pacific, Organized by FAO Regional Office for Asia and the Pacific, Khon Kaen, Thailand, 31 May–2 June 2012; pp. 168–174.
241. Meghwal, P.R.; Singh, A.; Singh, S.K. Maru Gaurav: New karonda variety. *Indian Hortic.* **2020**, *23*, 30–31.
242. Srivastava, A.; Sarkar, P.K.; Bishnoi, S.K. Value addition in under-exploited fruits of Karonda (*Carissa carandus* L.): An earning opportunity for rural communities in India. *Rashtriya Krishi* **2017**, *12*, 161–163.
243. Samadia, D.K. *Thar Sobha: Grow for Horticultural Exploitation*; Technical Folder; ICAR—Central Institute for Arid Horticulture: Bikaner, India, 2015.
244. Samadia, D.K. Thar Sobha: New khejri varieity. *Indian Hortic.* **2016**, *61*, 1–2.
245. Singh, S.; Singh, A.K.; Rao, V.V.A.; Bhargawa, R. Thar Rituraj: A new khirni variety. *Indian Hortic.* **2015**, *60*, 14–15.
246. Meghwal, P.R.; Singh, A.; Singh, D. Enhancing the productivity of lasora (*Cordia myxa* L.) through genetic improvement and production management. In Proceedings of the National Conference on Arid Horticulture for Enhancing Productivity and Economic Empowerment, Organized by Indian Society for Arid Horticulture and ICAR-Central Institute for Arid Horticulture, Bikaner, India, 27–29 October 2018.
247. Meghwal, P.R.; Singh, A. Maru Samridhi: New lasora variety. *Indian Hortic.* **2019**, *64*, 32–33.
248. Meghwal, P.R.; Singh, A.; Kumar, P. Evaluation of selected gonda genotypes (*Cordia myxa* L.) on different root stocks. *Indian J. Hortic.* **2014**, *71*, 415–418.
249. Singh, S.; Singh, A.K.; Rao, V.V.A.; Bhargawa, R. Thar Madhu: A new mahua variety. *Indian Hortic.* **2016**, *61*, 27–29.
250. Keerthika, R.; Rajangam, J.; Arumugam, T.; Venkatesan, K. Standardization of Propagation Techniques in Manila Tamarind (*Pithecellobium dulce* (Roxb.) Benth.) var. PKM 1. *Int. J. Curr. Microbiol. App. Sci.* **2020**, *9*, 269–277.
251. Goyal, P.; Kachhwaha, S.; Kothari, S.L. Micro propagation of *Pithecellobium dulce* (Roxb.)—A multipurpose leguminous tree and assessment of genetic fidelity of micropropagated plants using molecular markers. *Physiol. Mol. Biol. Plants* **2012**, *18*, 169–176. [CrossRef]
252. Lal, N.; Nath, V. *Sweet Tamarind; Minor fruits: Nutraceutical importance and cultivation*; Jaya Publishing House: Delhi, India, 2017; Volume 43, pp. 901–912.
253. Krishnaswami, S. *Mulberry Cultivation in South India*; Central Silk Board: Bangalore, India, 1986; pp. 1–19.
254. Begum, N.; Kiran, B.R.; Purushotham, R. Mulberry cultivation practices and diseases: An Overview. *Int. J. Curr. Eng. Sci. Res.* **2018**, *5*, 61–68.
255. Singh, S.; Mishra, D.S.; Singh, A.K. Tamarind (*Tamarindus indica* L.). In *Tropical Fruit Crops Theory to Practical Chapter: 16*; Jaya Publishing House: Delhi, India, 2021; pp. 610–631.
256. Bhore, D.P. Studies on propagation of tamarind (*Tamarindus indica*). *Maharastra J. Hort.* **1986**, *3*, 55–58.
257. Awasthi, O.P.; Shukla, N. Effect of time on success of soft wood grafting in tamarind (*Tamarindus indica* L.). *Range Manag. Agrof.* **2003**, *24*, 31–34.
258. Yadav, V.; Hiwale, S.S.; Singh, A.K.; Rao, V.V.A. Thar Gaurav: High-yielding wood apple variety for dry land. *Indian Hortic.* **2020**, *65*, 36–40.
259. Angadi, S.G.; Raghavendra, V.N.; Allolli, T.B. Propagation of wood apple (*Feronia limonia* L.) by budding. *Acta Hortic.* **2011**, *890*, 175–182. [CrossRef]

Review

Neglected and Underutilized Plant Species (NUS) from the Apulia Region Worthy of Being Rescued and Re-Included in Daily Diet

Aurelia Scarano [1], Teodoro Semeraro [2], Marcello Chieppa [3] and Angelo Santino [1,*]

[1] Institute of Science of Food Production, C.N.R. Unit of Lecce, 73100 Lecce, Italy; aurelia.scarano@ispa.cnr.it
[2] Department of Environmental and Biological Sciences and Technologies, University of Salento, S.P. 6 Lecce-Monteroni, 73100 Lecce, Italy; teodoro.semeraro@unisalento.it
[3] National Institute of Gastroenterology 'S. De Bellis', Institute of Research, 70013 Castellana Grotte, Italy; marcello.chieppa@irccsdebellis.it
* Correspondence: angelo.santino@ispa.cnr.it

Abstract: Neglected and underutilized species (NUS) are cultivated, semi-domesticated, or wild plant species, not included in the group of the major staple crops, since, in most cases, they do not meet the global market requirements. As they often represent resilient species and valuable sources of vitamins, micronutrients, and other phytochemicals, a wider use of NUS would enhance sustainability of agro-systems and a choice of nutritious foods with a strategic role for addressing the nutritional security challenge across Europe. In this review, we focused on some examples of NUS from the Apulia Region (Southern Italy), either cultivated or spontaneously growing species, showing interesting adaptative, nutritional, and economical potential that can be exploited and properly enhanced in future programs.

Keywords: NUS; sustainable food supply; nutritional security; Apulia Region

Citation: Scarano, A.; Semeraro, T.; Chieppa, M.; Santino, A. Neglected and Underutilized Plant Species (NUS) from the Apulia Region Worthy of Being Rescued and Re-Included in Daily Diet. *Horticulturae* **2021**, *7*, 177. https://doi.org/10.3390/horticulturae7070177

Academic Editors: Rosario Paolo Mauro, Carlo Nicoletto and Leo Sabatino

Received: 4 June 2021
Accepted: 28 June 2021
Published: 3 July 2021

Publisher's Note: MDPI stays neutral with regard to jurisdictional claims in published maps and institutional affiliations.

1. Introduction

Often considered a central argument in the scientific debates at a local or global scale, the biodiversity loss issue is becoming a critical challenge that needs to be carefully considered in future years. Following this debate, the newly launched EU Biodiversity strategy has put forward measures to address the biodiversity loss across the European Union [1]. Within this issue, a lively interest has been addressed towards the agro-biodiversity, which includes cultivated species and landraces, wild flora, soil microorganisms, pollinators, and the relative interconnections between plant and environment or genetic resources and agricultural management/practices [2]. Furthermore, local knowledge and culture also have an important role and should be considered part of agro-biodiversity [2].

Localized in the central part of the Mediterranean area, Italy offers a wide variety of ecological, pedoclimatic, and orographic conditions. The Italian flora is characterized by rare and endemic plants, with many domesticated crops and vegetables showing high genetic and phenotypic variability [3]. In the Italian territory, particularly South Italy, small family-owned farms and rural areas are rich in vegetable germplasm, represented by wild flora, different landraces, and plant species closely linked to the local historical memory [3].

The neglected and underused plant species (NUS) are cultivated varieties, semi-domesticated, or wild plant species that tend to be underutilized locally or globally, due to their relatively low value for the global production and marketplace, since they most often do not meet the modern standards of uniformity [4] as major cultivated varieties [4–7]. Indeed, starting from the Green Revolution, we assisted the decline of many local/traditional species and varieties, which were less competitive compared with commercial cultivars, and, therefore, they have been replaced by high-yielding and uniform cultivars developed by modern breeding programs [2,7]. This genetic erosion has been also amplified by urban spreading, changes in socio-economic conditions, and destruction of natural environments due to increased human activities [3].

As a source of vitamins, micronutrients, and other phytochemicals, NUS have the potential to play a strategic role for addressing nutritional security challenges [6]. A wider use of NUS would also enhance adaptability and resilience to biotic and abiotic stress factors and ultimately might lead to a more sustainable supply of diverse and nutritious foods [8]. In fact, many autochthonous plant species are characterized by a high nutritional value compared to cultivars or similar species belonging to the same family.

Furthermore, landraces and wild relatives can provide genetic traits that are useful for increasing biotic resistance and tolerance to abiotic stress in future breeding programs, especially when creating more sustainable and resilient production systems [9–12].

In this review, we focus on some examples of NUS from the Apulia Region (Southern Italy), either cultivated landraces or spontaneously growing as a part of the local flora, that are worthy of being rescued and enhanced for their interesting nutritional properties and economical potential.

1.1. Multicolored Carrots

Carrot (*Daucus carota* L.) is one of the most popular and consumed root vegetables worldwide and it is especially known in Western dietary regimes as an important source of dietary carotenoids, such as α-carotene and β-carotene, which are also known as provitamin A [13]. In fact, the popularity of carrots is mainly linked to their nutritional value, which makes them an economically important horticultural crop.

Carrot is consumed as a fresh vegetable, used in many traditional dishes or soups, commercially transformed into juices and concentrates, canned, or dried powdered [14]. Although the most common genotypes are orange-colored, in some countries, such as those in southern Europe, Turkey, China, or India, multicolored carrots are also well known [15]. In fact, the primary genotypes of carrots were yellow or purple and they originally spread from Afghanistan across the Middle East, North Africa, Europe, and China. During the domestication processes, yellow carrots have been preferred, leading to the final development and cultivation of orange carrots, the most prevalent at present [15–17]. On the other hand, black/purple carrots (*Daucus carota* ssp. *sativus* var. *atrorubens* Alef.), deriving from the primary domestication center and pigmented both in the epidermis and the inner central core of the taproots, are still cultivated and highly appreciated in some countries and represent one of the most used anthocyanin sources as food colorant, due to the high stability of the processing conditions and storage [18].

In Italy, some documents report the presence of multicolored carrots through 13th and 14th centuries [16]. In the Apulia region, multicolored carrot landraces (Figure 1) are cultivated from local farmers in different villages, and they have been officially inserted in the list of species at risk of genetic erosion, according to the Apulian Rural Development Program (2007/2013). In particular, only three different landraces related to the area of production (Polignano, Tiggiano, and Zapponeta) have been described. In the case of Polignano landrace, these carrots are currently cultivated in an area of about 20 ha, with cultivation practices at risk due to the age of elder farmers and the difficulty for farmers to collect reproductive material/seeds. The Polignano and Tiggiano carrots have been the subject of several studies in recent years, particularly due to the anthropic cultural heritage associated with them and their high nutritional value [16–21]. In fact, their typical yellow-purple color has been associated with increased levels of some classes of polyphenols compared to the commercial orange varieties. Among the multicolored carrots, the yellow carrots have showed a slight reduction in the content of carotenoids and phenolic compounds, whereas the purple-yellow and purple-orange carrots ensure high levels of polyphenols, mainly chlorogenic acid and anthocyanins, maintaining, at the same time, a carotenoids content similar to orange carrots (Table 1) [16,18,22–24]. Due to the presence of high levels of phenolic compounds, the extracts from yellow-purple carrots have shown to be high in vitro antioxidant capacities compared to the orange carrots, but their nutritional significance can be also extended to other molecular properties, since a body of evidence has associated polyphenols dietary administration to anti-inflammatory, anti-aging, and anti-tumoral effects, thus providing a preventive effect against chronic and inflammatory human diseases [23,24]. Based on these nutraceutical features, multicolored carrots represent important horticultural species that can be valorized in breeding programs aimed at biodiversity preservation and sustainable agriculture.

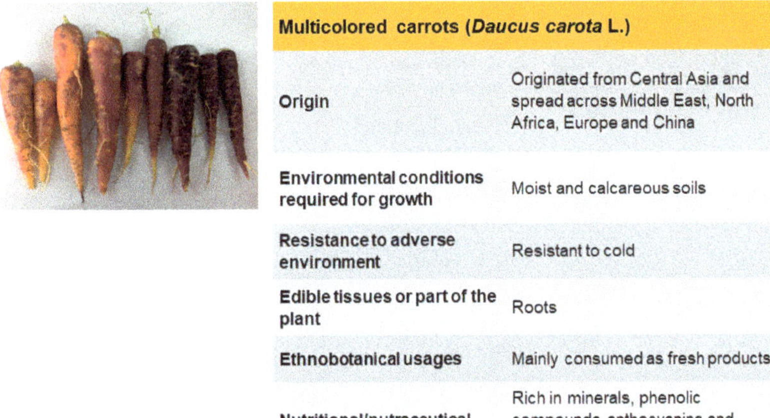

Figure 1. Scheme of the main features of multicolored carrots.

Table 1. Some examples of phenolic compounds, anthocyanins, and carotenoid content in multicolored carrots.

Multicolored Carrots (Different Accessions/Landraces)	Total Phenolics	Total Anthocyanins	Total Carotenoids
Yellow- or orange-purple carrots	15.04–38.69 mg GAE [1]/g DW [22]	17.3–17.9 µmol/g DW [22]	0.334–0.771 mg/g DW [22]
Polignano carrots from Apulia	0.676 mg GAE/g [16] 4.5 mg CGA [2]/g DW [18]	5.06–7.82 mg KE [3]/g DW (Blando 2021)	0.433 mg/g [16] 0.332 mg/g DW [18]
Tiggiano carrots from Apulia	~2.6 mg CGA/g DW [24]	~1 mg C3GE [4]/g FW [24]	~0.400 mg/g DW [24]

[1] GAE: Gallic acid equivalents; [2] CGA: Chlorogenic acid; [3] KE: Kuromanin equivalents; [4] C3GE: cyanidin-3-glucoside equivalents.

1.2. Roquette

Roquette (Figure 2), also known as arugula, belongs to the Brassicaceae family and is an important leafy salad worldwide. In the Mediterranean region, the main cultivated roquette species are: *Eruca vesicaria* L. Cav. (formerly *E. sativa* Mill.) [25], which is prevalently cultivated in rich soils, or alternatively can be found mixed with ruderal flora in marginal areas; *Diplotaxis tenuifolia* L., which has succulent leaves and is well adapted to harsh and poor soils and is mostly collected as a wild species [26]. In Apulia, *Eruca vesicaria* is currently suffering a strong genetic erosion, due to the growing attention focused on the wild *Diplotaxis tenuifolia*, which is preferred for culinary preparations. However, *E. vesicaria* is still cultivated in small gardens in the area of Bari but with rare cases of global marketing placement.

Roquette is characterized by a pungent and bitter taste, provided by a range of beneficial compounds (Vitamin C, carotenoids, phenolics, and glucosinolates) (Table 2) that contribute to its antioxidant capacity. Conversely, it can also accumulate anti-nutrients (e.g., nitrates) and heavy metals [27–31]. Besides the culinary uses, roquette has interesting medicinal properties, such as diuretic and depurative effects [26], and its extracts have antimicrobial properties [32], antigenotoxic properties in *D. melanogaster* [33], and cytotoxic effects in tumoral cell lines [34].

Roquette (*Eruca vesicaria* L.)	
Origin	It is widespread around the world
Environmental conditions required for growth	Rich soils in marginal areas
Resistance to adverse environment	Tolerance to nitrates and heavy metals
Edible tissues or part of the plant	Leaves
Ethnobotanical usages	Prevalently used as leafy/green salad
Nutritional/nutraceutical properties	Rich in vitamin C, carotenoids, phenolic compounds and glucosinulates; antioxidant properties

Figure 2. Scheme of the main features of roquette. (Source photo: biodiversitàpuglia.it [35]).

Table 2. Some examples of phytochemical compounds in roquette.

Roquette Leaves	Total Phenolics	Total Carotenoids	Vitamin C
"Nature" and "Naturelle" genotypes	0.446–1.024 mg GAE [1]/g FW [28]	0.076.2–0.137 mg/g FW [28]	0.0256–0.079 mg/g FW [28]
Italian *E. sativa*	3.62 mg GAE/g DW [36]		
Bulgarian *E. sativa*	4.45 mg GAE/g DW [36]		
D. tenuifolia (wild rocket)	Quercetin: 0.0189–0.0774 mg/g FW [37]	0.0847 mg/g DW [38] Lutein: 0.0455–0.0545 mg/g FW [37]	0.2078–0.8174 mg/g FW [37]
E. vesicaria (garden rocket)	1.93 mg GAE/g FW [39] 9.20 µmol/g DW [40]	0.13 mg/g DW [38]	0.02967 mg/g FW [41] 0.15 mg/g FW [42]

[1] GAE: Gallic acid equivalents.

1.3. Salicornia spp.

Salicornia spp. (Figure 3) is a group of edible halophytes able to grow in high salt soil conditions, commonly named glasswort, pickle-weed, or sea asparagus [43]. *Salicornia* spp. can be found as a wild species in transition zones between permanently flooded muds and perennial vegetation, characterized by a winter flooding period and dry summer. The geographical distribution of the wild species is very wide, since it can be found in USA, Mexico, Canada, Europe (e.g., Britain, Ireland, France, Spain, Italy), India, Iran, Korea, and some Africa regions [43].

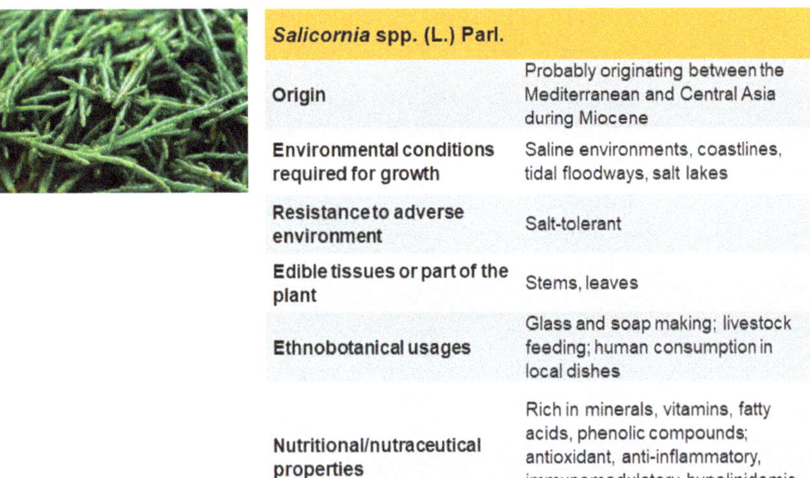

Figure 3. Scheme of the main features of *Salicornia* spp. (Source photo: biodiversitàpuglia.it [35]).

Apart from the historic usage as a source of sodium carbonate for glass making and an additive for soap production [43], some *Salicornia* spp. are utilized for culinary purposes.

The natural adaptation to saline environments, as well as the salt tolerant traits and the contextual content of bioactive compounds, makes *Salicornia* spp. interesting for many landscapes due to its cultivation in adverse, harsh environments and its contribution to human nutrition. In fact, *Salicornia* cultivation could represent a valid option in the context of global warming, in which edible plants with high salt tolerance are needed. *Salicornia* spp. can be also good candidates for reclamation of barren lands, salt flats, and seashores [43]. Its use has also been proposed in heavy metal removal and phytoremediation, but these applications are not compatible with nutritional purposes, as they can be a source of toxic metal ions and antinutrients. In fact, it is important to keep in mind that some species can accumulate high contents of oxalic acid and iatrogenic iodine and excessive content of salt, heavy metals, and saponines (as in the case of *S. bigelovii*) [44].

In the Apulia region, wild *Salicornia* spp. gathering is quite common, linked to ancient culinary uses, even though some cultivation practices (as in the case of *Salicornia patula*, belonging to the *S. europea* group) are also consolidated, especially in the northern area of Gargano, close to the areas of the Lesina and Varano salt lakes [44,45]. The first attempts of *Salicornia* cultivation have been reported along the Lesina lagoon, which occupies an area of about 51 km^2, with a length of 22 km, an average width of 2.4 km, and a depth of about 0.7 m [45]. Other scattered coastal sites suitable for *Salicornia* spp. growth and gathering are present in the southern parts of the Apulia region, such as "Torre Guaceto" coastal lagoon (province of Brindisi), "Le Cesine" (province of Lecce), and "Salina dei monaci" (province of Taranto).

S. patula can be generally cultivated from February–March to August–September in a soil that is typically black, sandy, acidic, and very rich in organic matter. The harvest of fresh and tender parts can be repeated depending on the level of development of the plant, with a final yield that can reach 10–15 tons per hectar [44]. The propagation can be carried out by gamic or agamic techniques, and in the case of gamic techniques, seeds need strategies for dormancy under hypersaline conditions and germination at low salt levels. Furthermore, germination is affected either by the type of salt or its concentration [45].

Regarding the nutritional value, *Salicornia* spp. L. contains essential amino acids, vitamins (mainly vitamins A and C), dietary fibers, and, as expected, a large diversity of minerals, including sodium, potassium, calcium, magnesium, iron, and iodine [46]. Going deeper into the phytochemistry of *Salicornia* spp., some studies have also evidenced the presence of: (i) saponins (in *S. europea* and *S. bigelovii*); (ii) lipids, with a prevalence of palmitic acid (e.g., in *S. ramosissima*) or α-linolenic acid (e.g., in *S. europea*) [46]; (iii) steroid compounds, such as spinasterol and stigmasterol (in *S. europea*, *S. herbacea*, *S. fruticosa*, and *S. bigelovii*); (iv) alkaloid derivatives, saliherbine, and salicornin [47]; (v) flavonoids (mainly favanones and flavone derivatives) and phenolic acids in methanolic extracts from *S. europea* [47]. Due to the presence of sterols, triterpenoids saponins, and polyphenolic compounds, beneficial properties have been associated with *Salicornia* extracts, such as antioxidant, anti-inflammatory, immunomodulatory, hypolipidemic, and hypoglycemic effects [46–50].

Phytochemical analyses on *S. patula* have focused on fatty acids content, with a percentage of saturated fatty acids reaching 80% and phenolic content ranging from 2.989 to 4.209 mg GAE/g DW, with the major components represented by salicylic and transcinnamic acids [51] (Table 3).

Table 3. Some examples of polyphenol and fatty acid content in *Salicornia* spp.

Salicornia spp.	Polyphenol Content	Fatty Acid Content
Salicornia (two ecotypes from Israel)	1.05–1.53 mg GAE [1]/g FW [52]	2.24–2.41 mg/g FW (Total FAs [2]) [52]
Salicornia herbacea	0.78 mg/g FW [53]	
Salicornia patula	2.989–4.209 mg GAE/g DW [51]	80% total FAs 2–3% MUFAs [3] 6–13% PUFAs [4] [51]
Salicornia ambigua	0.813–0.1252 mg/g FW [54]	1.2–1.6 mg/g FW 60–61% [5] SFAs 4–4.5% MUFAs 17–18% PUFAs [54]

[1] GAE: Gallic acid equivalents; [2] FAs: fatty acids; [3] MUFAs: monounsaturated fatty acids; [4] PUFAs: polyunsaturated fatty acids; [5] SFA: saturated fatty acid.

1.4. Purslane

Purslane (*Portulaca oleracea* L.) (Figure 4) is a very common spontaneous plant in gardens, lawns, vineyards, cultivated fields, eroded slopes, and bluffs, where it is considered one of the most common weeds. It is a very common plant in the temperate and subtropical regions, but it also grows in the tropics and at higher latitudes [55]. *P. oleracea* is a synanthropic species that can tolerate mechanical disturbance and can be derived from anthropic activities. It has fleshy, succulent, and very branched leaves and stems. The origin of *P. oleracea* is uncertain, but it has been suggested that it comes from India, even though it was also found in America in pre-Columbian times [56]. Purslane has a broad physiological adaptability and high morphological variability (highly polymorphic); therefore, the taxonomy of *P. oleracea* is still under debate [56]. This is quite important because the Italian peninsula and adjacent islands provide fragmentary information on the infraspecific diversity of *P. oleracea*. However, a recent elucidation about the distribution of various *P. oleracea* morphotypes has been provided [56,57]. Thus, in the *P. oleracea* complex, the *P. trituberculata* morphotype has been identified in the Apulia region. This morphotype is one of the most common in continental Italy since the Roman period [57].

Purslane (*Portulaca oleracea* L.)	
Origin	Uncertain
Environmental conditions required for growth	Temperate, subtropical and tropical areas
Resistance to adverse environment	Synanthropic species
Edible tissues or part of the plant	Stems, leaves
Ethnobotanical usages	Consumption of fresh or cooked leaves in local dishes
Nutritional/nutraceutical properties	Rich in minerals, phenolics, omega-3 fatty acids, tocopherol and Vitamin C; antioxidant activities

Figure 4. Scheme of the main features of *P. oleracea* (Source photo: biodiversitàpuglia.it [35]).

In the Apulia region, purslane has always been traditionally harvested, and recently, it has been officially recognized as a traditional food product [58]. Due to its sour and salty taste, similar to fresh spinach, purslane is generally served raw in salads to give flavor and freshness or cooked to prepare soups. In the past, it was used as a medicinal herb due to its purifying, diuretic, and anti-diabetic properties. Purslane is a good source of omega-3 fatty acids, tocopherols, and vitamin C (Table 4) and contains minerals, such as magnesium, manganese, potassium, iron, and calcium. Flavonoids and polyphenols have also been extracted from purslane leaves, particularly with oleracein A and C, found as major components in leaves, reaching 8.2–103.0 mg and 21.2–143 mg/100 g dried weight [59]. Concerning the biological activities, purslane has shown antioxidant and lipid oxidation inhibiting capacities [60–63] and provides protection against DNA damage in in vitro studies [61]. Di Cagno et al. [64] have also tested purslane juice obtained by lactic acid bacteria fermentation, finding that the fermented juice strongly decreased the levels of pro-inflammatory mediators and reactive oxygen species in the $CaCo_2$-cell line.

Table 4. Some examples of polyphenol, vitamin C, tocopherol, and fatty acid content in purslane.

P. oleracea	Polyphenols	Vitamin C	Tocopherols	Fatty Acids
P. oleracea accessions	0.96–9.12 mg GAE/g DW [63] 3.6 mg GAE/g DW [60] Oleracein A: 8–1.03 mg/g DW [59] Oleracein C: 21–1.43 mg/g DW 4.418–23.77 mg GAE/g DW) [62]	2.40–9.73 µg/g FW [65]	3.02–4.81 µg/g FW [59]	Total SFA [2] (% of total FA [1]): 27–55% MUFA [3] (% of total FA): 5–12% PUFA [4] (% of total FA): 38–66% [59]
Raw purslane juice	85 mg GAE/100 mL [64]	22 mg/100 mL [64]	2.5 mg/100 mL [64]	

[1] FA: fatty acids; [2] SFA: saturated fatty acid; [3] MUFAs: monounsaturated fatty acids; [4] PUFAs: polyunsaturated fatty acids.

1.5. Leopoldia comosa L.

Leopoldia comosa (L.) Parl., (Figure 5), previously named *Muscari comosum* (L.) Mill, is a perennial bulb, belonging to the Hyacinthaceae family and originating from South-East Europe, Turkey, and Iran, naturalized elsewhere and eaten in some Mediterranean countries. It is called the tassel of hyacinth or tassel grape hyacinth. It is a wild species, but it can also be properly cultivated. The wild specimens can be found in rocky ground or cultivated lands, cornfields, or vineyards. The cut bulbs transude mucilages, sugars, latex, tannins, salts, triterpenes, homoisoflavones, and muscarosides [66].

Leopoldia comosa (L.) Parl.	
Origin	Turkey and Mediterranean area
Environmental conditions required for growth	Temperate areas, rocky grounds
Resistance to adverse environment	Resistant to cold
Edible tissues or part of the plant	Bulbs
Ethnobotanical usages	Consumption of bulbs in local dishes
Nutritional/nutraceutical properties	Bulbs are rich in flavonoids, phenolic acids and fatty acids; antioxidant and hypoglycemic properties, prevention of obesity-related disorders

Figure 5. Scheme of the main features of *L. comosa*. (Source photo: biodiversitàpuglia.it [35]).

The bulbs of *L. comosa* are characterized by a typical strong sour and bitter taste and, in the culinary uses of the Apulia region, are traditionally boiled and consumed with olive oil, vinegar, and salt, or they can be fried. Additionally, they can be part of the preparation of other traditional local dishes [67]. Other popular usages of *L. comosa* bulbs include the cure of toothache and skin spots [68].

L. comosa bulbs are rich in several classes of phytochemicals, including flavonoids, phenolic acids, and fatty acids [69] (Table 4). Among the fatty acid fraction, palmitic acid has been reported as the major component, followed by linoleic, linolenic, and stearic acids [69] (Table 5).

Phytochemicals in *Leopoldia comosa* bulbs have shown metal chelating, antioxidant properties, pancreatic lipase inhibitory activity, and hypoglycemic activity via the inhibition of carbohydrate digestive enzymes, such as α-amilase and α-glucosidase [70]. Furthermore, enzyme-inhibitory effects and in vitro antitumoral activities in breast adenocarcinoma cells have also been reported [71].

In a comparative study of extracts deriving from wild and cultivated bulbs of *L. comosa*, Marrelli et al. [69] have shown higher radical scavenging activity and good in vitro pancreatic lipase inhibitory activity from the wild bulb extracts compared to the cultivated bulb extracts. In light of these data, the extracts from wild *L. comosa* bulbs have been suggested to be considered for subsequent in vivo studies and the activity could be attributed to phenolic compounds [69]. Accordingly, Casacchia et al added *L. comosa* extracts (20 or 60 mg/die) to a high-fat diet in rats fed for 2 weeks. Following these conditions, *L. comosa* extracts inhibited lipase and pancreatic amylase activities, counteracting abdominal obesity, dyslipidemia, liver steatosis, and improving glucose tolerance, suggesting an important effect of prevention of obesity-dependent metabolic disorders [72]. In another study, Casacchia et al. [71] used raw bulbs or bulbs cooked with two different methods (boiled or steam-cooked), confirming higher antioxidant activities and inhibition of pancreatic lipase and α-amylase, especially in the raw bulbs, relating these in vitro activities mainly to the phenolic compounds and suggesting that the traditional cooking methods can partially deplete the observed biological activities.

Table 5. Some examples of polyphenol and fatty acid content in *L. comosa*.

Leopoldia comosa	Total Polyphenols	Total Flavonoids	Total Fatty Acids
Wild raw bulbs	264.33 mg/g FW [69] 92.47 mg CAE [1]/g FW [71] 56.6 mg CAE/g extract [70] 102.89 mg CAE/g of FW [72]	10.40 mg/g FW [69] 4.57 mg QE [2]/g FW [71] 23.4 mg QE/g extract [70] 28.07 mg QE/g of FW [72]	Palmitic acid 16.2 mg/g of fatty acids fraction [69] Palmitic acid 15.5% of fatty acid composition [70]
Cultivated raw bulbs	42 mg/g FW [69]	5.74 mg/g FW [69]	Palmitic acid 17.5 mg/g of fatty acids fraction [69]
Boiled bulbs	39.53 mg CAE/g FW [71]	0.64 mg QE/g FW [71]	
Steam-cooked bulbs	49.80 mg CAE/g FW [71]	1.63 mg QE/g FW [71]	

[1] CAE: Chlorogenic acid equivalents; [2] QE quercetin equivalents.

1.6. Milk Thistle

Milk thistle (*Sylibum marianum* L.) (Figure 6) is a member of the Asteraceae family and is native to the Mediterranean basin, although it is widespread in Northern Africa, Asia, North and South America, and South Australia [73,74]. It can be cultivated as an ornamental plant, but it often grows widely as a proper weed in yields and roadsides, in warm environments and dry soils. Flowering season is between July and August. It is also considered a heavy metals tolerant species [73]. Milk thistle fruits, sometimes confused as seeds, have been used for medical purposes since ancient Greek civilization, especially for the treating of liver diseases for its hepatoprotective activities. A recent study has evidenced that wild accessions of *S. marianum* in Italy can be identified in three different stable chemotypes, based on the biochemical profile of these accessions. Two of these chemotypes have been reported from different Italian regions, including Apulia, with no clear correlation between the chemical profile and geographic features [75].

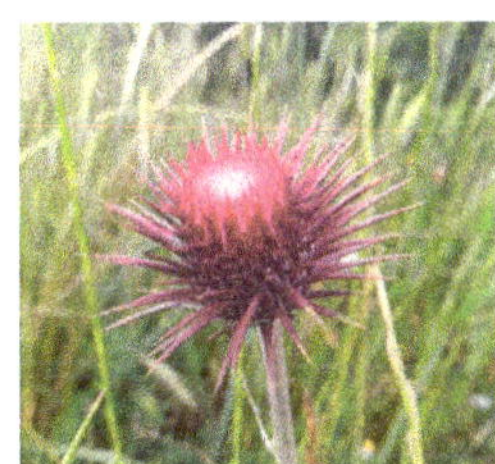

Milk thistle (*Silybum marianum* L.)	
Origin	Mediterranean basin, then widespread over the world
Environmental conditions required for growth	Warm environment, dry soils
Resistance to adverse environment	Tolerant to dry soils or polluted with heavy metals
Edible tissues or part of the plant	Stems, leaves, fruits
Ethnobotanical usages	Oils and tea from fruits; consumption of cooked leaves and stems in local dishes
Nutritional/nutraceutical properties	Rich in flavolignans (the most abundant is silybin), flavonoids, tocopherol; antioxidant, antinflammatory and hepatoprotective activities

Figure 6. Scheme of the main features of *S. marianum*.

Apulian traditional culinary usages included the leaves and the tender stems of the milk thistle, together with other well-known and appreciated species, *Cynara cardunculus* L. and *Scolymus hispanicus* L. (golden thistle). However, the main problem that has greatly limited its uses in recent years is represented by the first cleaning phase, which consists of eliminating leaf blade, which is exceedingly spiny. However, once cooked the milk thistles can be used to prepare very tasty dishes rich in beneficial compounds.

The most important biological activities of milk thistle are related to silymarin, a mixture of flavonoid complexes and flavolignans. In silymarin, many compounds have been reported, among which are silybin, isosilybin, silychristin, isosilychristin, sylidianin, and silimonin [76–78]. Apart these compounds, some flavonoids (quercetin, kaempferol, apigenin, naringenin, eriodyctiol, and taxifolin), tocopherol, sterols, sugars, and proteins have been reported [76], even though silybin is the most abundant compound in the extracts [74] (Table 6).

Milk thistle extracts, from the heads, leaves, and stems, have shown several biological properties [79], including strong antioxidant and anti-inflammatory properties and anti-tumoral activities [79–81]. In an experimental model of nonalcoholic steatohepatitis, the administration of *S. marianum* extract reduced the severity of steatohepatitis and the levels of alanine amino transferase and aspartate amino transferase and improved the levels of glutathione [82]. The hepatoprotective activities were also observed in human hepatocytes and human liver microsomes by inhibiting cytochrome-P450 isoenzymatic activities [83].

Table 6. Some examples of bioactive compounds in *S. marianum*.

Milk Thistle Organs	Polyphenols	Flavonoids	Sylibin
Leaves	14–17 mg GAE [1]/g DW [73]	~11 mg GAE/g DW [73]	
Heads	11–12 mg GAE/g DW [73]	~5 mg GAE/g DW [73]	
Seeds (fruits)	24–35 mg GAE/g [84]	16–29 mg QE [2]/g [84]	3–311 mg/g [85]

[1] GAE: Gallic acid equivalents; [2] QE quercetin equivalents.

2. Conclusions

In this review, we focused on some examples of NUS from the Apulia region that are worthy of being enhanced, with the intent to preserve the living heritage and biodiversity. The major staple crops, intensively cultivated because they ensure the standards of global market requirements, are preferred to NUS, thus hiding their great potentials to contribute to the process of adaptation to changing climates. Indeed, most of NUS are characterized by a high resilience to harsh and adverse environments and a rich source of nutrients. Further efforts need to address:

1. The molecular basis and genetic traits linked to the adaptation to harsh environmental conditions (with a special look at tolerance to heat/salt/heavy metals stresses);
2. The characterization of main nutrient classes and their biosynthesis pathways;
3. The quantification and characterization of the main antimetabolic factors/antinutrients;
4. A better knowledge of the biological activities in the prevention of human diseases.

Expanding our knowledge on these issues will increase awareness of the importance of NUS and the activities related to their recovery and enhancement. Investing in research on NUS, an inter/multi-disciplinary approach and shared scientific and traditional knowledge will help to fully realize the benefits of these crops.

Author Contributions: A.S. (Aurelia Scarano), T.S., M.C., and A.S. (Angelo Santino) wrote the paper. All authors have read and agreed to the published version of the manuscript.

Funding: This work was, in part, funded by the Apulia region SICURA project (KC3U5Y1) and CNR-DiSBA project NutrAge (project nr. 7022).

Institutional Review Board Statement: Not applicable.

Informed Consent Statement: Not applicable.

Acknowledgments: The authors acknowledge the Apulia Region project "BiodiverSO" and A. Signore, P. Santamaria, and G. Mastrosimini for the pictures.

Conflicts of Interest: The authors declare no conflict of interest.

References

1. Slow Food. The New EU Biodiversity strategy: The Most Important Questions Answered. Available online: https://www.slowfood.com/the-new-eu-biodiversity-strategy-the-most-important-questions-answered/ (accessed on 25 May 2021).
2. Renna, M.; Montesano, F.F.; Signore, A.; Gonnela, M.; Santamaria, P. BiodiverSO: A case study of integrated project to preserve the biodiversity of vegetable crops in Puglia (Southern Italy). *Agriculture* **2018**, *8*, 128. [CrossRef]
3. Laghetti, G.; Bisignano, V.; Urbano, M. Genetic resources of vegetable crops and their safeguarding in Italy. *Hort. Int. J.* **2018**, *2*, 72–74. [CrossRef]
4. Stamp, P.; Messmer, R.; Walter, A. Competitive underutilized crops will depend on the state funding of breeding programmes: An opinion on the example of Europe. *Plant Breed.* **2012**, *131*, 461–464. [CrossRef]
5. Ochatt, S.; Jain, S.M. *Breeding of Neglected and Under-Utilized Crops, Spices and Herbs*; Science Publishers Inc.: Enfield, NH, USA, 2007.
6. Ebert, A.W. Potential of underutilized traditional vegetables and legumes crops to contribute to food and nutritional security, income and more sustainable production systems. *Sustainability* **2014**, *6*, 319–335. [CrossRef]
7. Padulosi, S.; Thompson, J.; Rudebjer, P. *Fighting Poverty, Hunger and Malnutrition with Neglected and Underutilized Species (NUS): Needs, Challenges and the Way Forward*; Bioversity International: Rome, Italy, 2013.
8. Semeraro, T.; Scarano, A.; Buccolieri, R.; Santino, A.; Aarrevaara, E. Planning of urban green spaces: An ecological perspective on human benefits. *Land* **2021**, *10*, 105. [CrossRef]
9. Frison, E.A.; Cherfas, J.; Hodgkin, T. Agricultural biodiversity is essential for a sustainable improvement in food and nutrition security. *Sustainability* **2011**, *3*, 238–253. [CrossRef]
10. Jackson, L.E.; Pascual, U.; Hodgkin, T. Utilizing and conserving agrobiodiversity in agricultural landscapes. *Agric. Ecosyst. Environ.* **2007**, *121*, 196–210. [CrossRef]
11. McCouch, S. Feeding the future. *Nature* **2013**, *499*, 23–24. [CrossRef]
12. Scarano, A.; Olivieri, F.; Gerardi, C.; Liso, M.; Chiesa, M.; Chieppa, M.; Frusciante, L.; Barone, A.; Santino, A.; Rigano, M.M. Selection of tomato landraces with high fruit yield and nutritional quality under elevated temperatures. *J. Sci. Food Agric.* **2020**, *100*, 2791–2799. [CrossRef]
13. Zhang, D.; Hamauzu, Y. Phenolic compounds and their antioxidant properties in different tissues of carrots (*Daucus carota* L.). *J. Food Agric. Environ.* **2004**, *2*, 95–100.
14. Sharma, K.D.; Karki, S.; Thakur, N.S.; Attri, S. Chemical composition, functional properties and processing of carrot—A review. *J. Food Sci. Technol.* **2012**, *49*, 22–32. [CrossRef]
15. Arscott, S.A.; Tanumihardjo, S.A. Carrots of many colors provide basic nutrition and bioavailable phytochemicals acting as a functional food. *Comp. Rev. Food Sci. Food Saf.* **2010**, *9*, 223. [CrossRef]
16. Cefola, M.; Pace, B.; Renna, M.; Santamaria, P.; Signore, A.; Serio, F. Compositional analysis and antioxidant profile of yellow, orange and purple Polignano carrots. *Ital. J. Food Sci.* **2012**, *24*, 284–291.
17. Blando, F.; Marchello, S.; Maiorano, G.; Durante, M.; Signore, A.; Laus, M.N.; Soccio, M.; Mita, G. Bioactive compounds and antioxidant capacity in anthocyanin-rich carrots: A comparison between the black carrot and the Apulian landrace "Polignano" carrot. *Plants* **2021**, *10*, 564. [CrossRef] [PubMed]
18. Elia, A.; Santamaria, P. Biodiversity in vegetable crops, a heritage to save: The case of Puglia region. *Ital. J. Agron.* **2013**, *8*, 4. [CrossRef]
19. Signore, A.; Renna, M.; D'Imperio, M.; Serio, F.; Santamaria, P. Preliminary evidences of biofortification with iodine of "Carota di Polignano", an Italian carrot landrace. *Front. Plant Sci.* **2018**, *9*, 170. [CrossRef] [PubMed]
20. Scarano, A.; Gerardi, C.; D'Amico, L.; Accogli, R.; Santino, A. Phytochemical analysis and antioxidant properties of colored Tiggiano carrots. *Agriculture* **2018**, *8*, 102. [CrossRef]
21. Renna, M.; Serio, F.; Signore, A.; Santamaria, P. The yellow-purple Polignano carrot (*Daucus carota* L.): A multicolored landrace from the Puglia region (Southern Italy) at risk of genetic erosion. *Gen. Res. Crop. Evol.* **2014**, *61*, 1611–1619. [CrossRef]
22. Sun, T.; Simon, P.W.; Tanumihardjo, S.A. Antioxidant phytochemicals and antioxidant capacity of biofortified carrots (*Daucus carota* L.) of various colors. *J. Agric. Food Chem.* **2009**, *57*, 4142. [CrossRef]
23. Blando, F.; Calabriso, N.; Berland, H.; Maiorano, G.; Gerardi, C.; Carluccio, M.A.; Andersen, Ø.M. Radical scavenging and anti-inflammatory activities of representative anthocyanin groupings from pigment-rich fruits and vegetables. *Int. J. Mol. Sci.* **2018**, *19*, 169. [CrossRef]
24. Scarano, A.; Chieppa, M.; Santino, A. Plant polyphenols-biofortified foods as a novel tool for the prevention of human gut diseases. *Antioxidants* **2020**, *9*, 1225. [CrossRef] [PubMed]
25. Hassan, S.M.; Abshour, M.; Soliman, A.A.F.; Hassanien, H.A.; Alsanie, W.F.; Gaber, A.; Elshobary, M. The potential of a new commercial seaweed extract in stimulating morpho-agronomic and bioactive properties of *Eruca sativa* (L.) Cav. *Sustainability* **2021**, *13*, 4485. [CrossRef]
26. Padulosi, S.; Pignone, D. Rocket: A Mediterranean crop for the world—Report of a workshop. In Proceedings of the Project of Underutilized Mediterranean Species, Padova, Italy, 13–14 December 1996.
27. Cannata, M.G.; Carvalho, R.; Bertoli, A.C.; Augusto, A.S.; Bastos, A.R.R.; Carvalho, J.G.; Freitas, M.P. Effects of Cadmium and Lead on Plant Growth and Content of Heavy Metals in Arugula Cultivated in Nutritive Solution. *Comm. Soil Sci. Plant Anal.* **2013**, *44*, 952–961. [CrossRef]

28. Bonasia, A.; Lazzizera, C.; Elia, A.; Conversa, G. Nutritional, Biophysical and Physiological Characteristics of Wild Rocket Genotypes As Affected by Soilless Cultivation System, Salinity Level of Nutrient Solution and Growing Period. *Front. Plant Sci.* **2017**, *8*, 300. [CrossRef] [PubMed]

29. Al Jassir, M.S.; Shaker, A.; Khaliq, M.A. Deposition of heavy metals on green leafy vegetables sold on roadsides of Riyadh city, Saudi Arabia. *Bull. Environ. Contam. Toxicol.* **2005**, *75*, 1020–1027. [CrossRef]

30. Ali, M.H.H.; Al-Qahtani, K.M. Assessment of some heavy metals in vegetables, cereals and fruits in Saudi Arabia markets. *Egypt. J. Acq. Res.* **2012**, *38*, 31–37. [CrossRef]

31. Bantis, F.; Kaponas, C.; Charalambous, C.; Koukounaras, A. Strategic successive harvesting of rocket and spinach baby leaves enhanced their quality and production efficiency. *Agriculture* **2021**, *11*, 465. [CrossRef]

32. Rani, I.; Akhund, S.; Suhail, M.; Abro, H. Antimicrobial potential of seed extract of *Eruca sativa*. *Pakistan J. Bot.* **2010**, *42*, 2949–2953.

33. Villatoro-Pulido, M.; Font, R.; Saha, S.; Obregón-Cano, S.; Anter, J.; Muñoz-Serrano, A.; De Haro-Bailón, A.; Alonso-Moraga, A.; Del Río-Celestino, M. In vivo biological activity of rocket extracts (*Eruca vesicaria* subsp. *sativa* (Miller) Thell) and sulforaphane. *Food Chem. Toxicol.* **2012**, *50*, 1384–1392. [CrossRef]

34. Michael, H.N.; Shafik, R.E.; Ramsy, G.E. Studies on the chemical constituents of fresh leaf of *Eruca sativa* extract and its biological activity as anticancer agent in vitro. *J. Med. Plants Res.* **2011**, *5*, 1184–1191.

35. BiodiverSO. Available online: https://biodiversitapuglia.it (accessed on 25 May 2021).

36. Matev, G.; Dimitrova, P.; Petkova, N.; Ivanov, I.; Mihaylova, D. Antioxidant activity and mineral content of rocket (*Eruca sativa*) plant from Italian and Bulgarian origins. *J. Microbiol. Biotechnol. Food Sci.* **2018**, *8*, 756–759. [CrossRef]

37. Durazzo, A.; Azzini, E.; Lazzè, M.C.; Raguzzini, A.; Pizzala, R.; Maiani, G. Italian wild rocket [*Diplotaxis tenuifolia* (L.) DC.]: Influence of agricultural practices on antioxidant molecules and on cytotoxicity and antiproliferative effects. *Agriculture* **2013**, *3*, 285–298. [CrossRef]

38. Žnidarčič, D.; Ban, D.; Šircelj, H. Carotenoid and chlorophyll composition of commonly consumed leafy vegetables in Mediterranean countries. *Food Chem.* **2011**, *129*, 1164–1168. [CrossRef] [PubMed]

39. Li, Z.; Lee, H.W.; Liang, X.; Liang, D.; Wang, Q.; Huang, D.; Ong, C.N. Profiling of phenolic compounds and antioxidant activity of 12 cruciferous vegetables. *Molecules* **2018**, *23*, 1139. [CrossRef]

40. Degl'Innocenti, E.; Pardossi, A.; Tattini, M.; Guidi, L. Phenolic compounds and antioxidant power in minimally processed salad. *J. Food Biochem.* **2008**, *32*, 642–653. [CrossRef]

41. Marchioni, I.; Martinelli, M.; Ascrizzi, R.; Gabbrielli, C.; Flamini, G.; Pistelli, L.; Pistelli, L. Small functional foods: Comparative phytochemical and nutritional analyses of five microgreens of the *Brassicaceae* family. *Foods* **2021**, *10*, 427. [CrossRef] [PubMed]

42. Mieszczakowska-Frąc, M.; Celejewska, K.; Płocharski, W. Impact of innovative technologies on the content of Vitamin C and its bioavailability from processed fruit and vegetable products. *Antioxidants* **2021**, *10*, 54. [CrossRef] [PubMed]

43. Patel, S. Salicornia: Evaluating the halophytic extremophile as a food and a pharmaceutical candidate. *3Biotech* **2016**, *6*, 104. [CrossRef]

44. Loconsole, D.; Cristiano, G.; De Lucia, B. Glassworts: From wild salt marsh species to sustainable edible crops. *Agriculture* **2018**, *9*, 14. [CrossRef]

45. Urbano, M.; Tomaselli, V.; Bisignano, V.; Veronico, G.; Hammer, K.; Laghetti, G. *Salicornia patula* Duval-Jouve: From gathering of wild plants to some attempts of cultivation in Apulia region (Southern Italy). *Genet. Resour. Crop. Evol.* **2017**. [CrossRef]

46. Lopes, M.; Cavaleiro, C.; Ramos, F. Sodium reduction in bread: A role for glasswort (*Salicornia ramosissima* J. Woods). *Compreh. Rev. Food Sci. Food Saf.* **2017**, *16*, 1056–1071. [CrossRef]

47. Isca, V.M.S.; Seca, A.M.L.; Pinto, D.C.G.A.; Silva, A.M.S. An overview of *Salicornia* gnus: The phytochemical and pharmacological profile. In *Natural Products: Research Reviews*; Gupta, V.K., Ed.; Daya Publishing House: New Delhi, India, 2014; Volume 2, pp. 145–176.

48. Kim, S.; Lee, E.-Y.; Hillman, P.F.; Ko, J.; Yang, I.; Nam, S.-J. Chemical strucutre and biological activities of secondary metabolites from *Salicornia europea* L. *Molecules* **2021**, *26*, 2252. [CrossRef] [PubMed]

49. Gouda, M.; Elsebaie, E.M. Glasswort (*Salicornia* spp.) as a source of bioactive compounds and its health benefits: A review. *Alex. J. Food Sci. Technol.* **2016**, *13*, 1–7. [CrossRef]

50. Essaidi, I.; Brahmi, Z.; Snoussi, A.; Koubaier, H.B.H.; Casabianca, H.; Abe, N.; El Omri, A.; Chaabouni, M.M.; Bouzouita, N. Phytochemical investigation of Tunisian *Salicornia herbacea* L., antioxidant, antimicrobial and cytochrome P450 (CYPs) inhibitory activities of its methanol extract. *Food Control* **2013**, *32*, 125–133. [CrossRef]

51. Sánchez-Gavilán, I.; Ramírez, E.; de la Fuente, V. Bioactive compounds in Salicornia patula Duval-Jouve: A mediterranean edible euhalophyte. *Foods* **2021**, *10*, 410. [CrossRef]

52. Ventura, Y.; Wuddineh, W.A.; Myrzabayeva, M.; Alikulov, Z.; Khozin-Goldberg, I.; Shpigel, M.; Samocha, T.M.; Sagi, M. Effect of seawater concentration on the productivity and nutritional value of annual *Salicornia* and perennial *Sarcocornia* halophytes as leafy vegetable crops. *Sci. Hortic.* **2011**, *128*, 189–196. [CrossRef]

53. Cho, J.-Y.; Kim, J.Y.; Lee, Y.G.; Lee, H.J.; Shim, H.J.; Lee, J.H.; Kim, S.-J.; Ham, K.-S.; Moon, J.-H. Four new dicaffeoylquinic acid derivatives from glasswort (*Salicornia herbacea* L.) and their antioxidative activity. *Molecules* **2016**, *21*, 1097. [CrossRef]

54. Labronici Bertin, R.; Valdemiro Gonzaga, L.V.; da Silva Campelo Borges, G.; Stremel Azevedo, M.; Maltez, H.F.; Heller, M.; Micke, G.A.; Benathar Ballod Tavares, L.; Fett, R. Nutrient composition and, identification/quantification of major phenolic compounds in *Sarcocornia ambigua* (*Amaranthaceae*) using HPLC-ESI-MS/MS. *Food Res. Int.* **2014**, *55*, 404–411. [CrossRef]

55. Matthews, J.F.; Ketron, D.W.; Zane, S.F. The biologiy and taxonomy of the *Portulaca oleracea* L. (*Portulaceae*) complex in North America. *Rhodora* **1993**, *95*, 166–183.
56. Danin, A.; Bundrini, F.; Bandini Mazzanti, M.; Bosi, G.; Caria, M.C.; Dandria, D.; Lanfranco, E.; Mifsud, S.; Bagella, S. Diversification of *Portulaca oleracea* L. complex in the Italian peninsula and adjacent islands. *Bot. Lett.* **2016**. [CrossRef]
57. Galasso, G.; Conti, F.; Peruzzi, L.; Ardenghi, N.M.G.; Banfi, E.; Celesti-Grapow, L.; Albano, A.; Alessandrini, A.; Bacchetta, G.; Ballelli, S.; et al. An updated checklist of the vascular flora alien to Italy. *Plant Biosyst.* **2018**, *152*, 556–592. [CrossRef]
58. BiodiverSO. Available online: https://biodiversitapuglia.it/fbpost/la-portulaca-portulaca-oleracea-l-e-una-pianta-spontanea-molto-comune-in-orti-e-campi-dove-e-con/ (accessed on 25 May 2021).
59. Petropoulos, S.A.; Fernandes, Â.; Dias, M.I.; Vasilakoglou, I.B.; Petrotos, K.; Barros, L.; Ferreirs, I.C.F.R. Nutritional value, chemical composition and cytotoxic properties of common purslane (*Portulaca oleracea* L.) in relation to harvesting stage and plant part. *Antioxidants* **2019**, *8*, 293. [CrossRef] [PubMed]
60. Uddin, M.K.; Juraimi, A.S.; Ali, M.E.; Ismail, M.R. Evaluation of antioxidant properties and mineral composition of purslane (*Portulaca oleracea* L.) at different growth stages. *Int. J. Mol. Sci.* **2012**, *13*, 10257–10267. [CrossRef]
61. Erkan, N. Antioxidant activity and phenolic compounds of fractions from *Portulaca oleracea* L. *Food Chem.* **2012**, *133*, 775–781. [CrossRef]
62. Silva, R.; Carvalho, I.S. In vitro antioxidant activity, phenolic compounds and protective effect against DNA damage provided by leaves, stems and flowers of *Portulaca oleracea* (purslane). *Nat. Prod. Comm.* **2014**, *9*, 45–50. [CrossRef]
63. Alam, M.A.; Juraimi, A.S.; Rafii, M.Y.; Hamid, A.A.; Aslani, F.; Hasan, M.M.; Zainudin, M.A.M.; Uddin, M.K. Evaluation of antioxidant compounds, antioxidant activities, and mineral composition of 13 collected purslane (*Portulaca oleracea* L.) accessions. *BioMed Res. Int.* **2014**. [CrossRef]
64. Di Cagno, R.; Filannino, P.; Vincentini, O.; Cantatore, V.; Cavoski, I.; Gobbetti, M. Fermented *Portulaca oleracea* L. juice: A novel functional beverage with potential ameliorating effects on the intestinal inflammation and epithelial injury. *Nutrients* **2018**, *11*, 248. [CrossRef] [PubMed]
65. Dewanti, F.D.; Pujiasmanto, B.; Sudendah, S.; Yunus, A. Analysis of ascrobic acid content (vitamin C) of purslane (*Portulaca oleracea* L.) at various altitudes in East Java, Indonesia. *IOP Conf. Ser. Earth Environ. Sci.* **2021**, *637*, 012074. [CrossRef]
66. Adinolfi, M.; Barone, G.; Corsaro, M.M.; Lanzetta, R.; Mangoni, L.; Parrilli, M. Glycosides from Muscari comosum. 7. Structure of three novel muscarosides. *Can. J. Chem.* **1987**, *65*, 2317–2326. [CrossRef]
67. Renna, M.; Rinaldi, V.A.; Gonnella, M. The Mediterranean diet bewteen traditional foods and human health: The culinary example of Puglia (Southern Italy). *Int. J. Gastron. Food Sci.* **2014**. [CrossRef]
68. Motti, R.; Antignani, V.; Idolo, M. Traditional plant use in the Phlegraean Fields Regional Park (Campania, Southern Italy). *Hum. Ecol.* **2009**, *37*, 775–782. [CrossRef]
69. Marrelli, M.; La Grotteria, S.; Araniti, F.; Conforti, F. Investigation of the potential health benefits as lipase inhibitor and antioxidant of *Leopoldia comosa* (L.) Parl.: Variability of chemical composition of wild and cultivated bulbs. *Plant Foods Hum. Nutr.* **2017**, *72*, 274–279. [CrossRef]
70. Loizzo, M.R.; Tundis, R.; Menichini, F.; Pugliese, A.; Bonesi, M.; Solimene, U.; Menichini, F. Chelating, antioxidant and hypoglycaemic potential of *Muscari comosum* (L.) Mill. Bulb extracts. *Int. J. Food Sci. Nutr.* **2010**, *61*, 780–791. [CrossRef]
71. Casacchia, T.; Sofo, A.; Casaburi, I.; Marrelli, M.; Conforti, F.; Statti, G.A. Antioxidant, enzyme-inhibitory and antitumor activity of the wild dietary plant *Muscari comosum* (L.) Mill. *Int. J. Plant. Biol.* **2017**, *8*, 6895. [CrossRef]
72. Casacchia, T.; Scavello, F.; Rocca, C.; Granieri, M.C.; Beretta, G.; Amelio, D.; Gelimini, F.; Spena, A.; Mazza, R.; Toma, C.C.; et al. *Leopoldia comosa* prevents metabolic disorders in rats with high-fat diet-induced obesity. *Eur. J. Nutr.* **2019**, *58*, 965–979. [CrossRef] [PubMed]
73. Sulas, L.; Re, G.A.; Bullitta, S.; Piluzza, G. Chemical and productive properties of two Sardinian milk thistle (*Silybum marianum* (L.) *Gaertn.*) populations as sources of nutrients and antioxidants. *Genet. Resour. Crop. Evol.* **2016**, *63*, 315–326. [CrossRef]
74. Bijak, M. Silybin, a major bioactive component of milk thistle (*Silybum marianum* L. Gaernt.)—Chemistry, bioavailability, and metabolism. *Molecules* **2017**, *22*, 1942. [CrossRef]
75. Martinelli, T.; Fulvio, F.; Pietrella, M.; Focacci, M.; Lauria, M.; Paris, R. In *Silybum marianum* Italian wild populations the variability of silymarin profiles results from the combination of only two stable chemotypes. *Fitoterapia* **2021**, *148*, 104797. [CrossRef]
76. Abenavoli, L.; Capasso, R.; Milic, N.; Capasso, F. Milk thistle in liver diseases: Past, present, future. *Phytother. Res.* **2010**, *24*, 1423–1432. [CrossRef] [PubMed]
77. Kren, V.; Walterova, D. Silybin and silymarin—New effects and applications. *Biomed. Pap. Med. Fac. Univ. Palacky Olomouc Czechoslov. Repub.* **2005**, *149*, 29–41. [CrossRef]
78. Lee, J.I.; Narayan, M.; Barrett, J.S. Analysis and comparison of active constituents in commercial standardized silymarin extracts by liquid chromatography-electrospray ionization mass spectrometry. *J. Chromatogr. B Anal. Technol. Biomed. Life Sci.* **2007**, *845*, 9–103. [CrossRef] [PubMed]
79. Porwal, O.; Ameen, M.S.M.; Anwer, E.T.; Uthirapathy, S.; Ahamad, J.; Tahsin, A. *Silybum marianum* (milk thistle): Review on its chemistry, morphology, ethno medical uses, phytochemistry and pharmacological activities. *J. Drug Deliv. Therap.* **2019**, *9*, 199–206. [CrossRef]
80. Le, Q.-U.; Lay, H.-L.; Wu, M.-C.; Joshi, R.K. Phytoconstituents and pharmacological activities of *Silybum marianum* (milk thistle): A critical review. *Am. J. Ess. Oils Nat. Prod.* **2018**, *6*, 41–47.

81. Chambers, C.S.; Holečková, V.; Petrásková, L.; Biedermann, D.; Valentová, K.; Buchta, M.; Křen, V. The silymarin composition . . . and why does it matter? *Food Res. Int.* **2017**, *100*, 339–353. [CrossRef] [PubMed]
82. Aghazadeh, S.; Amini, R.; Yazdanparast, R.; Ghaffari, S.H. Anti-apoptotic and anti-inflammatory effects of *Silybum marianum* in treatment of experimental steatohepatitis. *Exp. Toxicol. Pathol.* **2011**, *63*, 569–574. [CrossRef]
83. Doehmer, J.; Weiss, G.; McGregor, G.P.; Appel, K. Assessment of a dry extract from milk thistle (*Silybum marianum*) for interference with human liver cytochrome-p450 activities. *Toxicol. Vitr.* **2011**, *25*, 21–27. [CrossRef]
84. Aziz, M.; Saeed, F.; Ahmad, N.; Ahmad, A.; Afzaal, M.; Hussain, S.; Mohamed, A.A.; Alamri, M.S.; Anjum, F.M. Biochemical profile of milk thistle (*Silybum marianum* L.) with special reference to silymarin content. *Food Sci. Nutr.* **2021**, *9*, 244–250. [CrossRef]
85. Elwekeel, A.; Elfishawy, A.; Abouzid, S. Silymarin content in *Silybum marianum* fruits at different maturity stages. *J. Med. Plants Res.* **2013**, *7*, 1665–1669. [CrossRef]

Review

Solanum aethiopicum: The Nutrient-Rich Vegetable Crop with Great Economic, Genetic Biodiversity and Pharmaceutical Potential

Mei Han [1], Kwadwo N. Opoku [1], Nana A. B. Bissah [2] and Tao Su [1,3,*]

[1]   Co-Innovation Center for Sustainable Forestry in Southern China, College of Biology and the Environment, Nanjing Forestry University, Nanjing 210037, China; sthanmei@njfu.edu.cn (M.H.); opokunketiah38@gmail.com (K.N.O.)
[2]   Department of Crop and Soil Sciences, Kwame Nkrumah University of Science and Technology, Kumasi AK-449, Ghana; bissawaa.bissah@gmail.com
[3]   Key Laboratory of State Forestry Administration on Subtropical Forest Biodiversity Conservation, Nanjing Forestry University, Nanjing 210037, China
*   Correspondence: sutao@njfu.edu.cn

Abstract: *Solanum aethiopicum* is a very important vegetable for both rural and urban communities in Africa. The crop is rich in both macro- and micronutrients compared with other vegetables and is suitable for ensuring food and nutritional security. It also possesses several medicinal properties and is currently employed in the treatment of high blood pressure, diabetes, cholera, uterine complaints as well as skin infections in humans. The crop is predominantly cultivated by traditional farmers and plays an important role in the subsistence and economy of poor farmers and consumers throughout the developing world. It also holds potential for dietary diversification, greater genetic biodiversity and sustainable production in Africa. Despite the numerous benefits the crop presents, it remains neglected and underutilized due to the world's over-dependence on a few plant species, as well as the little attention in research and development it has received over the years. This review highlights the importance of *S. aethiopicum*, its role in crop diversification, reducing hidden hunger, the potential for nutritive and medicinal benefits, agricultural sustainability and future thrusts for breeding and genetic improvement of the plant species.

Keywords: nutraceuticals; genetic diversity; breeding; *Solanum aethiopicum*; neglected and underutilized; phytochemicals

Citation: Han, M.; Opoku, K.N.; Bissah, N.A.B.; Su, T. *Solanum aethiopicum*: The Nutrient-Rich Vegetable Crop with Great Economic, Genetic Biodiversity and Pharmaceutical Potential. *Horticulturae* **2021**, 7, 126. https://doi.org/10.3390/horticulturae7060126

Academic Editors: Rosario Paolo Mauro, Carlo Nicoletto and Leo Sabatino

Received: 1 May 2021
Accepted: 26 May 2021
Published: 28 May 2021

Publisher's Note: MDPI stays neutral with regard to jurisdictional claims in published maps and institutional affiliations.

1. Introduction

Solanum aethiopicum is a horticultural crop species consumed in Africa and it contains a great amount of nutrients [1]. The crop is an important food source for many people. The fruits and leaves are employed in the preparation of stews and soups. It is used to treat certain diseases because of the phytochemicals it contains and serves as a rich source of important macro and micronutrients [2]. It is thus a potential curative to hidden hunger [3–5]. The crop is one of the most widely grown vegetables in Africa and was brought to Brazil through the slave trade. The crop belongs to the *Solanaceae* family with a chromosome number of 24. It is commonly called African eggplant, scarlet eggplant, garden eggs and bitter tomato. The species is believed to have originated from Africa and was domesticated from the wild *Solanum anguivi* Lam., via the semi-domesticated *Solanum distichum* Schumach. & Thonn., both of which are found throughout tropical Africa [6,7]. The crop is cultivated in the humid zones of West Africa for its immature fruit, in the savanna area frequently for both its leaves and immature fruits (often called 'djakattou'), and in East Africa, especially Uganda, mainly as a leaf vegetable (called 'nakati'). African eggplant is a good and reliable crop as it yields a harvest of about 25 tons per hectare [8]. The species *Solanum melongena* and *Solanum macrocarpon* are the close relatives of *S. aethiopicum* and both species are commonly grown in Africa. Mwinuka et al. describe this crop as a neglected and underutilized horticultural species, despite possessing the ability to significantly contribute to nutrition and food security. It has been overlooked as there is limited knowledge on the plant and research efforts about it are limited, especially in terms of climate adaptation and sustainable agricultural growth in Africa and particularly in East Africa [9].

Although African eggplants possess several important properties, knowledge on the genetic improvement potential, possibility of providing medicinal or nutritional benefits, such as disease prevention and treatment and its potential for agricultural sustainability is still limited. This is because the crop has not received adequate research attention for years. There is therefore a need to promote the dynamic use, documentation, conservation and evaluation of genetic resources of this important but neglected crop [10]. Additionally, due to the overdependence on the three major food crops (wheat, maize and rice) and their subsequent shortages, the conservation, improvement and utilization of underutilized plant species such as African eggplant is of utmost importance. The cultivation of underutilized crops provides greater genetic biodiversity, and can potentially improve food security [11]. The purpose of this review is to give an overview of research carried out on *S. aethiopicum* for the detection of avenues that will contribute to promoting the plant species, mainly in its potential for biodiversity, nutritive and medicinal benefits, and agricultural sustainability.

2. Physiological and Genetic Properties of *S. aethiopicum*

2.1. Characterization and Ecophysiology

S. aethiopicum grows to about 2.5 m in height and is an often a branched deciduous shrub. On the stems, the leaves are alternately arranged and have smooth or lobed margins. Leaf-blades can reach a length of up to 30 cm and a width of 21 cm. The leaves' petioles are oval or elliptical, reaching a length of up to 11 cm [12]. The species is hermaphrodite (has both male and female organs) and mainly self-pollinated, but insect pollinators complement self-pollination as they provide cross-pollen, which augments gene exchange and hybrids in natural population and lessens inbreeding depression. The inflorescence is a five-flowered lateral, racemose cyme [13]; peduncle often short or even absent, rachis short to long. The flowers develop into egg- or spindle-shaped berries, which are red to orange in color with a smooth or grooved surface depending on the variety. The crop becomes ready for harvesting from 100 to 120 days after planting and the fruit should be plucked before it changes color from white to pale yellow. African eggplant seeds are flattened, 2–5 mm in diameter, lenticular to reinformed and pale brown or yellow in color. Germination is epigeal with cotyledons being thin and leafy [14]. Multiplication of the species is carried out mainly through the seeds.

S. aethiopicum is a complex species consisting of groups that are very distinct morphologically and were formerly regarded as four separate species [15] (Figure 1). It is also described to be a hypervariable species as it is made up of many forms and types that differ morphologically, with hundreds of local varieties [16]. The species *S. aethiopicum* can be classified with respect to its use into four distinct groups. They are the Gilo, Shum, Kumba and Aculeatum group [17]. The Gilo group gives rise to several differently shaped edible fruit (ranging from a spherically depressed form to elliptic in outline); the Kumba group possess a stout main stem with large hairless leaves that can be picked as a green vegetable, and later produces very large grooved fruit that is picked green or even red; the Shum group is a short much-branched plant with small hairless leaves and shoots that are plucked frequently as a leafy green vegetable but the small (1.5 cm across) very bitter fruits are not eaten; the Aculeatum group produces flat-shaped fruit [18]. The plant is also sometimes grown as an ornamental one.

Being a tropical plant species, the African eggplant is intolerant to low temperatures and very cold or water-logged conditions. Some level of tolerance to irrigation-induced salinity has been reported from Senegal [19]. The edible groups of African eggplant (Gilo, Kumba and Shum) are adapted to diverse areas depending on the climate. The Gilo group is commonly found in humid areas all over tropical Africa where its members grow best at the full sun of woodland savanna on fairly deep and well-drained soils of pH 5.5–6.8, with 25–35 °C and 20–27 °C day and night temperatures, respectively [20]. The Kumba group, which is cultivated chiefly in semi-arid areas from Occidental Sahel to the north of Nigeria, is able to put up with environments that are hotter than normal (even to 45 °C day temperature) with an occasionally low air humidity of about 20%, especially after irrigation [20]. The Shum group is typically seen to shed its leaves during plant water loss under the warm, humid conditions. In Africa, the group is found at high altitude and very humid areas mainly in Uganda and the southeast of Nigeria [20]. In Uganda, it is grown in swamps during the dry season [21,22].

Figure 1. Fruit vegetable pictures for (**a**) Kumba, (**b**) Aculeatum, (**c**) Shum and (**d**) Gilo group of *Solanum aethiopicum*.

The hybrids emanating from the cross between *S. aethiopicum* and *S. anguivi* Lam (wild ancestor) are fertile [23]. Characterization of a plant species gives a description of its germplasm and determines the expression of highly heritable traits ranging from morphological or agronomical features to seed proteins or molecular markers. The characterization of cultivated eggplants and their wild relatives has commonly been analyzed by employing conventional morphological descriptors that are highly heritable and simple to assess [15,24]. Studies on the morphological characterization of scarlet eggplant have shown it to be a highly variable crop. The most extensive research was performed by Lester in 1986 [16]. The authors characterized 108 accessions of the scarlet eggplant complex using morphological and taxonomically relevant characters (e.g., fruit size, number of locules, number of flowers per inflorescence, prickliness) and observed that the four cultivar groups could be differentiated by a set of traits (including leaf shape, fruit shape, prickly leaves and stem, bitter taste) detected in a single group. A number of the accessions were also observed to be intermediate between *S. anguivi* and *S. aethiopicum*. According to Osei et al. [25], a wide variation exists among species *S. aethiopicum*, *S. anguivi* and *S. macrocarpon*. They also observed a lot of similarities between the lines of *S. aethiopicum* and *S. anguivi*.

2.2. Propagation, Cultivation Techniques and Systems

The African eggplant is often grown as an annual, but generally a perennial, plant with stems that become more or less woody and persist. It grows well in deep, well-drained soils. Cultivation of African eggplant is mainly dependent on the rain, but irrigation can be applied during the dry seasons. The African eggplant requires a pH of 5.5–6.8, and thrives well at daytime temperatures ranging between 20–30 °C, but it can tolerate 10–40 °C. It cannot tolerate very cold or water-logged conditions [26]. It is more prudent to grow the crop on a minimum to low salinity soil to promote maximum growth and development, since salinity plays a part in disturbing the anatomical and morphological features of the crop [27]. *S. aethiopicum* is propagated by seed. The seeds for planting are obtained from fully ripe fruits that should not be exposed to direct sunlight. Seeds remain viable for a long time when stored in a cool, dry place. Seeds also store well inside air-dried fruits, which is the traditional form of seed storage by farmers. The 1000-seed weight of African eggplant is 2–4 g. Its germination takes 5–9 days for the Gilo and Shum groups, but only 3–5 days for the Kumba group, although the latter may express seed dormancy and end up having few seeds per fruit. Seeds are sown in sandy soil in nursery beds or containers. The seedlings are transplanted to the field after 30–35 days, when they have 5–7 leaves and are 15–20 cm tall. Plants of the Kumba group grown in dry savanna regions are often planted at a distance of 1 m × 1 m, whereas those of the Gilo group can be spaced at 50–100 cm in the row and 75–100 cm between rows, depending on the cultivar [14]. They can be grown either on flat land or ridges. The cultivation of Shum Group is somewhat different. Cultivars of the Shum group are grown for their young shoots, which are frequently harvested, and the crop can thus be spaced at 20–30 cm in the row and 60–75 cm between rows. An alternative is to broadcast seeds of Shum group, for thinned plants to be used as the first harvest. Seeds are sometimes broadcast together with amaranths (*Amaranthus* spp.) and spider plant (*Cleome gynandra* L.), where the latter two crops are harvested early by uprooting and the plants of *S. aethiopicum* Shum group remain.

Production of African eggplant is created using a range of cultivation techniques. These techniques are aimed at the preservation and modification of the physical and chemical characteristics of the soil (soil preparation and tillage, irrigation, fertilization), while improving plant production (training, pruning, fruiting, production and pesticide treatments). Some examples of cultivation techniques employed by farmers include intercropping, mixed cultures and monocultures. On the farm, it can be intercropped with crops such as cowpea, sorghum, *Ziziphus mauritiana* Lam, *Solanum lycopersicum*, *Capsicum annuum*, *Corchorus olitorius* and *Abelmoschus esculentus* [28,29]. In mixed cultures, African eggplant is the main crop, whereas *Zea mays* and *Manihot esculenta* are grown as secondary crops. Monoculture is also practiced where an optimal crop density is cultivated on the farm and sometimes under irrigation during the dry season. Most of the farmers in Africa cultivate using conventional and organic systems. From the study of Aguessy et al., the majority of the farmers who were interviewed acknowledged the use of NPK and urea fertilizers a few days before transplanting as compensation for soil deficiency conditions. The other proportions of farmers are into the use of organic manures [30]. Animal manure and compost are being employed in the cultivation of African eggplant. They are helpful in soil amendment due to the needed organic matter and sequester carbon they add, as well as the reduced reliance on chemical pesticides and fertilizers. Compost from false yam (*Icacina oliviformis*) tuber was applied to a field planted with African eggplant and it showed support for plant growth and yield characteristics [31].

Efforts are also underway to optimize the yield of African eggplant. Mwinuka and others in a recent report assessed how irrigation water and nitrogen affects the growth parameters of African eggplant, as well as the yield, fruit quality, water use efficiency (WUE) and nitrogen use efficiency (NUE). They observed that the African eggplant growth variables (plant height and Leaf Area Index) correlated well with fruit yield, and when 100% water was added to a 75% (187 kg/ha) nitrogen treatment, the quality of the fruit was at maximum. Superlative WUE and NUE were obtained at 80% and 100% water supply, respectively, in addition to a nitrogen concentration of 75%. It was further suggested that in soils with a mixture of sand, clay and loam, under sub humid circumstances that are tropical, the optimal application for African eggplant will likely be about 80% of the total irrigation requirement and 75% of the nitrogen requirement, in order to minimize the rate of exchange between the various indicators [9].

The crop is affected by several diseases and pests. Common diseases of African eggplant include *Chilli veinal* mottle virus (ChiVMV) borne by the vector green peach aphid (*Myzus persicae*) [32], Stemphylium disease caused by *Stemphylium solani* and other soil-borne severe diseases such as wilt caused by *Ralstonia solanacearum*, collar rot and wilting caused by *Sclerotium rolfsii* and *Verticillium dahliae*, and root-knot nematodes (*Meloidogyne* spp.) [33]. Pests of the plant include green peach aphid *Myzus persicae*, grasshoppers (*Zonocerus* sp.), fruit and flower borers (*Leucinodes* and *Scrobipalpa*), leafhopper (*Jacobiasca lybica*) and caterpillars (*Selepa docilis*) [34]. Spider mites (*Hemitarsonemus* and *Tetranychus*) are a serious problem in drier regions; acaricide sprays can normally control them [35].

2.3. Biodiversity and Conservation

Today, the production of crops is based on just a small circle of plant species [36]. Despite this fact, several undermined crops are grown and collected, thereby adding to biodiversity [37]. Although *S. aethiopicum* is one of the less recognized crops, it shows a distinct diversity in the cultivated types of varieties in Ghana. As well as contributing to biodiversity intraspecifically, it also helps to maintain a wide intraspecies biodiversity through local cultivation. Limited scale producers keep up the genetic diversity of this crop. Not only is there scattering of cultivars across locations, their naming is often also set differently. Moreover, the cultivars could be equivalent, yet exhibit differences in their phenotype due to diverse biotic and abiotic factors. *S. aethiopicum* has noticeable qualities such as taste, size, color, shape and others, but these attributes differ widely. For instance, its color could be deep green, green, white, cream, or even yellow and this could be used as the basis of freshness. The sizes range from small to big with a massive demand for the latter. The shape could be round, elongated, with ridged or smooth surfaces. Taste is additionally a significant quality property pursued by buyers and very frequently identified with the shape. Most often than not, the bitter ones are the round shaped ones, and this characteristic is genotype-specific but environmental factors could also play a part in it [37]. In Ghana, research scope to determine the reason for this diversity is very little, though not for lack of diversity in African eggplant germplasm, and due to this myopic research level, to our knowledge, nothing has been implemented to conserve diversity.

Accurate information on the analysis of genetic diversity, which is expressed as variants in a linear sequence of nucleotides in DNA, is relevant to know which genes are responsible for which trait expressed in *S. aethiopicum*. African eggplant differs severely when it comes to agronomic traits, some of which include time of flowering, fruit yield, fruit maturity, branching habits [25]. Taking the physiology, morphology and biochemical properties of the eggplant's related species and wild types into consideration, there exists a wide diversity morphologically [38,39]. For any breeding program to be effective, the detail in the genetic magnitude of diversity within the crop species is dire. This is because informed knowledge on genetic diversity could go a long way in sustaining selection gain in the long term [40]. From Sharma and Jana's [41] point of view, a priority for starting an efficient breeding program is the assessment of how species vary genetically, since it gives a platform for cutting desirable genes to size. Additionally, knowing how genetically diverse a huge population of African eggplant germplasm is aids in the decision-making of breeding techniques and management schedules for current and future use [42]. The study of genetic diversity allows for the selection of African eggplant parents that are genetically diverse to obtain in the segregating generations, recombinants that are desirable [43]. This choice of parental selection using biodiversity studies is valuable because it showcases how useful variations can be expressed in later offspring. Thus, with African eggplant, diversity studies at their early stages would amount to the creation of cultivars and varieties that are shared across the board and known for their stability and adaptability in terms of performance (consistency in tolerance of harsh weather conditions and disease resistance) based on evidential characteristics.

Since there is a lack of large genomic resources for conservation, this implies that *S. aethiopicum* breeding is lagging behind other vegetable crops, especially of the same species. It is thus highly important to evaluate and safeguard *S. aethiopicum*, as it is highly underrepresented in the global conservation system of plant genetic resources and may have hidden genes that are of relevance when it comes to tolerance or resistance of biotic or abiotic stresses [8]. The World Vegetable Center (WorldVeg) secures some amount of public germplasm collection of eggplant, which includes the major cultivated species (*S. melongena*, *S. aethiopicum* and *S. macrocarpon*), and more than 30 eggplant wild relatives, with more than 3200 accessions collected from 90 countries of which *S. aethiopicum* is included and considered as important. Over the last 15 years, more than 10,000 seed samples from the Center's eggplant collection have been shared with public and private sector entities, including other genebanks. An analysis of the global occurrences and genebank holdings of cultivated eggplants and their wild relatives reveals that the WorldVeg genebank holds the world's most extensive public collection of the three cultivated eggplant species with *S. aethiopicum* being the third. However, comparing this with other vegetable crops (with huge genomic resources for conservation), there is still a significant deficiency in the conservation of *S. aethiopicum*. The main cluster of *S. aethiopicum* is in West Africa, with a total of 1288 occurrences based on the literature of previous studies and characterization data available at the WorldVeg [8]. From 1981–1986, the International Plant Genetic Resources Institute (IPGRI) performed a collection of genetic resources of African eggplant. In 2001, efforts were made to regenerate and evaluate the collections as part of the EGGNET project, an international project for managing the genetic resources of eggplant. A rich germplasm collection is maintained at INRA in Montfavet, France. A collection is kept at AVRDC in Arusha, Tanzania, and plant breeders in Ghana, Côte d'Ivoire, Nigeria, Senegal and elsewhere in Africa also maintain some accessions of African eggplant [14].

2.4. Genetics and Breeding

The breeding of African eggplant has received little attention, although germplasm collections of it exist. Most of the cultivated species came about as a result of selection by farmers and consumers' preference for fruit size, color and taste (sweet or bitter), skin toughness, shelf life, as well as yield potential, disease resistance, earliness and duration of the harvest season, plant architecture and fruit location, and in the case of the Shum group, ease of leaf harvest. *S. aethiopicum* for a long time has been neglected by formal programs involved in the improvement of crops except in breeding programs where some of its specific traits are exploited to improve the resistance of *S. melongena* to diseases. Domesticated and wild relatives of African eggplant possess important traits that remain to be explored. *S. aethiopicum* has been reported to express low susceptibility to pests and diseases than *S. melongena* [44], leading to the selection of *S. aethiopicum* (Aculeatum group) in Japan as a rootstock for tomato and garden egg due to its resistance to wilt. It also shows a higher level of tolerance to drought and heat than tomato and eggplant in the field. Ano et al. [45] reported on *S. aethiopicum* in a breeding program used as a source of the disease resistance gene. From this study, *S. melongena* was crossed with *S. aethiopicum* and the hybrids subsequently backcrossed to *S. melongena*. It was likely to derive families with a high resistance level to bacterial wilt from the second backcross, as well as vast differences in the shape and fruit color. Studies into the molecular mechanism underlying double fertilization between self-crossed *S. melongena* and that hybridized with *S. aethiopicum* have been carried out using comparative transcriptome analysis. A number of differentially expression genes (DEGs) were found which are involved in plant hormone transduction, cell senescence, metabolism, and biosynthesis pathways. The findings of this study provide insights into the regulatory mechanisms underlying variations between ovaries of self-crossed and hybrid eggplants and a basis for future studies of crossbreeding *Solanum* [46].

In a study by Hamidou et al. [47], 12 scarlet eggplant accessions were assessed via morphological and cytomolecular characterization. This study established the genetic closeness of *S. aethiopicum* and *S. melongena* not only in terms of their chromosome number but also in terms of their genome size. Similarities and contrasts among the 12 scarlet eggplant accessions were observed for 27 quantitative descriptors. The cluster analysis was able to classify the accessions into four distinct groups. A great deal of variability was found among the 12 scarlet eggplants accessions, most of which was associated with flower and fruit-related traits. However, no significant differences among the accessions for the GC content, for seedling and stomatal traits evaluated were found. Additionally, there was no significant difference ($\alpha = 0.05$), with the exception of one accession (PI 420226), among the scarlet eggplants evaluated for their nuclear DNA content.

Availability of the genome information of a neglected crop such as *S. aethiopicum* will hasten a more productive and exact selection of exceptional accessions of the crop and breeding of it, as well as other crops within the *Solanaceae* family. A recently published draft genome sequence of the African eggplant has afforded some insights into disease resistance, drought tolerance, and the evolution of the genome [48]. The 1.02 Gb draft genome assembly of *S. aethiopicum* contained largely repetitive sequences (78.9%). Gene models were annotated including 34,906 protein-coding genes. Expansion of disease resistance genes was found through two rounds of amplification of long terminal repeat retrotransposons, which may have happened ~1.25 and 3.5 million years ago. The resequencing of *S. aethiopicum* and *S. anguivi* genotypes resulted in identification of 18,614,838 single-nucleotide polymorphisms (SNPs), of which 34,171 were positioned within disease resistance genes [48]. Furthermore, the investigation into the domestication and demographic history showed active selection in both the Gilo and Shum groups for genes involved in drought tolerance. A pan-genome of *S. aethiopicum* was generated containing 51,351 protein-coding genes, 7069 of which were missing from the reference genome [48].

More recently, few researchers have focused their attention on improving the agronomic characters of farmer's selection. In Senegal, very hairy, mite-resistant cultivars of the Kumba group have been bred. Some promising cultivars for high yield were obtained when *S. anguivi* was crossed with 'Dwomo', a cultivar whose fruits resembles an egg in its size and shape. This was achieved by collaboration between the Crops Research Institute in Ghana and the Natural Resources Institute (United Kingdom) [49]. A seed company in Senegal (Technisem Seed Company) is into the commercialization of improved cultivars of African eggplant. The most popular cultivars of the Kumba group are 'Ndrowa' (flat, ribbed, green-yellow fruits of 70–80 g, diameter 5 cm, with a mild taste, plant height 60–100 cm, harvestable 50–70 days after planting, yield potential 25 t/ha, resistant to mites), 'Ngalam' (flat, strongly ribbed, pale green to white fruits of 120–180 g, diameter 7 cm, slightly bitter taste, plant height 60 cm, harvestable 50–60 days after planting, yield potential 10 fruits per plant, resistant to mites) and 'Jaxatu Soxna' (flat, ribbed, pale green to white fruits of 40–50 g, diameter 5–6 cm, bitter taste, plant height 50 cm, harvestable 40–60 days after planting, yield 20–25 fruits per plant or 30 t/ha, or used as a leafy vegetable, resistant to mites, drought, high rainfall and high temperatures).

The inheritance of some important traits in *S. aethiopicum* has also been studied intensively. Most wild-type characters such as prickles, stellate hairs and long racemose cymes are often dominant. Imperfect morphogenesis, indicating loss of genetic regulation, was involved in many of the recessive domesticated traits. The F_1 hybrids between cultivars

(*S. anguivi* and *S. aethiopicum*) also showed significant heterosis and are recommended for crop production [23]. Research for yield of fruit and yield component traits on the mode of inheritance, genetic control, heritability and heterosis has been implemented. Traits such as fruits per cluster, the cluster per plant and length and width, are the parameters of fruit yield in fruit vegetable. Out of the components of fruit yield studied in an experiment for *S. aethiopicum*, fruit length and fruit clusters per plant were both affected by the additive gene action, while on the other hand, diameter and fruits per cluster required the dominance gene action when it comes to their inheritance. Dominance and dominance by dominance digenic interaction expressed the duplicate type of epistasis for fruit clusters per plant [50]. This genetic information is very imperative for efficient breeding strategy and development of hybrids and open-pollinated varieties.

In vitro regeneration from cotyledonary and true leaf explants (direct organogenesis) of *S. aethiopicum* and *S. macrocarpon* have been reported by Carmina et al. [51]. A higher level of regeneration and the average of shoots per explant has been demonstrated by using low concentrations (0.1 and 0.2 µM) of thidiazuron (TDZ) in the media. The success allows for further genetic transformation and improvement of *S. aethiopicum* through in vitro techniques. Somaclonal variation occurring in tissue culture media has proven to be a potential source of variability for crop improvement [52,53]. Studies have shown that in vitro regeneration of plants through indirect organogenesis (i.e., via callus phase) could induce more variants as compared to direct organogenesis. Hence, tissue culture avenues could be exploited to induce somaclonal variants of African eggplant, so as to broaden the genetic base of the crop for future breeding programs. Mutants of African eggplant demonstrate differences in traits that are agro-morphological and yield attributes with varying doses of Gamma irradiation. The effect of the mutagen on *S. aethiopicum* is assessed in terms of its influence on growth improvement by inducing changes in cells and tissues of the plant that are cytological, genetical, biochemical, physiological and morphogenetic. Low doses of gamma irradiation (40 Gy and 60 Gy) caused significant variation in some traits that are agro-morphological, like plant height, leaf characteristics, days to first flowering, number of leaves per plant and number of branches per plant in *S. aethiopicum* [54]. Therefore, mutation breeding by gamma irradiation and other mutagens can be employed as an effective means of inducing genetic variability in African eggplant to improve and select desirable mutants for breeding purposes.

2.5. Phytochemical and Pharmacological Activities

The African eggplant fruits are usually used in the preparation of soups, in stews as a vegetable, and they can sometimes be eaten raw. It is also possible to eat the shoots and leaves in a cooked form. African eggplant contains many protein, minerals, vitamins, carbohydrate, fat, crude fiber, ash and water substances that are relevant and massively helpful in nutrient supplement and health promotion. The fruits mostly possess high moisture content and low dry matter [55]. Several fundamental mineral elements, including calcium, magnesium, potassium, sodium, manganese, iron, copper, zinc and phosphorus are also contained in the fruits of *S. aethiopicum*. These minerals are involved in functions such as maintenance of heart rhythm, muscle contractility, formation of bones and teeth, acid-base balance, regulation of cellular metabolism and enzymatic reactions [56]. Moreover, *S. aethiopicum* has been shown to be rich in vitamins (such as vitamin A, B, C, D and E) [57]. Analysis of the nutritional and mineral composition of the fruits and leaves of *S. aethiopicum* brought to light that the crop is rich in major and minor nutrients and all the nutrients are known to be vital for the proper functioning of the body [1]. Of note, the exocarp of both ripe and unripe *S. aethiopicum* fruit bears no incidence of toxic metals (e.g., chromium, cadmium, lead, etc.), which makes it a better source of health products. The nutrient, mineral and vitamin composition of *S. aethiopicum* fruits is presented in Table 1 as adopted from previous studies [55–57]. Chinedu et al. in their study on the proximate and phytochemical analyses of the fruits of *S. aethiopicum* also identified nutrients and mineral elements [58] that had similar values to the ones presented above.

Table 1. Proximate composition of *Solanum aethiopicum* L. fruits.

Nutrient	Composition (per 100 g of Fresh Fruit)
Moisture content	91.20 ± 0.34%
Crude protein	1.07 ± 0.01%
Crude fat	0.38 ± 0.03%
Crude fiber	2.44 ± 0.04%
Ash content	0.73 ± 0.03%
Carbohydrate	4.18 ± 0.08%
Dry matter	8.80 ± 0.19%
Mineral Element	**Concentration (mg/g Dry Weight Basis)**
Calcium	0.310~0.360
Magnesium	0.595~0.625
Iron	0.025~1.125
Potassium	4.475~9.525
Sodium	0.865~1.005
Manganese	0.005
Copper	0.007~0.008
Zinc	0.077~2.938
Phosphorus	1.091~1.245
Vitamins	**Content (mg/100 g Fresh Fruit)**
Vitamin A (retinol)	53.550 ± 0.55
Vitamin B1 (thiamine).	0.037 ± 0.00
Vitamin B2 (riboflavin)	0.034 ± 0.00
Vitamin B (niacin)	0.700 ± 0.00
Vitamin C (ascorbic acid)	2.300 ± 0.00
Vitamins D (calciferol)	0.010 ± 0.00
Vitamin E (tocopherol)	0.310 ± 0.00

Phytochemical screening of the crop showed a copious presence of alkaloids, flavonoids, phytosterols, saponins and vitamin C, moderate presence of cardiac glycosides, steroids and tannins, and a trace amount of terpenoids in the fruits [55,58]. Alkaloidal extracts of *Solanum* species have been reported to express analgesic effects and central nervous system depression [59]. The presence of alkaloids, mainly glycoalkaloids, gives the bitterness in eggplants and the relative bitterness determines to a great extent their edibility or otherwise. The incidence of toxic glycoalkaloids leads to poisoning in some *Solanum* species by causing diarrhea or carcinogenic glycosides which bring about excessive deposition of calcium in tissues. Researchers have cautioned that care should be taken when using the fruit and insist that its consumption should be in small quantities [60]. African eggplant is of great medicinal importance and thus has many functional food properties. Its saponins are important dietary supplements and nutraceuticals. They possess antimicrobial properties and protect plants from microbial pathogens. Reports have shown that saponins present in traditional medicine preparations cause hydrolysis of glycosides from terpenoids which avert the toxicity associated with the intact molecule [61,62]. Ascorbic acid and flavonoids have been found to be present in the fruits and stalk of the plant, which contain high antioxidant potential [63–65]. Traditionally, many tropical African countries use the fruits of bitter cultivars as medicine to cure certain ailments [66]. The roots and fruits of African eggplant are used as a carminative and sedative, and to treat colic and high blood pressure; leaf juice as a sedative to treat uterine complaints; an alcoholic extract of leaves as a sedative, anxiolytic, anti-emetic and to treat tetanus after an abortion [67,68]. The crushed and softened fruits are also used as a purgative. The juice of boiled roots is used to treat hookworms, while the crushed leaves are used for gastric ailments [19]. Several parts of the plant are used, as powder or ash, to treat diseases such as diabetes, otitis, cholera, toothache, bronchitis, skin infections, dysuria, asthenia, dysentery and haemorrhoids in decoction. Moreover, antiviral, anticancer, anticonvulsant and anti-infective effects have been reported in eggplants due to the phytochemicals they contain. Narcotic, anti-asthmatic and antirheumatic effects are also ascribed to eggplants [69].

A previous study identified four groups of phenolic acids in the fruit of *S. aethiopicum*. Group 1 comprised chlorogenic acid isomers and 3-O-trans, 5-O-trans, 5-O-cis isomers of caffeoylquinic acid (3CQA, neochlorogenic acid; 5-CQA, chlorogenic acid; 4-CQA, cryptochlorogenic acid and 5-(Z)-CQA, cis-chlorogenic acid). Chlorogenic acid was the predominant compound in Group 1, with average levels 176-, 28.9-, and 92.5-fold higher than levels of neochlorogenic acid, cryptochlorogenic acid and cis-chlorogenic acid, respectively. Group 2 consisted of two phenolics: 3,5-diCQA and 4,5-diCQA. Group 3 comprised four compounds: N,N′-dicaffeoylspermidine (predominated); N-caffeoylputrescine and their isomers; and Group 4 comprised 3-O-actyl-5-O-caffeoylquinic acid (3-acetyl-5CQA), 3-O-actyl-4-O-caffeoylquinic acid (3-actetyl-4-CQA) [70]. The leaves of African eggplant possess oxalate and alkaloids such as solasodine, with reported glycocorticoid effects [14]. Its characteristic bitter taste is a result of the furostanol glycosides [44,67].

Anosike et al. [71] in their study reported that *S. aethiopicum* fruit had anti-inflammatory properties. Compared to the control group, the fruit extract substantially reduced the fresh egg albumin-15-mediated rat paw oedema and reduced the granuloma tissue formation in the fruit treated groups. According to Tunwagun et al. [72], phytochemicals such as alkaloids, saponin, tannin, volatile oil, phenol and flavonoids can be found in different anatomical parts of ripe African eggplant fruits. The presence of these phytochemicals in the different parts of the fruit indicates that the fruit might be pharmacologically active against a number of diseases. Another study by Emiloju and Chinedu [73] observed a dose-dependent, weight-reducing effect on rats in their study of the effects of dietary supplements of *S. aethiopicum* and *S. macrocarpon* on weight gain and glucose metabolism in rats. They further concluded that these underutilized crops may be used to alleviate the challenge of obesity.

2.6. Postharvest Factors Affecting their End-Use Quality

African eggplant production is characterized by poor post-harvest handling practices, a short shelf life of approximately three to four days and fruit spoilage losses estimated at 25% depending on the fruit harvesting stage and storage environment [37,74]. The lack of adequate management practices for post-harvest storage is estimated to cause up to 50 percent of vegetable crop yield losses between harvesting and consumption [75]. The most critical issue affecting the quantity, quality and thus the market value of fruits is poor harvesting and storage practices [76]. It is estimated that about 40–50% of horticultural crops produced are lost even before consumption in developing countries, mainly due to poor post-harvest handling which results in bruising, water loss and eventual decay during post-harvest handling [77]. A number of factors will lead to a reduction in the quality of fruits after harvest when not properly handled. Postharvest factors influencing postharvest quality of fruits include maturity stage, method of harvesting, time of harvesting, sorting and grading, packaging and packaging materials, storage, type of storage, temperature and relative humidity during storage.

The maturity stage at harvest is an essential factor that influences flavor and shelf-life qualities of fruits [78]. Failure to harvest fruits on time causes them to over-mature and ripen while still on the plant. On the other hand, harvesting fruits at the immaturity stage gives rise to increased shriveling and fruits susceptible to mechanical damages [79]. In Tanzania, African eggplant fruits are harvested by farmers based on their size and regularly comprise immature, mature and over-mature fruits. It is reported that cultivating the fruits at the immature level (fruits with non-shiny peel) is suitable for increasing yield in number of fruits per hectare and increasing vitamins and minerals content while harvesting at the mature fruits stage (by fruits with shinny peel) is recommended for increasing fruit beta carotene content, carbohydrate and fiber content [80]. An appropriate harvesting method is necessary to prevent bruises or injuries during harvesting as they may later express black or brown patches that make them unattractive. Injury to the peel may serve as an entry point for microorganisms, causing rotting. In Africa, harvesting of *S. aethiopicum* is carried out primarily with the hand by pulling or twisting the fruit pedicel or harvesting individual fruits or fruit bunch with the help of fruit clippers/secateurs/scissors. Time for harvesting also influences quality. Fruits harvested and transported early in the morning for sorting, grading, and packing at the packing house are long-lasting and of better quality. To minimize the risk of heat injury and sunburn, it is advisable that the fruits be harvested during the cooler periods of the day [81]. Harvesting fruits early in the morning facilitates faster pre-cooling and yields better quality.

Sorting and grading after harvest are one of the most critical postharvest practices. This is carried out mainly to remove diseased and defective fruits from the lot. This activity is conducted in a pack house or farmer's field and an assurance of a quality produce depends on a proper sorting and grading. Currently, grading is performed manually by farmers, but mechanical graders can also be employed. Most farmers and traders in Africa neither grade nor sort their produce after harvest. According to Apolot et al. [81], 16.6% of traders in Uganda are generally involved with sorting and grading of their vegetables as against 83.4% who are not. Packaging of fruits is needed in order to protect them from mechanical injuries. The packages should be well-ventilated and care should also be taken to prevent compression damages on the fruit during storage and transportation. A previous report revealed that storing fruits in perforated polyethylene bags reduced water loss and had the longest shelf life, although they faced the highest incidence of decay [82]. In Africa, the most commonly used storage facilities are air-cooled storage houses which rely on natural cold air. Proper management of temperature, ventilation and relative humidity are the key factors that affect the postharvest quality and storage life of horticultural product. Presently, mechanical refrigerators are employed in the storage of African eggplant, but they are energy intensive, unaffordable for local farmers and require constant supplies of electricity which is unavailable in many rural parts of Africa. In a study by Sekulya et al. [83], the shelf life of Shum (leafy vegetable) saw an increase (from one day to four days) when it was submerged into portable water intermittently for 2 to 3 s after every one hour during the day for sample in ambient storage both with roots intact and with roots cut-off. Samples stored in perforated polyethylene and meshed perforated polyethylene maintained more moisture and showed a minimal percentage of weight loss with the highest chlorophyll content when stored in the charcoal cooler, making it the best-tested packaging material.

The most efficient way to preserve quality has been found to be proper temperature control between the time of harvesting and consumption [84]. Since the fruits are alive after harvesting, all physiological processes, such as respiration and transpiration (water loss), begin after harvesting, and the supply of nutrients and water is not possible since the product is no longer attached to the parent plant. Respiration leads to product deterioration, including loss of nutritional value, changes in texture and taste, and weight loss through transpiration. Storing harvested fruits at low temperatures of approximately 20 °C can slow down many metabolic activities that lead to ripening, hence allowing more time for all postharvest handling. The loss of water from the harvested fruit product is primarily due to the amount of humidity present in the ambient air, expressed as relative humidity [85]. Harvested fruits at high levels of relative humidity retain their nutritional quality, weight, appearance and flavor, while reducing the rate at which wilting, softening and juiciness takes place. Fruits of African eggplant are vulnerable to shrinkage after harvest so any small loss of moisture and fruit shrinkage will become apparent. Enzymes are vital in the postharvest degradation of fruits of African eggplant. In order to avoid their degradation, enzymatic activities responsible for hydrolysis and oxidation activities in the fruits have been studied. Out of the enzymatic activities experimented, phosphatases, N'Acethylglucosaminidase, dopamineoxydase and pyrocatecholoxydase were the principal enzymatic proteins. Being able to control the activities of these enzymes will be very helpful in the postharvest preservation of the fruits of *S. aethiopicum* [86].

3. Product and Process Innovations

African eggplant is very beneficial to human health, but due to the short shelf life of about three to five days, it results in postharvest losses when the harvesting season is at its peak. Since in its fresh state, it has a relatively limited postharvest life, African eggplant can be changed through processing to forms that are shelf-stable [87]. African eggplant would usually require cold chain systems for the process following harvesting (handling, transport and distribution) to maintain its quality. Unluckily, these facilities are not enough and even sourly established in developing countries where this crop is predominantly located. One of the most recent innovations has been in drying of *S. aethiopicum* as a processing technology [88]. Certain initiatives for the processing of African eggplant, such as canning, have also been put underway. With the drying technology, a plausible way of producing food products that are shelf-stable is assured.

Drying technology for *S. aethiopicum* is simple, safe and easy to learn. Water activity that promotes microbial activities is reduced with this technology thereby reducing its weight and volume, effectively making packaging and transportation costs minimal [89]. The most reliable drying methods include osmotic, freeze, vacuum, Ohmic, solar, hot air, and microwave [90]. But Sagar and Kumar [91] reported in their review that, freeze-drying, osmotic dehydration and vacuum drying are too expensive for a broad-scale production of African eggplant. The freeze-drying technique is crucial in water removal as it stops deterioration activities due to enzyme and microbial activity, thus generating the highest quality of the final product [91]. According to Mbondo [92], drying of *S. aethiopicum* using freeze-drying retains a high amount of phenolics and beta carotene after the process. The drying of African eggplant through sun drying and hot air cabinet drying has been reported [88]. Another study recommended slicing the vegetable, dipping them in plain water and drying using a cabinet dryer as the best method for producing dried *S. aethiopicum* powder [88]. Osmotic dehydration of African eggplant using sodium chloride (NaCl) solution can be employed after harvest to preserve the fruits by reducing the water activity of the fruits [93]. Processing of *S. aethiopicum* for the purpose of export is being carried out by the Food Processing International as well as the Nsawam Cannery Company in Ghana. This is a practicable and achievable attempt to deal with its poor shelf life [94]. It could be seen by private initiatives as a means of positive return if they invest, which in turn could encourage further research [37].

4. Landscape Protection and Restoration

African eggplant is well known for its ability to yield so much from limited spaces. They can produce a worthwhile harvest while being grown on the tiniest of plots or even in garden pots. This characteristic partiality for small spaces of the crop enables it to be used for city gardens squeezed in modern structures like factories, roads, train tracks, high-rise buildings and factories, allowing for conservation of soil fertility while maximizing landscape spaces. Known also for its aesthetic property, African eggplant can turn many eyes just with their looks. As a result, it is a very eligible crop for ornamental and beautification purposes [95]. Due to its tendency to tolerate shade, the crop is usually employed to cover bare soil found between main crops grown on the farm, to promote soil conservation activities (nutrients and moisture), thus keeping the soil enriched and fertile. This makes them likely candidates for infertile and complex soils, thereby putting to use and reducing the numerous agricultural wastelands present. Moreover, despite its profuse branching, this propensity allows the crop to shade out unwanted competitors as it makes weeding difficult [95].

National Research Council suggested that, the crops' resistance to diseases that are soil-borne and caused by pathogens such as *Fusarium oxysporum* and *Verticillium dahlia* gives the potential for it to curb soil sickness. It has also been claimed that the crop hosts several commercial crop-affecting pests, bacteria and fungi. Hence, it can be used as bait to lure these pests and pathogens away from the main crops. In addition, the crop has also shown some molluscicidal abilities, making it beneficial in the control of garden snails, slugs and schistosomiasis parasite-harboring water snails.

5. Conclusions

S. aethiopicum, although an underutilized crop species, is a very important vegetable that can contribute to improving food security and provide sustainable productions. Since fruits are culinarily and aesthetically attractive, farmers can export dried or canned fruits to European or Asian countries as an alternative food or garnish. Alongside its agricultural importance and ornamental value, African eggplant also has pharmaceutical uses. The crop is dense in secondary metabolites, which can be used in medicine and in crop protection. Cultivating this plant species should be a better business for local farmers to sell these products to a chemical company that can start investigations on natural medicines and pesticide. In order to increase the consumption and cultivation of this vegetable, the vital genetic resource should receive much research attention and improvement. In vitro and ex situ conservation of the plant's genetic resource should be maintained, as well as a genomic resource database created to store the crop's sequenced genome information. Currently, the crop faces huge postharvest losses due to its short shelf life. Fruits are not durable, therefore, they should be quickly dried or processed. Solar drying could be employed as the cheapest drying method, and one can use UV filters if necessary, to protect those precious compounds in the fruits. Breeding regimes for African eggplant should develop fruits that stay longer on the shelf. Furthermore, promoting the use of this vegetable will stimulate dietary diversification among the people in Africa.

Author Contributions: K.N.O. gathered the information and prepared the initial draft of the manuscript. M.H. and T.S. developed the concept and revised the final draft of the manuscript. N.A.B.B. assisted with literature search and proofreading of the manuscript. All authors have read and agreed to the published version of the manuscript.

Funding: This research was funded by the National Natural Science Foundation of China (NSFC), grant number (31870589; 31700525); the Scientific Research Foundation for High-Level Talents of Nanjing Forestry University (SRFNFU) (GXL2017011; GXL2017012); the Priority Academic Program Development of Jiangsu Higher Education Institutions.

Institutional Review Board Statement: Not applicable.

Informed Consent Statement: Not applicable.

Data Availability Statement: Data sharing not applicable.

Conflicts of Interest: The authors declare no conflict of interest.

References

1. Kamga, R.T.; Kouamé, C.; Atangana, A.R.; Chagomoka, T.; Ndango, R. Nutritional Evaluation of Five African Indigenous Vegetables. *J. Hortic. Res.* **2013**, *21*, 99–106. [CrossRef]
2. Hossein Aminifard, M.; Aroiee, H.; Fatemi, H.; Ameri, A.; Karimpour, S. Responses of eggplant (*Solanum melongena* L.) to different rates of nitrogen under field conditions. *J. Cent. Eur. Agric.* **2010**, *11*, 453–458. [CrossRef]
3. Bisamaza, M.; Banadda, N. Solar drying and sun drying as processing techniques to enhance the availability of selected African indigenous vegetables, *Solanum aethiopicum* and *Amaranthus lividus* for nutrition and food security in Uganda. *Afr. J. Food Sci. Technol.* **2017**, *8*, 1–6. [CrossRef]
4. Padulosi, S.; Thompson, J.; Rudebjer, P. *Fighting Poverty, Hunger and Malnutrition with Neglected and Underutilized Species (NUS): Needs, Challenges and the Way Forward*; Biodiversity International: Rome, Italy, 2013.
5. Cernansky, R. The rise of Africa's super vegetables. *Nature* **2015**, *522*, 146–148. [CrossRef] [PubMed]
6. Anaso, H.U. Comparative cytological study of Solanum aethiopicum Gilo group, Solanum aethiopicum Shum group and Solanum anguivi. *Euphytica* **1991**, *53*, 81–85. [CrossRef]
7. Lester, R.N. Genetic resources of capsicum and eggplants. In Proceedings of Xth Meeting on Genetics and Breeding of Capsicum and Eggplant, Avignon, France, 7–11 September 1998; Palloix, A., Daunay, M.C., Eds.; Eucarpia: Wageningen, The Netherlands, 1998; pp. 25–30.
8. Taher, D.; Solberg, S.Ø.; Prohens, J.; Chou, Y.; Rakha, M.; Wu, T. World Vegetable Center Eggplant Collection: Origin, Composition, Seed Dissemination and Utilization in Breeding. *Front. Plant Sci.* **2017**, *8*, 1484. [CrossRef] [PubMed]

9. Mwinuka, P.R.; Mbilinyi, B.P.; Mbungu, W.B.; Mourice, S.K.; Mahoo, H.F.; Schmitter, P. Optimizing water and nitrogen application for neglected horticultural species in tropical sub-humid climate areas: A case of African eggplant (*Solanum aethiopicum* L.). *Sci. Hortic.* **2021**, *276*, 109756. [CrossRef]
10. Padulosi, S.; Hodgkin, T.; Williams, J.T. Underutilized crops: Trends, challenges and opportunities in the 21st Century. In *Managing Plant Genetic Resources*; Engels, J.M.M., Ramanatha Rao, V., Brown, A.H.D., Jackson, M.T., Eds.; CAB International: Wallingford, UK; IPGRI: Rome, Italy, 2002; pp. 323–338. [CrossRef]
11. Ebert, A.W. Potential of Underutilized Traditional Vegetables and Legume Crops to Contribute to Food and Nutritional Security, Income and More Sustainable Production Systems. *Sustainability* **2014**, *6*, 319–335. [CrossRef]
12. Bukenya, Z.R.; Carasco, J.F. Ethnobotanical aspects of *Solanum* L. (Solanaceae) in Uganda. In *Solanaceae IV: Advances in Biology and Utilization*; Royal Botanic Gardens: Kew, UK, 1999; pp. 345–360.
13. Prasad, D.N.; Prakash, R. Floral biology of brinjal (*Solanum melongena* L.). *Indian J. Agric. Sci.* **1968**, *38*, 1053.
14. Lester, R.N.; Seck, A. Solanum aethiopicum L. In *Plant Resources of Tropical Africa/Ressources Végétales de l'Afrique Tropicale*; Grubben, G.J.H., Denton, O.A., Eds.; Backhuys Publishers: Wageningen, The Netherlands, 2004; pp. 530–536.
15. Polignano, G.; Uggenti, P.; Bisignano, V.; Gatta, C. Della Genetic divergence analysis in eggplant (*Solanum melongena* L.) and allied species. *Genet. Res. Crop Evol.* **2010**, *57*, 171–181. [CrossRef]
16. Lester, R.N. Taxonomy of Scarlet Eggplants, *Solanum aethiopicum* L. *Acta Hort.* **1986**, *182*, 125–132. [CrossRef]
17. Lester, R.N.; Daunay, M.C. Diversity of African vegetable Solanum species and its implications for a better understanding of plant domestication. *Schr. Genet. Ressour.* **2003**, *22*, 137–152.
18. Plazas, M.; Andújar, I.; Vilanova, S.; Gramazio, P.; Herraiz, F.J.; Prohens, J. Conventional and phenomics characterization provides insight into the diversity and relationships of hypervariable scarlet (*Solanum aethiopicum* L.) and gboma (*S. macrocarpon* L.) eggplant complexes. *Front. Plant Sci.* **2014**, *5*, 1–13. [CrossRef] [PubMed]
19. Lim, T.K. Edible medicinal and non medicinal plants. *Springer* **2015**, *6*, 310–317. [CrossRef]
20. Kouassi, A.; Béli-Sika, E.; Tian-Bi, T.Y.-N.; Alla-N'Nan, O.; Kouassi, A.; N'Zi, J.-C.; N'Guetta, A.S.-P.; Tio-Touré, B. Identification of Three Distinct Eggplant Subgroups within the *Solanum aethiopicum* Gilo Group from Côte d'Ivoire by Morpho-Agronomic Characterization. *Agriculture* **2014**, *4*, 260–273. [CrossRef]
21. Schippers, R.R. *African Indigenous Vegetables: An Overview of the Cultivated Species*; Natural Resources Institute: Chatham, UK, 2000; p. 221.
22. Nyadanu, D.; Lowor, S.T. Promoting competitiveness of neglected and underutilized crop species: Comparative analysis of nutritional composition of indigenous and exotic leafy and fruit vegetables in Ghana. *Genet. Resour. Crop Evol.* **2015**, *62*, 131–140. [CrossRef]
23. Lester, R.N.; Thitai, G.N.W. Inheritance in *Solanum aethiopicum*, the scarlet eggplant. *Euphytica* **1989**, *40*, 67–74.
24. Prohens, J.; Plazas, M.; Raigón, M.D.; Seguí-Simarro, J.M.; Stommel, J.R.; Vilanova, S. Characterization of interspecific hybrids and backcross generations from crosses between two cultivated eggplants (*Solanum melongena* and *S. aethiopicum* Kumba group) and implications for eggplant breeding. *Euphytica* **2012**, *186*, 517–538. [CrossRef]
25. Osei, M.K.; Banful, B.; Osei, C.K.; Oluoch, M.O. Characterization of African eggplant for morphological characteristics. *J. Agric. Sci. Tech.* **2010**, *4*, 33–37.
26. James, B.; Atcha-Ahowe, C.; Godonou, I.; Baimey, H.; Goergen, G.; Sikirou, R.; Toko, M. *Integrated Pest Management in Vegetable Production: A Guide for Extension Workers in West Africa*; International Institute of Tropical Agriculture (IITA): Ibadan, Nigeria, 2010; p. 120.
27. Mensah, S.I.; Ekeke, C.; Udom, M. Effect of Salinity on the Growth Characteristics of *Solanum aethiopicum* L. (Solanaceae). *Asian Res. J. Agric.* **2020**, *12*, 17–22. [CrossRef]
28. Ofori, K.; Gamedoagbao, D.K. Yield of scarlet eggplant (*Solanum aethiopicum* L.) as influenced by planting date of companion cowpea. *Sci. Hortic.* **2005**, *105*, 305–312. [CrossRef]
29. Sidibé, D.; Sanou, H.; Bayala, J.; Teklehaimanot, Z. Yield and biomass production by African eggplant (*Solanum aethiopicum*) and sorghum (*Sorghum bicolor*) intercropped with planted Ber (*Ziziphus mauritiana*) in Mali (West Africa). *Agrofor. Syst.* **2017**, *91*, 1031–1042. [CrossRef]
30. Aguessy, S.; Laura, Y.; Loko, E. Ethnobotanical Characterization of Scarlet Eggplant (*Solanum aethiopicum* L.) Varieties Cultivated in Benin Republic. *Res. Sq.* **2021**, 1–27, pre-print. [CrossRef]
31. Agyapong, R.B.; Quainoo, A.K. Effects of False yam (*Icacina oliviformis*) tuber compost manure on the growth performance of garden eggs (*Solanum aethiopicum* L.). *Ghana J. Sci. Tech. Dev.* **2021**, *7*, 20–28. [CrossRef]
32. Ahabwe, S. Role of Seed and Aphid Vectors in Proliferation of Tomato Virus Diseases on Small Holder Farms in Kasese District, Uganda. Master's Dissertation, Makerere University, Kampala, Uganda, June 2019.
33. Sabatino, L.; Iapichino, G.; D'Anna, F.; Palazzolo, E.; Mennella, G.; Rotino, G.L. Hybrids and allied species as potential rootstocks for eggplant: Effect of grafting on vigour, yield and overall fruit quality traits. *Sci. Hortic.* **2018**, *228*, 81–90. [CrossRef]
34. Amengor, N.E.; Boamah, E.D.; Akrofi, S.; Gamedoagbao, D.K.; Egbadzor, K.F.; Davis, H.E.; Owusu, E.O.; Kotey, D.A. Reconnaissance Study of Insect-Pests Infestation of Garden Eggs in the Volta Region of Ghana. *Int. J. Agric. Ext.* **2017**, *5*, 1–10.
35. Taher, D.; Schafleitner, R. Sources of Resistance for Two-spotted Spider Mite (*Tetranychus urticae*) in Scarlet (*Solanum aethiopicum* L.) and Gboma (*S. macrocarpon* L.) Eggplant Germplasms. *Hort. Sci.* **2019**, *54*, 240–245. [CrossRef]
36. Prescott-Allen, R.; Prescott-Allen, C. How Many Plants Feed the World? *Conserv. Biol.* **1990**, *4*, 365–374. [CrossRef]

37. Horna, D.; Gruère, G. *Marketing Underutilized Crops for Biodiversity: The Case of African Garden Egg (Solanum ethiopicum) in Ghana*; Global Facilitation Unit for Underutilized Species (GFU): Rome, Italy, 2007.
38. Daunay, M.C.; Lester, R.N.; Laterrot, H. *The Use of Wild Species for the Genetic Improvement of Brinjal Egg-Plant (Solanum melongena) and Tomato (Lycopersicon esculentum)*; The Royal Botanic Gardens: Kew, UK, 1991.
39. Collonnier, C.; Fock, I.; Kashyap, V.; Rotino, G.L.; Daunay, M.C.; Lian, Y.; MariskaI, K.; Rajam, M.V.; Servaes, A.; Ducreux, G.; et al. Applications of biotechnology in eggplant. *Plant CellTissue Organ Cult.* **2001**, *65*, 91–107. [CrossRef]
40. Chowdhury, M.A.; Vandenberg, B.; Warkentin, T. Cultivar identification and genetic relationship among selected breeding lines and cultivars in chickpea (*Cicer arietinum* L.). *Euphytica* **2002**, *127*, 317–325. [CrossRef]
41. Sharma, T.R.; Jana, S. Random amplified polymorphic DNA (RAPD) variation in Fagopyrum tataricum Gaertn. accessions from China and the Himalayan region. *Euphytica* **2002**, *127*, 327–333. [CrossRef]
42. Kumar, R.; Venuprasad, R.; Atlin, G.N. Genetic analysis of rainfed lowland rice drought tolerance under naturally-occurring stress in eastern India: Heritability and QTL effects. *Field Crop. Res.* **2007**, *103*, 42–52. [CrossRef]
43. Kubie, L.D. Evaluation of Genetic Diversity in Garden Egg (*Solanum aethiopicum*) Germplasm in Ghana. Doctoral Dissertation, University of Ghana, Accra, Ghana, July 2013.
44. Lock, M.; Grubben, G.J.H.; Denton, O.A. Plant Resources of Tropical Africa 2. Vegetables. *Kew Bull.* **2004**, *59*, 650. [CrossRef]
45. Ano, G.; Hebert, Y.; Prior, P.; Messiaen, C. A new source of resistance to bacterial wilt of eggplants obtained from a cross: *Solanum aethiopicum* L. × *Solanum melongena* L. *Agronomie* **1991**, *11*, 555–560. [CrossRef]
46. Li, D.; Li, S.; Li, W.; Liu, A.; Jiang, Y.; Gan, G.; Li, W.; Liang, X.; Yu, N.; Chen, R.; et al. Comparative transcriptome analysis provides insights into the molecular mechanism underlying double fertilization between self- crossed *Solanum melongena* and that hybridized with *Solanum aethiopicum*. *PLoS ONE* **2020**, *15*, 1–20. [CrossRef] [PubMed]
47. Sakhanokho, H.F.; Islam-faridi, M.N.; Blythe, E.K.; Smith, B.J.; Rajasekaran, K.; Majid, M.A. Morphological and Cytomolecular Assessment of Intraspecific Variability in Scarlet Eggplant (*Solanum aethiopicum* L.). *J. Crop Improv.* **2014**, *28*, 437–453. [CrossRef]
48. Song, B.; Song, Y.; Fu, Y.; Kizito, B.; Kamenya, S.N.; Kabod, P.N.; Liu, H.; Muthemba, S.; Kariba, R.; Njuguna, J.; et al. Draft genome sequence of *Solanum aethiopicum* provides insights into disease resistance, drought tolerance, and the evolution of the genome. *GigaScience* **2019**, *8*, 1–16. [CrossRef]
49. Grubben, G.J.H. *Plant Resources of Tropical Africa 2. Vegetables*; PROTA Foundation: Wageningen, The Netherlands; Backhuys Publishers: Leiden: The Netherlands; CTA: Wageningen, The Netherlands, 2004; p. 668.
50. Adeniji, O.T.; Kusolwa, P.M.; Reuben, S.W.O.M. Quantitative analysis of fruit size and fruit number in *Solanum aethiopicum* group. *AGRIEAST* **2019**, *13*, 27. [CrossRef]
51. Gisbert, C.; Prohens, J.; Nuez, F. Efficient regeneration in two potential new crops for subtropical climates, the scarlet (*Solanum aethiopicum*) and gboma (*S. macrocarpon*) eggplants. *N. Z. J. Crop Hortic. Sci.* **2006**, *34*, 55–62. [CrossRef]
52. Larkin, P.J.; Scowcroft, W.R. Somaclonal variation—A novel source of variability from cell cultures for plant improvement. *Theor. Appl. Genet.* **1981**, *60*, 197–214. [CrossRef]
53. Anil, V.S.; Lobo, S.; Bennur, S. Somaclonal variations for crop improvement: Selection for disease resistant variants in vitro. *Plant Sci. Today* **2018**, *5*, 44–54. [CrossRef]
54. Titus, S.D.; Falusi, O.A.; Abdulhakeem, A.; Liman, M. Effects of gamma irradiation on the agro-morphological traits of selected Nigerian eggplant (*Solanum aethiopicum* L.) accessions. *GSC Biol. Pharm. Sci.* **2018**, *2*, 23–30. [CrossRef]
55. Eletta, O.A.A.; Orimolade, B.O.; Oluwaniyi, O.O.; Dosumu, O.O. Evaluation of proximate and antioxidant activities of Ethiopian eggplant (*Solanum aethiopicum* L.) and Gboma Eggplant (*Solanum macrocarpon* L.). *J. Appl. Sci. Environ. Manag.* **2017**, *21*, 967–972. [CrossRef]
56. Michael, U.; Banji, A.; Abimbola, A.; David, J.; Oluwatosin, S.; Aderiike, A.; Ayodele, O.; Adebayo, O. Assessment of variation in mineral content of ripe and unripe African eggplant fruit (*Solanum aethiopicum* L.) Exocarps. *J. Pharmacog. Phytochem.* **2017**, *6*, 2548–2551.
57. Offor, C.E.; Igwe, S.U. Comparative Analysis of the Vitamin Composition of Two Different Species of Garden Egg (*Solanum aethiopicum* and *Solanum macrocarpon*). *World J. Med. Sci.* **2015**, *12*, 274–276. [CrossRef]
58. Chinedu, S.N.; Olasumbo, A.C.; Eboji, O.K.; Emiloju, O.C.; Olajumoke, K.; Dania, D.I. Proximate and Phytochemical Analyses of *Solanum aethiopicum* L. and *Solanum macrocarpon* L. Fruits. *Res. J. Chem. Sci.* **2011**, *1*, 65–71.
59. Vohora, S.B.; Kumar, I.; Khan, M.S.Y. Effect of alkaloids of *Solanum melongena* on the central nervous system. *J. Ethnopharmacol.* **1984**, *11*, 331–336. [CrossRef]
60. Bello, S.O.; Muhammad, B.Y.; Gammaniel, K.S.; Abdu-Aguye, I.; Ahmed, H.; Njoku, C.H.; Pindiga, U.H.; Salka, A.M. Preliminary Evaluation of the Toxicity and Some Pharmacological Properties of the Aqueous Crude Extract of *Solanum melongena*. *Res. J. Agric. Biol. Sci.* **2005**, *1*, 1–9.
61. Xu, R.; Zhao, W.; Xu, J.; Shao, B.; Qin, G. Studies on Bioactive Saponins from Chinese Medicinal Plants. *Adv. Exp. Med. Biol.* **1996**, *404*, 371–382. [CrossRef] [PubMed]
62. Asl, M.N.; Hosseinzadeh, H. Review of Pharmacological Effects of Glycyrrhiza sp. and its Bioactive Compounds. *Phytother. Res.* **2008**, *22*, 709–724. [CrossRef] [PubMed]
63. Vinson, J.A.; Hao, Y.; Su, X.; Zubik, L. Phenol Antioxidant Quantity and Quality in Foods: Vegetables. *J. Agric. Food Chem.* **1998**, *46*, 3630–3634. [CrossRef]

64. Bagchi, M.; Milnes, M.; Williams, C.; Balmoori, J.; Ye, X.; Stohs, S.; Bagchi, D. Acute and chronic stress-induced oxidative gastrointestinal injury in rats, and the protective ability of a novel grape seed proanthocyanidin extract. *Nutr. Res.* **1999**, *19*, 1189–1199. [CrossRef]

65. Diatta, K.; Diatta, W.; Fall, A.D.; Ibra, S.; Dieng, M.; Mbaye, A.I.; Sarr, A.; Adegbindin, M.A. Evaluation of the Antioxidant Activity of Stalk and Fruit of *Solanum aethiopicum* L. (Solanaceae). *Asian J. Res. Biochem.* **2020**, *6*, 6–12. [CrossRef]

66. Sodipo, O.A.; Abdulrahman, F.I.; Alemika, T.E.; Gulani, I.A. Chemical Composition and Biological Properties of the Petroleum Ether Extract of *Solanum macrocarpum* L. (Local Name: Gorongo). *Br. J. Pharm. Res.* **2012**, *2*, 108–128. [CrossRef]

67. Tchoupang, E.N.; Reder, C.; Ateba, S.B.; Zehl, M.; Kählig, H.; Njamen, D.; Holler, F.; Glasi-Tazreiter, S.; Krenn, L. Acetylated Furostene Glycosides from *Solanum gilo* Fruits. *Planta Med.* **2017**, *83*, 1227–1232. [CrossRef] [PubMed]

68. Abubakar, A.R.; Sani, I.H.; Malami, S.; Yaro, A.H. Anxiolytic and Sedative Activities of Methanol Extract of *Solanum aethiopicum* (Linn.) Fruits in Swiss Mice. *Trop. J. Nat. Prod. Res.* 2021; 5, 353–358. [CrossRef]

69. Yang, R.-Y.; Ojiewo, C. African Nightshades and African Eggplants: Taxonomy, Crop Management, Utilization, and Phytonutrients. In *African Natural Plant Products: Discoveries and Challenges in Chemistry, Health, and Nutrition*; Juliani, H.R., Simon, J.E., Ho, C.-T., Eds.; ACS Symposium Series; American Chemical Society: Washington, DC, USA, 2013; Volume 2, pp. 137–165. [CrossRef]

70. Stommel, J.R.; Whitaker, B.D. Phenolic Acid Content and Composition of Eggplant Fruit in a Germplasm Core Subset. *J. Am. Soc. Hortic. Sci.* **2003**, *128*, 704–710. [CrossRef]

71. Anosike, C.A.; Obidoa, O.; Ezeanyika, L.U.S. egg albumin-induced oedema and granuloma tissue formation in rats. *Asian Pac. J. Trop. Med.* **2012**, *5*, 62–66. [CrossRef]

72. Tunwagun, D.A.; Akinyemi, O.A.; Ayoade, T.E.; Oyelere, S.F. Qualitative Analysis of the Phytochemical Contents of Different Anatomical Parts of Ripe *Solanum aethiopicum* Linneaus Fruits. *J. Complement. Altern. Med. Res.* **2020**, *9*, 14–22. [CrossRef]

73. Emiloju, O.C.; Chinedu, S.N. Effect of *Solanum aethiopicum* and *Solanum macrocarpon* Fruits on Weight Gain, Blood Glucose and Liver Glycogen of Wistar Rats. *World* **2016**, *4*, 1–4. [CrossRef]

74. Chadha, M.L. *AVRDC's Experiences within Marketing of Indigenous Vegetables–A Case Study on Commercialization of African Eggplant*; AVRDC-Regional Center for Africa Duluti: Arusha, Tanzania, 2003.

75. Lumpkin, T.A. A Perspective from the World Vegetable Center. *Acta Hortic.* **2009**, *841*, 245–248. [CrossRef]

76. Bachmann, J.; Earles, R. *Postharvest Handling of Fruits and Vegetables*; ATTRA: Ozark Mountains, University of Arkansas, Fayetteville, NC, USA, 2000; pp. 1–19.

77. Ray, R.C.; Ravi, V. Post-Harvest Spoilage of Sweet potato in Tropics and Control Measures. *Crit. Rev. Food Sci. Nutr.* **2005**, *45*, 623–644. [CrossRef]

78. Kader, A.A. Flavor quality of fruits and vegetables. *J. Sci. Food Agric.* **2008**, *88*, 1863–1868. [CrossRef]

79. Kader, A.A. Fruit Maturity, Ripening, and Quality Relationships. *Acta Hortic.* **1999**, *485*, 203–208. [CrossRef]

80. Msogoya, T.J.; Majubwa, R.O.; Maerere, A.P. Effects of harvesting stages on yield and nutritional quality of African eggplant (*Solanum aethiopicum* L.) fruits. *J. Appl. Biosci.* **2014**, *78*, 6590–6599. [CrossRef]

81. Apolot, M.G.; Acham, H.; Ssozi, J.; Namutebi, A.; Masanza, M.; Kizito, E.; Jagwe, J.; Kasharu, A.; Deborah, R. Postharvest practices along supply chains of *Solanum aethiopicum* (shum) and *Amaranthus lividus* (linn) leafy vegetables in Wakiso and Kampala Districts, Uganda. *Afr. J. Food Agric. Nutr. Dev.* **2020**, *20*, 15978–15991. [CrossRef]

82. Majubwa, R.O.; Msogoya, T.J.; Maerere, A.P. Effects of local storage practices on deterioration of African eggplant (*Solanum aethiopicum* L.) fruits. *Tanzan. J. Agric. Sci.* **2015**, *14*, 106–111.

83. Sekulya, S.; Nandutu, A.; Namutebi, A.; Ssozi, J.; Masanza, M.; Jagwe, J.N.; Kasharu, A.; Rees, D.; Kizito, E.B.; Acham, H. Effect of Post-Harvest Handling Practices, Storage Technologies and Packaging Material on Post-Harvest Quality and Antioxidant Potential of *Solanum aethiopicum* (Shum) Leafy Vegetable. *Am. J. Food Technol.* **2018**, *6*, 167–180. [CrossRef]

84. Kitinoja, L.; AlHassan, H.Y. Identification of Appropriate Postharvest Technologies for Small Scale Horticultural Farmers and Marketers in Sub-Saharan Africa and South Asia—Part 1. Postharvest Losses and Quality Assessments. *Acta Hortic.* **2012**, *934*, 31–40. [CrossRef]

85. Hong, M.N.; Lee, B.C.; Mendonca, S.; Grossmann, M.V.E.; Verhe, R. Effect of infiltrated calcium on ripening of tomato fruits. *LWT J. Food Sci.* **1999**, *33*, 2–8.

86. Bedel, F.J.; Yolande, A.; Yves, D.; Alban, N.K.; Lucien, K. Screening of some catalytic activities of mature berries of *Solanum aethiopicum* cultivar "Klongbo". *Magna Sci. Adv. Biol. Pharm.* **2021**, *1*, 35–41. [CrossRef]

87. Vincente, A.R.; Manganaris, G.A.; Ortiz, C.M.; Sozzi, G.O.; Crisosto, C.H. Nutritional Quality of Fruits and Vegetables. In *Postharvest Handling*; Academic Press: Cambridge, MA, USA, 2014; pp. 69–122. [CrossRef]

88. Sam, A.; Hedwig, A.; Gaston, T.; Namutebi, A.; Masanza, M.; Jagwe, J.N.; Apolo, K.; Kizito, B.E.; Rees, D. Effect of different processing conditions on proximate and bioactive contents of *Solanum aethiopicum* (Shum) powders, and acceptability for cottage scale production. *Am. J. Food Nutr.* **2018**, *6*, 46–54.

89. Swanson, M.A.; McCurdy, S.M. *Drying Fruits & Vegetables*; A Pacific Northwest Extension Publication: Moscow, ID, USA; Corvallis, OR, USA; Pullman, DC, USA, 2009.

90. George, S.D.; Cenkowski, S.; Muir, W.E. A review of drying technologies for the preservation of nutritional compounds in waxy skinned fruit. In Proceedings of the 2004 North Central ASAE/CSAE Conference, Winnipeg, MB, Canada, 24–25 September 2004; pp. 4–104.

91. Sagar, V.R.; Suresh Kumar, P. Recent advances in drying and dehydration of fruits and vegetables: A review. *J. Food Sci. Technol.* **2010**, *47*, 15–26. [CrossRef] [PubMed]
92. Mbondo, N.N. Characteristic Effects of Drying Processes on Bioactive Compounds in African Eggplant (*Solanum aethiopicum* L.). Master's Dissertation, Jomo Kenyatta University of Agriculture and Technology, Juja, Kenya, 2019.
93. Adekeye, D.K.; Aremu, O.; Popoola, O.K.; Fadunmade, E.O.; Adedotun, I.S.; Araromi, A.A. Osmotic Dehydration of Garden Egg (*Solanum aethiopicum*): The Shrinkage and Mass Transfer Phenomena. *J. Anal. Tech. Res.* **2020**, *2*, 1–17. [CrossRef]
94. Danquah-Jones, A. Variation and Correlation among Agronomic Traits in Garden Egg (Solanum gilo Radii). Master's Dissertation, University of Ghana, Legon, Ghana, 2000.
95. National Research Council. *Lost Crops of Africa: Volume II: Vegetables*; National Academies Press: Washinton, DC, USA, 2006; pp. 1–352. [CrossRef]

MDPI

St. Alban-Anlage 66

4052 Basel

Switzerland

Tel. +41 61 683 77 34

Fax +41 61 302 89 18

www.mdpi.com

Horticulturae Editorial Office

E-mail: horticulturae@mdpi.com

www.mdpi.com/journal/horticulturae

www.ingramcontent.com/pod-product-compliance
Lightning Source LLC
LaVergne TN
LVHW071137230726
PP19538200001B/10